Fractal Growth Phenomena

(Second Edition)

Fractal Growth Phenomena

(Second Edition)

Tamás Vicsek

Department of Atomic Physics
Eötvös University
Budapest, Puskin u.5-7
H-1088 Hungary

World Scientific
Singapore • New Jersey • London • Hong Kong

Published by

World Scientific Publishing Co. Pte. Ltd.
P O Box 128, Farrer Road, Singapore 9128
USA office: Suite 1B, 1060 Main Street, River Edge, NJ 07661
UK office: 73 Lynton Mead, Totteridge, London N20 8DH

Library of Congress Cataloging-in-Publication Data
Vicsek, Tamás.
Fractal growth phenomena / Tamás Vicsek. --- 2nd ed.
p. cm.
Includes bibliographical references and index.
ISBN 9810206682. -- ISBN 9810206690 (pbk.)
1. Fractals. I. Title.
QA614.86.V53 1992
514'.74 -- dc20 92-11311
CIP

Printed in Singapore by General Printing Services Pte. Ltd.

To *Mária,*
Lilla and András

FOREWORD

The discovery of fractal growth phenomena has been, among all the scientific developments of recent years, one of the most rewarding. As a complete surprise, rules of random growth which are so simple as to seem trivial are in fact capable of generating an unimaginably rich variety of fractal shapes! The finding has immediately fired many skillful and ardent investigators worldwide, and soon proved to involve a fascinating interface between the roughness associated with fractals and the smoothness associated with the Laplace equation. By now, we have moved beyond the necessary and even unavoidable first stage, when the main task was to sort out the newly discovered riches. For example, some critical features of diffusion limited aggregates have now been explained. But the topic is beautifully alive, and will stay with us for a long time. Thus, a study born in the context of carbon deposits on the cylinder walls of Diesel engines promises to become part of a great new stream of mathematical physics.

What has been lacking is a broad survey of the field. It is our very good fortune that this gap is being beautifully filled by the present book. Professor Vicsek has taken full advantage of the unusual combination of circumstances that surrounds his life and work. He is in close professional interaction with all other workers in the field, therefore the book is very much up to date. On the other hand, he has achieved a degree of objectivity, detachment and balance that was indispensable, but might have been hard to achieve in the scientific "downtowns". It is good, also, that the book is self contained, thanks to its introductory chapters. But the author stresses, and I agree, that the surveys in these chapters do not claim to provide a complete introductory treatise.

Personally, I have taken no direct part in this great adventure, though, of course, I have provided in advance its tools, namely the fractal sets and the multifractal measures! Hence, I feel free to express my enthusiasm.

The present book will hasten the pace of progress in the study of fractal aggregates, and will help interest in this topic to spread everywhere.

Benoit B. Mandelbrot

Physics Department, IBM T.J. Watson Research Center,
Yorktown Heights, NY 10598 and
Mathematics Department, Yale University, New Haven, CT 06520

PREFACE

During the last couple of years the field of fractal growth phenomena has *continued to develop fast* and a large number of important new results have been published since the manuscript of the first edition has been completed. Based on this fact, and on the Publisher's evaluation of the success of the first edition, we arrived at the conclusion that the need for a second edition containing several new features is justified.

The main extra features of the present edition are the following: i) Perhaps most important, a *new part* (Part IV) has been added to provide a review of the recent developments in the field of fractal growth phenomena in order to make the book completely up to date. Particular stress is made on treating the most rapidly evolving subfield of growing fractal surfaces. The three chapters of this new part represent natural extensions of the earlier ones; they are built on the information given previously. Since the material presented in the first three parts can still be used effectively as it is, it did not seem necessary to change those parts of the book.

Next, ii) A number of *colour plates* has been included in this edition with the purpose of making the book more enjoyable for readers who like to learn about physics by inspecting beautiful pictures. iii) Misprints which were detected in the first edition are now *corrected.* iv) Finally, but not last, together with my wife we have developed a *software* package demonstrating fractal growth models on personal computers. This *Fractal Growth* software is now available together with the second edition and represents a complementary source for those who are interested in fractal growth phenomena.

I would like to thank again all of my colleagues listed in the Preface to the first edition. With most of them I continued to work together. In addition I am grateful to my *new* collaborators, A.-L. Barabási, S. Buldyrev, R. Burbonnais, M. Cserző, S. Havlin, H. Herrmann, B. B. Mandelbrot, S. Manna, and H. E. Stanley. Critical remarks on the first edition by colleagues including F. Family, B. B. Mandelbrot and J. Kertész were very helpful when writing the present edition.

Budapest, August 1991 Tamás Vicsek

PREFACE TO THE FIRST EDITION

Even bearing in mind that we live in an era of explosive advances in various areas of science, the investigation of phenomena involving fractals has gone through a spectacular development in the last decade. Many physical, technological and biological processes have been shown to be related to and described by objects with non-integer dimensions – an idea which was originally proposed and beautifully demonstrated by Benoit B. Mandelbrot in his classic books on fractals.

The physics of far-from-equilibrium *growth phenomena* represents one of the main fields in which fractal geometry is widely applied. During the past couple of years considerable experimental, numerical and theoretical information has accumulated about such processes, and it seemed reasonable to bring together most of this knowledge into a *separate book*, in addition to the numerous conference proceedings and reviews devoted to irreversible growth.

My intention was to provide a book which would summarize the basic concepts born in the studies of fractal growth as well as to present some of the most important new results for more specialized readers. Thus, it is hoped that the book will be able to serve as a textbook on the *geometrical aspects of fractal growth* and will also treat this area in sufficient depth to make it useful as a reference book. It follows from the nature of this approach that the emphasis is on presenting results in a reproducible manner rather than on briefly reviewing a large number of contributions. Obviously, the field of fractal growth phenomena is too broad to enable all of the related topics to be included. Among the important aspects not treated are, for example, cellular automata or the physical properties (elasticity, conductivity, etc.) of growing fractals.

Collaboration with many colleagues has greatly helped me in gaining an insight into the processes discussed in this book. For the last ten years my closest colleague and friend János Kertész and I have worked together at the Institute for Technical Physics of the Hungarian Academy of Sciences on a number of problems related to fractals. We have had many stimulating discussions in the past and we have a wealth of interesting new ideas to study together in the future. A considerable amount of my activity in the field of aggregation was realized during my visit to Emory University, Atlanta, where I was working with Fereydoon Family. With him, and with Paul Meakin of du Pont, Wilmington, our fruitful cooperation has become regular and now spans the ocean. I am also grateful to Á. Buka, D. Grier, V. Horváth, L. J. Montag, H. Nakanishi, D. Platt, Z. Rácz, G. Radnóczy, L. M. Sander, A. Szalay, B. Taggett, T. Tél and Y. Zhang for their kind collaboration.

I should also like to thank a number of colleagues who greatly stimulated my work by showing interest in my investigations. Discussions with Gene Stanley and Dietrich Stauffer have helped me to be involved in the most interesting current problems of the physics of fractals. I have learned much about fractals from long conversations with Benoit Mandelbrot. At the suggestion of György Marx and Péter Szépfalusy I became involved into the teaching of growth phenomena. I thank Tamás Geszti and Tivadar Siklós for their numerous helpful suggestions.

It was Len Sander who encouraged me to write this book. János Kertész and Tamás Tél read parts of the preliminary version and I am grateful for their useful comments and suggestions. My thanks to Harvey Shenker for a number of last-minute linguistic corrections. Many of the figures were reproduced from works by other authors and here I thank these colleagues for granting me the necessary permissions and for providing the corresponding originals. The rest of the illustrations were made and reproduced by Mihály

Hubai and Sára Tóth.

Finally, this Preface provides me with a good opportunity to express my gratitude to my wife, Mária Strehó. She, a specialist in numerical analysis, has helped me in many ways in the writing of this book.

Budapest, September 1988 Tamás Vicsek

CONTENTS

Part III. FRACTAL PATTERN FORMATION

9. Computer Simulations

10. Experiments on Laplacian Growth

Part IV. RECENT DEVELOPMENTS

11. Cluster Models of Self-similar Growth

12. Dynamics of Self-affine Surfaces

13. Experiments

Appendices

Chapter 1

INTRODUCTION

During the last decade it has widely been recognized by physicists working in diverse areas that many of the structures common in their experiments possess a rather special kind of geometrical complexity. This awareness is largely due to the activity of Benoit Mandelbrot (1977, 1979, 1982, 1988), who called attention to the particular geometrical properties of such objects as the shore of continents, the branches of trees, or the surface of clouds. He coined the name *fractal* for these complex shapes to express that they can be characterized by a *non-integer* (fractal) *dimensionality*. With the development of research in this direction the list of examples of fractals has become very long, and includes structures from microscopic aggregates to the clusters of galaxies.

An important field where fractals are observed is that of far-from-equilibrium growth phenomena which are common in many fields of science and technology. Examples for such processes include dendritic solidification in an undercooled medium, viscous fingering which is observed when a viscous fluid is injected into a more viscous one, and electrodeposition of ions onto an electrode. Figure 1 demonstrates the complexity of possible patterns growing under a wide variety of experimental conditions. In the experiments leading to the structures shown in Fig. 1 quasi two-dimensional samples were used

Figure 1. Examples for complex geometrical structures observed in various experiments on growth of unstable interfaces. The three major types of patterns found in the experiments on i) crystallization (a, b and c), ii) viscous fingering (d, e and f), and iii) electrodeposition of zinc (g, h and i) are grouped in separate columns. The fractal dimension of the structures shown in the middle column is close to 1.7. (This set of pictures is reproduced from Vicsek and Kertész (1987). The individual pictures are from: (a) Ben-Jacob et al (1986), (b) Radnóczy et al (1987), (c) Bentley and Humpreys (1962), (d) Buka et al (1986), (e) Daccord et al (1986), (f) Ben-Jacob et al (1985), (g and i) Sawada et al (1986) and (h) Matsushita et al (1984); For details see references to Part III.)

and the motion of the interfaces was determined by the spatial distribution of a quantity which satisfies the Laplace equation with moving boundary conditions.

In addition to interfacial growth, *aggregation* of similar particles represents another important class of growth phenomena producing complicated geometrical objects. Aggregation may take place particle by particle, while in other cases (for example during the formation of aerogels) the aggregates themselves are also mobile and are joined together to form larger clusters during their motion.

A broad class of growing patterns is characterized by an open branching structure as is illustrated by the middle column of Fig. 1. Such objects can be described in terms of fractal geometry. In the present case this means that the growing structures are *self-similar* in a statistical sense and the volume $V(R)$ of the region bounded by the interface scales with the increasing linear size R of the object in a non-trivial way

$$V(R) \sim R^D . \qquad (1)$$

Here $D < d$ is typically a non-integer number called the *fractal dimension* and d is the Euclidean dimension of the space the fractal is embedded in. Naturally, for a real object the above scaling holds only for length scales between a lower and an upper cutoff.

There are a number of reasons for the recent rapid development in the research of fractal growth. The interest is greatly motivated by the fact that fractal growth phenomena are closely related to many processes of practical importance. Here we shall mention only two examples. The internal texture of alloys due to the dendritic structures developing during their solidification is largely responsible for most of their mechanical properties. Another area of application is secondary oil recovery, where water pumped into the ground through one well is used to force the oil to flow to the neighbouring wells. The effectiveness of this method is influenced by the fractal structure of viscous fingers corresponding to the water-oil interface.

The internal evolution of physics as a discipline has also given rise to an increased interest in the investigation of structures growing under far-from-equilibrium conditions. In the 1970's most of the researchers working in the field of statistical mechanics were involved in problems related to *phase transitions* in equilibrium systems. These studies led to many important theories and methods including renormalization based on the scale invariance of thermodynamical systems at their critical point.

Since growing fractals are also *scale-invariant* objects (this property is equivalent to their self-similarity), the knowledge which had accumulated during the investigations of second order phase transitions was particularly useful in making a step forward and investigating scaling in growth processes. Thus the fields involving fractals were developing fast and it has become evident that *multifractal scaling*, which is the generalization of simple scaling, represents an important characteristic of many growth phenomena.

Most of the already large amount of new results on fractal growth (and fractals in general) can be found in conference and school proceedings (Family and Landau 1984, Shlesinger 1984, Stanley and Ostrowsky 1985, Pynn and Skjeltorp 1985, Boccara and Daoud 1985, Pietronero and Tosatti 1986, Engelman and Jaeger 1986). Some aspects of growth phenomena are discussed in the recent books by Jullien and Botet (1987) on aggregation, and by Feder (1988) on fractals. Finally, a number of review papers have been published recently about processes related to fractal growth (e.g., Herrmann 1986, Sander 1986, Witten and Cates 1986, Meakin 1987a, Meakin 1987b, Jullien 1987).

Here, I concentrate on the geometrical aspects of fractal growth. My intention was to give a balanced account of the most important results in a pedagogical style. The material is divided into four parts, viz. I: Fractals, II: Cluster Growth Models; III: Fractal Pattern Formation and IV: Recent Developments. References are provided at the end of each part.

Part I introduces the basic definitions and concepts related to *fractal geometry* in general. The major types of fractals are discussed and a few useful rules for the estimation of fractal dimensions are given. Fractal measures are treated in a separate chapter since many recent results demonstrate their important role in physical processes. The last chapter of the first part contains a collection of methods which are commonly used to determine fractal dimensions of various objects including experimental samples and computer generated clusters. Throughout this part examples are given to illustrate the principles introduced in the text.

Computer models based on growing clusters made of identical particles have proven to be a particularly useful tool in the investigation of fractal growth. The main advantage of such models is that they provide a possibility to determine the most relevant factors affecting the geometrical properties of objects developing in a given kind of growth phenomenon. Accordingly, in Part II important results concerning a wide variety of *cluster growth models* are discussed. First those models (called local) are examined in which the probability of adding a particle to the growing cluster depends only on the immediate environment of the given position. In the models of diffusion-limited growth (Chapter 6) the probability of adding a particle to the cluster is determined by the structure of the whole cluster; consequently, these processes are truly non-local. In some cases both local and non-local models may lead to compact structures with self-affine surfaces which are treated in Chapter 7. The last chapter of Part II discusses results obtained in the numerical and experimental studies of cluster-cluster aggregation. An important aspect of the aggregation of clusters is that the time is well defined in such processes. This fact allows for the development of a dynamic scaling theory for the cluster size distribution.

Part III deals with *fractal pattern formation*, where the term pattern formation is used for interfacial growth phenomena in which the motion of the unstable interfaces is dominated by the surface tension. Diffusion-limited growth processes may lead to a variety of structures (see Fig. 1), and in a number of cases it is still not known which are the conditions for the development of a given type of the possible interfaces. The answer to the questions about the relevance of the factors affecting the growth of complicated patterns could unambiguously be provided if it were possible to solve directly the corresponding non-linear equations. However, this approach does not seem to be feasible at present, because of the instability and the extreme complexity of the solutions. Thus, fractal pattern formation has mainly been studied by numerical simulations and model experiments which are reviewed in the two chapters of this part.

The purpose of Part IV is to review the most recent developments in the field of fractal growth phenomena in order to make the book up to date. The chapters in this part represent natural extensions of the first three parts; they are built on the information given previously. The new results are complementing primarily Chapters 6, 7 and 10. However, developments related to other chapters will be discussed as well.

The field of fractal growth phenomena is still growing quickly, and there are many new results which could not be included into this book. Those readers who are interested in the developments not treated here are advised to consult the already mentioned literature or to look for the numerous conference proceedings and reviews which are currently in a preparatory stage.

Part I

FRACTALS

Chapter 2

FRACTAL GEOMETRY

Our present knowledge of fractals is a result of an increasing interest in their behaviour. Some of the basic properties of objects with anomalous dimension were noticed and investigated at the beginning of this century mainly by Hausdorff (1919) and Besicovitch (1935). The relevance of fractals to physics and many other fields was pointed out by Mandelbrot, who demonstrated the richness of fractal geometry and presented further important results in his recent books on the subject (Mandelbrot 1975, 1977 and 1982). The purpose of this chapter is to give an introduction to the basic concepts, properties and types of fractals.

2.1. FRACTALS AS MATHEMATICAL AND PHYSICAL OBJECTS

One of the common features of fractal objects is that they are *self-similar* (scale invariant). This means that if we first cut out a part of them, and then we blow this piece up, the resulting object (in a statistical sense) will look the same as the original one. For example, if we took a picture of the shore of England from an airplane we would get a curve with an overall appearance rather similar to another picture which we would see when standing on the ground and looking at a rocky part of the shore. Analogously, a bough with lateral branches looks like the whole tree when looked at from a larger distance. For simpler shapes, self-similarity is not fulfilled or it is satisfied in a trivial way. A circle and its parts (the arcs) do not look the same, but naturally, a filled circle (a disc) and any smaller disc cut out of it are trivially similar (can be obtained from each other by reduction or extension). Another typical property of fractals is related to their volume with respect to their linear size. To demonstrate this we first need to introduce a few notions. We call *embedding dimension* the Euclidean dimension d of the space the fractal can be embedded in. In addition, d has to be the smallest such dimension. Obviously, the volume of a fractal (or any object), $V(l)$, can be measured by covering it with d dimensional balls of radius l. Then the expression

$$V(l) = N(l)l^d \tag{2.1}$$

gives an estimate of the volume, where $N(l)$ is the *number of balls needed to cover the object completely* and l is much smaller than the linear size L of the whole structure. The structure is regarded to be covered if the region occupied by the balls includes it entirely. The phrase "number of balls needed to cover" corresponds to the requirement that $N(l)$ should be the smallest number of balls with which the covering can be achieved. For ordinary objects $V(l)$ quickly attains a constant value, while for fractals typically $V(l) \to 0$ as $l \to 0$. On the other hand, the surface of fractals may be anomalously large with respect to L.

There is an alternative way to determine $N(l)$ which is equivalent to the definition given above. Consider a d-dimensional hypercubic lattice of lattice spacing l which occupies the same region of space where the object is located. Then the number of boxes (mesh units) of volume l^d which overlap with the structure can be used as a definition for $N(l)$ as well. This approach is called *box counting*.

Returning to the example of the shore of England we can say that it can be approximately embedded into a plane ($d = 2$). Measuring its total length (corresponding to the surface in a two-dimensional space) we would find that it tends to grow almost indefinitely with the decreasing length l of the measuring sticks. At the same time, the measured "area" of the shore (volume in $d = 2$) goes to zero if we determine it by using discs of decreasing radius. The reason for this is rooted in the extremely complicated, self-similar character of the shore. Therefore, such a curve seems to be definitely much "longer" than a line but having infinitely small area: it is neither a one- nor a two dimensional object.

We have seen on the example of the shore that the volume of a finite geometrical structure measured according to Eq. (1) may go to zero with the decreasing size of the covering balls while, simultaneously, its measured surface diverges. In general, we *call a physical object fractal, if measuring its volume, surface or length with d, $d-1$ etc. dimensional hyperballs it is not possible to obtain a well converging finite measure for these quantities when changing l over several orders of magnitude.*

It is possible to construct mathematical objects which satisfy the criterion of self-similarity exactly, and their measured volume depends on l even if l or (l/L) becomes smaller than any finite value. Figure 2.1 gives examples how one can construct such fractals using an iteration procedure. Usually one starts with a simple initial configuration of units (Fig. 2.1a) or with a geometrical object (Fig. 2.1b). Then, in the growing case this simple seed configuration (Fig. 2.1a, $k = 2$) is repeatedly added to itself in such a way that the seed configuration is regarded as a unit and in the new structure these units are arranged with respect to each other according to the same symmetry as the original units in the seed configuration. In the next stage the previous configuration is always looked at as the seed. The construction of Fig. 2.1b is based on division of the original object and it can be well followed how the subsequent replacement of the squares with five smaller squares leads to a self-similar, scale invariant structure.

One can generate many possible patterns by this technique; the fractal shown in Fig. 2.1 was chosen just because it has an open branching structure analogous to many observed growing fractals (Vicsek 1983). Only the first couple of steps (up to $k = 3$) of the construction are shown. *Mathematical fractals* are produced after *infinite number of such iterations.* In this $k \to \infty$ limit the fractal displayed in Fig. 2.1a becomes infinitely large, while the details of Fig. 2.1b become so fine that the picture seems to "evaporate" and

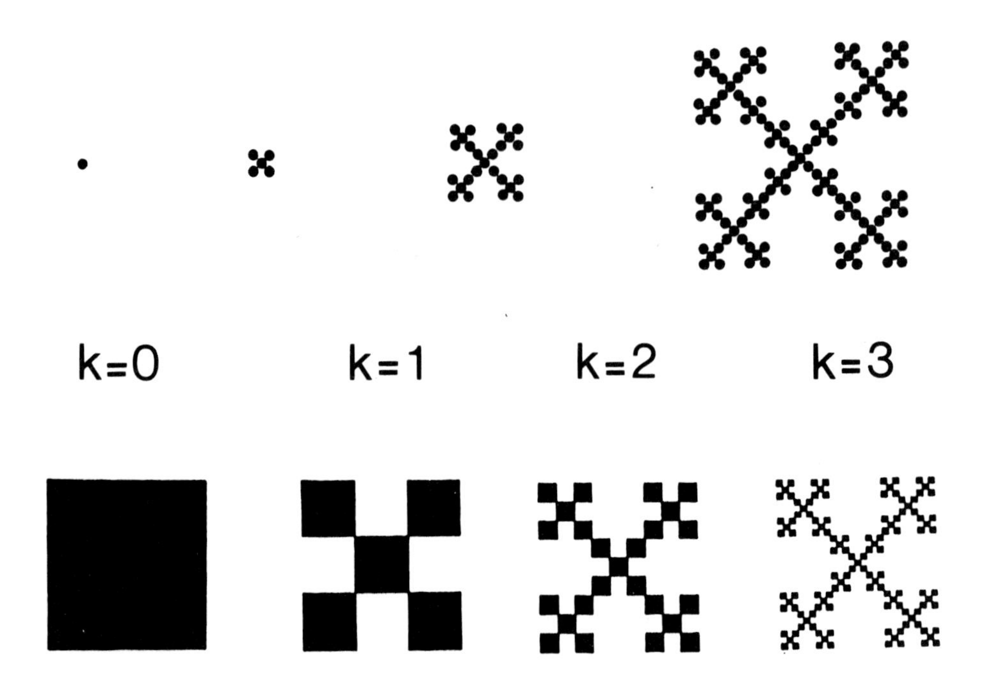

Figure 2.1. Example for the construction of a deterministic fractal embedded into two dimensions. Figure 2.1a demonstrates how one can generate a growing fractal using an iteration procedure. In Fig. 2.1b an analogous structure is constructed by subsequent divisions of the original square. Both procedures lead to fractals for $k \to \infty$ with the same dimension $D \simeq 1.465$.

can not be seen any more. For every finite k the structures in Fig. 2.1a can be scaled into each other, but this can not be done exactly in the $k \to \infty$ limit. Our example shows a connected construction, but disconnected objects distributed in a nontrivial way in space can also form a fractal.

There are a few important things to be pointed out in connection with Fig. 2.1 and fractal growth phenomena. Obviously, it is Fig. 2.1a which better approximates a real growth process. In a physical system there is always a lower cutoff of the length scale; in our case this is represented by the size of the particles. On the other hand a real object has a finite linear size which inevitably introduces an upper cutoff of the scale on which fractal scaling can be observed. This leads us to the conclusion that, in contrast to the mathematical fractals, for fractals observed *in physical phenomena the anomalous scaling of the volume can be observed only between two well defined length scales.* For growing fractals the volume is usually measured as a function of increasing linear size of structure.

There is a widely studied phenomenon in which the second type of fractals (Fig. 2.1b) play an essential role. It can be shown that in chaotic dissipative systems the trajectories in the phase space approach a fractal called strange attractor. In this case the fractal object is both finite and mathematical; it has infinitely fine details. It is a general belief that chaos and fractal growth are closely related, but the two fields have not been included into a unified picture yet.

2.2. DEFINITIONS

Since measuring the volume of fractals embedded into a d dimensional Euclidean space leads to the conclusion that they are objects having no integer dimension, we assume that the dimensionality of fractals is usually given by a noninteger number D that will be called fractal dimension. Because of the two main types of fractals demonstrated in Fig. 2.1, to define and determine D one typically uses two related approaches.

In the case of growing fractals, where there exists a smallest typical size a, one cuts out d-dimensional regions of linear size L from the object and the volume, $V(L)$, of the fractal within these regions is considered as a function of the linear size L of the object. When determining $V(L)$, the structure is covered by balls or boxes of unit volume ($l = a = 1$ is usually assumed), therefore $V(L) = N(L)$, where $N(L)$ is the number of such balls. In some cases this definition has to be modified to obtain consistent results (Tél and Vicsek 1987). According to the modification one requires that the size of the covering balls has to diverge as well, i.e., $a/l \to 0$ and $l/L \to 0$ should be satisfied when $L \to \infty$ (see Section 3.4).

For fractals having fixed L and details on very small length scale D is defined through the scaling of $N(l)$ as a function of decreasing l, where

$N(l)$ is the number of d dimensional balls of diameter l needed to cover the structure. The fact that an object is a mathematical fractal then means that $N(l)$ diverges as $L \to \infty$ and $l \to 0$, respectively, according to a non-integer exponent. Correspondingly,

$$N(L) \sim L^D \tag{2.2}$$

and

$$D = \lim_{L\to\infty} \frac{\ln N(L)}{\ln(L)} \tag{2.3}$$

for the *growing case*, where $l = 1$. Here, as well as in the following expressions the symbol $\sim$ means that the proportionality factor, not written out in (2.2), is independent of l. For fractals having a finite size and infinitely small ramifications we have

$$N(l) \sim l^{-D} \tag{2.4}$$

with

$$D = \lim_{l\to 0} \frac{\ln N(l)}{\ln(1/l)}\,. \tag{2.5}$$

Obviously, the above definitions for non-fractal objects give a trivial value for D coinciding with the embedding Euclidean dimension d. For example, the area (corresponding to the volume $V(L)$ in $d = 2$) of a circle grows as its squared radius which according to (2.3) results in $D = 2$. Similarly, the number of circles needed to cover a square diverges as the inverse of the squared radius of these circles leading again to $D = d = 2$ on the basis of (2.5).

Now we are in the position to calculate the dimension of the objects shown in Fig. 2.1. It is evident from the figure that for the growing case

$$N(L) = 5^k \qquad \text{with} \qquad L = 3^k\,, \tag{2.6}$$

where k is the number of iterations completed. From here using (2.3) we get the value $D = \ln 5/\ln 3 = 1.465...$ which is a number between $d = 1$ and $d = 2$ just as we expected. Analogously, for the fractal shown in Fig. 2.1b $N(l) = 5^k$ with $l = 3^{-k}$, leading to the same D.

Non-trivial self-similarity and the fractal value for the dimensionality of the objects are closely related. This can be seen from the fact that $D < d$ results in a negligible volume in the d dimensional embedding space. For growing fractals $V(L)/L^d \to 0$ as $L \to \infty$ means that the structure must possess large empty regions (holes) with diameters comparable to its actual linear size L. It is the presence of holes on every length scale which is the origin of non-trivial scale invariance.

In the previous paragraph we made use of the fact that the object we considered had a lower (the size of the particles) and an upper cutoff length scale (the size of the whole structure). This is, however, a property we can assume for all objects arising as a result of any physical process. The finite size ($L < \infty$) of physical fractals makes it possible to treat all of the physically relevant cases using a dimensionless quantity

$$\epsilon = \frac{l}{L} \tag{2.7}$$

which is the size of covering balls normalized by the linear size of the structure. In the case of growing fractals when the fractal dimension is investigated l (the size of the particles) is kept constant and L is increasing, while for fractals generated by subsequent divisions L is constant and l is decreasing. Therefore, (2.2) and (2.4) in terms of ϵ are recast into the same form

$$N(\epsilon) \sim \epsilon^{-D}, \tag{2.8}$$

where $\epsilon \ll 1$, $N(\epsilon)$ is the number of d dimensional balls of radius ϵL needed to cover the fractal, and D is the same fractal dimension as in (2.2) and (2.4). In the following when discussing properties of fractals in general, we shall use (2.8), while in the case of describing various particular growth processes it is more convenient to apply expression (2.2). A definition analogous to (2.8) was first used to determine a non-integer dimension for geometrically very complex objects by Hausdorff (1919) and later put into a more systematic framework by Besicovitch (1935). In fact, the definitions given by them are more general than the above expression and contain (2.8) as a special case. Equation (2.8) is more directly related to the Kolmogorov capacity (Kolmogorov and Tihomirov 1959).

The original intention of Hausdorff was to define a measure being independent of the resolution of the measurement, ϵ. We have seen that for non-trivially self-similar objects measuring the volume with balls of integer dimension, it goes either to zero or to infinity. This problem was avoided by Hausdorff who suggested that the volume should be measured covering the structure with ordinary balls, but assuming that the volume or measure of a ball is ϵ^D. Then the so called Hausdorff measure is calculated according to

$$F = N(\epsilon)\epsilon^D . \tag{2.9}$$

It can be seen easily from (2.8) that F is independent of ϵ then, and only then if D, the assumed dimension of the balls, coincides with the fractal dimension D of the object studied.

To conclude this section *ordinary fractals are defined as objects for which D determined from (2.8) is smaller than the embedding dimension d.* This definition, however, should be completed by a few remarks. For physical fractals (2.8) holds only within a few magnitudes of changing ϵ. In addition, as we shall see later, there are objects (called fat fractals) for which $D = d$, but share some of the properties of ordinary fractals. Namely, when measuring their finite volume one obtains a correction converging to zero very slowly (according to a power law, just as the total volume of an ordinary fractal).

USEFUL RULES

Before reviewing some of the most typical types of the rich botanic garden of fractals we mention a few rules which can be useful in predicting various properties related to the fractal structure of an object. Of course, because of the great variety of self-similar geometries the number of possible exceptions is not small and the rules listed below should be regarded, at least in part, as starting points for more accurate conclusions.

a) Many times it is the *projection* of a fractal which is of interest or can be experimentally studied (e.g., a picture of a fractal embedded into $d = 3$). In general, projecting a $D < d - m$ dimensional fractal onto a $d - m$ dimensional surface results in a structure with the same fractal dimension $D_p = D$. For $D \geq d - m$ the projection fills the surface, $D_p = d - m$.

b) It follows from a) that for $D < d - m$ the density correlations $c(r)$ (see the next section) within the projected image decay as a power law with an exponent $d - m - D$ instead of $d - D$ which is the exponent characterizing the algebraic decay of $c(r)$ in d.

c) Cutting out a $d-m$ dimensional slice (*cross section*) of a D dimensional fractal embedded into a d dimensional space usually leads to a $D - m$ dimensional object. This seems to be true for self-affine fractals as well, with D being their local dimension (see Section 2.3.2).

d) Consider two sets A and B having fractal dimensions D_A and D_B, respectively. *Multiplying* them together results in a fractal with $D = D_A + D_B$. As a simple example, imagine a fractal which is made of parallel sticks arranged in such a way that its cross section is the fractal shown in Fig. 2.1b. The dimension of this object is $D = 1 + \ln 5 / \ln 3$.

e) The *union* of two fractal sets A and B with $D_A > D_B$ has the dimension $D = D_A$.

f) The fractal dimension of the *intersection* of two fractals with D_A and D_B is given by $D_{A \cap B} = D_A + D_B - d$. To see this, consider a box of linear size L within the overlapping region of two growing stochastic fractals. The density of A and B particles is respectively proportional to L^{D_A}/L^d and L^{D_B}/L^d. The number of overlapping sites $N \sim L^{D_{A \cap B}}$ is proportional to these densities and to the volume of the box which leads to the above given relation. The rule concerning intersections of fractals with smooth hypersurfaces is a special case of the present one.

g) The distribution of empty regions (holes) in a fractal of dimension D scales as a function of their linear size with an exponent $-D-1$. The following heuristic argument supporting the above result is here applied to the one-dimensional case. The statement is essentially the following

$$n(\epsilon, \Delta\epsilon) \sim \epsilon^{-D-1}\Delta\epsilon\,,$$

where $n(\epsilon, \Delta\epsilon)$ is the number of gaps (empty regions) of length between ϵ and $\epsilon - \Delta\epsilon$. This can be seen by noting that the total length covered with intervals ϵ is

$$L(\epsilon) \sim \epsilon^{-D+1}\,.$$

The increase of the uncovered part when ϵ is decreased to $\epsilon - \Delta\epsilon$ is

$$\frac{dL(\epsilon)}{d\epsilon}\Delta\epsilon \sim \epsilon^{-D}\Delta\epsilon\,.$$

This comes from the gaps of length between ϵ and $\epsilon - \Delta\epsilon$, because they will not be covered any more. Thus,

$$\epsilon^{-D}\Delta\epsilon \sim n(\epsilon, \Delta\epsilon)\epsilon$$

which is equivalent to the statement.

2.3. TYPES OF FRACTALS

One of the most fascinating aspects of fractals is the extremely rich variety of possible realizations of such geometrical objects. This fact raises the question of classification, and in the book of Mandelbrot (1982) and in the following publications many kinds of fractal structures have been described. Below we shall discuss a few important classes with some emphasis on their relevance to growth phenomena.

2.3.1. Deterministic and random fractals

Since fluctuations are always present in physical processes, they never lead to structures with perfect symmetry. Instead, physical fractals are more or less random with no high level of symmetry. Yet it is of interest to investigate simple, idealized fractal constructions, because the main features of fractal geometry can be effectively demonstrated using them as examples.

Figure 2.1a shows a typical fractal generated by a *deterministic rule.* In general when constructing such *growing mathematical fractals* one starts with an object (particle) of linear size a. In the first step ($k = 1$) $n - 1$ copies of this seed object are added to the original one so that the linear size of the resulting configuration becomes ra, where $r > 1$. Next ($k = 2$) each particle in the first configuration is substituted by the whole $k = 1$ configuration itself. In this way the number of particles and the linear size of the structure become n^2 and $r^2 a$, respectively. In the kth step the same rule is applied: each particle is replaced by the $k = 1$ configuration. Similarly, the kth configuration is made of n units being identical to the $k - 1$th cluster. In other words, when making the $k + 1$th step the n subunits of the kth configuration corresponding to the structure obtained in the $k - 1$th step are replaced by the structure generated in the kth stage of this iteration procedure. The $k \to \infty$ limit results in a deterministic mathematical fractal. At the end of this section a few examples are given for the types of fractals described above and in the following.

The fractal dimension for such objects can readily be obtained from (2.3). Taking into account that for $L = r^k a$ the volume (the number of particles) of the structure is $N(L) = n^k$, we get

$$D = \frac{\ln n}{\ln r} \tag{2.10}$$

which is an exact expression for D.

The construction of deterministic fractals generated by subsequent *divisions* of a starting object proceeds in an analogous manner. In the first step this object having a linear size $L_0 = 1$ is divided into n identical parts each of which is a reduced version of the original structure with the same factor $1/r$. During the next step n copies of the starting object reduced by a factor $(1/r)^2$ are arranged inside a part generated in the previous step. This is done in a way which exactly corresponds to the arrangement of the parts placed in the first step within the object. In this case each of the n^k objects obtained in the kth step is replaced by the $k = 1$ configuration reduced by a factor of $(1/r)^k$. It is obvious that the number of balls of radius $\epsilon = (1/r)^k$ needed to cover the structure grows with k as n^k which on the basis of (2.4) leads to the expression (2.10) for the fractal dimension.

As was mentioned earlier, the two methods of generating deterministic fractals are closely related. For every finite k the linear size of a growing fractal can be rescaled to the same value L_0 and the objects obtained in this way are the same as those generated by subsequent divisions using the appropriate rule. Thus, for the sake of simplicity, in the following we shall mainly use the language corresponding to the method of divisions.

The above described constructions lead to uniform fractals in the sense that we used the same reduction parameter for all of the copies made. An important generalization of these is represented by the case when the reduction factor $(1/r)$ is not identical for all of the n newly created copies within the parts generated in the previous step of the procedure. As before, the fractal is produced by dividing an original object into parts being reduced versions of it, but this time the factors $r_i > 1, (i = 1, 2, ..., n)$ can not be all identical. Such non-uniform fractals are obtained in the limit of repeating the iteration procedure infinitely many times (see the examples at the end of this section).

To determine the fractal dimension of a *non-uniform* mathematical fractal we first note that it can be divided into n parts each being a rescaled version of the complete fractal. Let $N_i(\epsilon)$ denote the number of balls of radius ϵ needed to cover the ith part. The number of balls needed to cover the whole fractal is

$$N(\epsilon) = \sum_{i=1}^{n} N_i(\epsilon)\,. \tag{2.11}$$

Since the fractal is self-similar,

$$N_i(\epsilon/r_i) = N(\epsilon) \tag{2.12}$$

expressing the fact that one needs the same number of balls of reduced radius ϵ/r_i to cover a smaller version of the fractal of size L_0/r_i, than one needs for covering the complete structure with balls of radius ϵ. Then substitution of (2.4) and (2.12) into (2.11) leads to

$$\sum_{i=1}^{n} \left(\frac{1}{r_i}\right)^D = 1 \tag{2.13}$$

which is an implicit equation for the fractal dimension of non-uniform fractals. For $r_i = r_1 = r_2 = ... = r_n$, (2.13) is equivalent to (2.10).

Although in the following chapters we shall concentrate on the study of *random fractals* growing in physical processes, here we first show that one can generate simple stochastic fractals in a way analogous to the above described constructions. To take an example let us consider the fractal shown in Fig. 2.1b. It is constructed by dividing the original square into 9 equal parts and deleting 4 of them selected randomly (i.e., keeping 5). In the following steps the same procedure is repeated with the remaining squares. Figure 2.2 shows the resulting structure after 3 iterations. Comparing the geometrical appearance of Fig. 2.1b and 2.2 we find that they are quite different, however, their fractal dimension is identical $D = \ln 5/\ln 3 = 1.465...$, because one needs the same number of balls to cover them.

Of course, this construction represents only a simple (perhaps the simplest) version of possible random fractals. For example, it is not only the position of the reduced parts which can be varied, but the number of such units and/or the reduction parameter can also fluctuate around their average

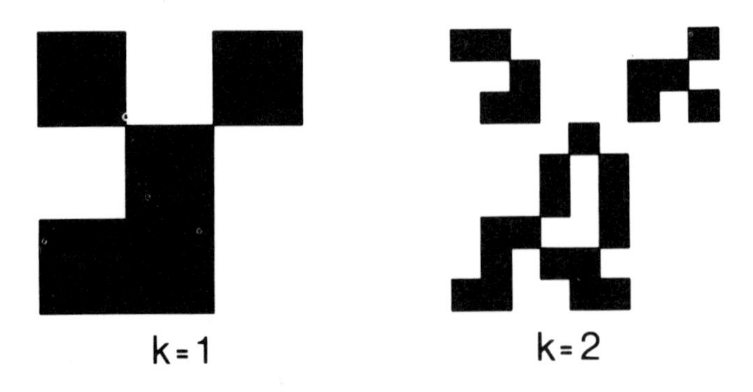

Figure 2.2. Construction of a stochastic fractal. Its fractal dimension is exactly the same as that of the structure shown in Fig. 2.1, despite the fact that they look quite different.

value. In general, for the fractal dimension of random fractals an explicit expression analogous to (2.10) does not exist and D has to be determined using various theoretical and numerical techniques which will be discussed in Chapter 4.

Self-similarity can be directly checked for a deterministic fractal constructed by iteration, but in the case of random structures one needs other methods to detect the fractal character of a given object. In fact, *random fractals are self-similar only in a statistical sense* (not exactly) and to describe them it is more appropriate to use the term scale invariance than self-similarity. Naturally, for demonstrating the presence of fractal scaling one can use the definition based on covering the given structure with balls of varying radii, however, this would be a rather troublesome procedure. It is more effective to calculate the so called *density-density or pair correlation function*

$$c(\mathbf{r}) = \frac{1}{V} \sum_{\mathbf{r}'} \rho(\mathbf{r} + \mathbf{r}')\, \rho(\mathbf{r}') \tag{2.14}$$

which is the expectation value of the event that two points separated by $\mathbf{r}$

belong to the structure. For growing fractals the volume of the object is $V = N$, where N is the number of particles in the cluster, and (2.14) gives the probability of finding a particle at the position $\mathbf{r} + \mathbf{r}'$, if there is one at $\mathbf{r}'$. In (2.14) ρ is the local density, i.e., $\rho(\mathbf{r}) = 1$ if the point $\mathbf{r}$ belongs to the object, otherwise it is equal to zero. Ordinary fractals are typically isotropic (the correlations are not dependent on the direction) which means that the density correlations depend only on the distance r so that $c(\mathbf{r}) = c(r)$.

Now we can use the pair correlation function introduced above as a criterion for fractal geometry. An object is non-trivially scale-invariant if its correlation function determined according to (2.14) is unchanged up to a constant under rescaling of lengths by an arbitrary factor b:

$$c(br) \simeq b^{-\alpha} c(r) \tag{2.15}$$

with α a non-integer number larger than zero and less than d. It can be shown that the only function which satisfies (2.15) is the power law dependence of $c(r)$ on r

$$c(r) \sim r^{-\alpha} \tag{2.16}$$

corresponding to an algebraic decay of the local density within a random fractal, since the pair correlation function is proportional to the density distribution around a given point. This fact can be used for expressing the fractal dimension through the exponent α. To show this for growing fractals, we calculate the number of particles $N(L)$ within a sphere of radius L from their density distribution

$$N(L) \sim \int_0^L c(r) d^d r \sim L^{d-\alpha}, \tag{2.17}$$

where the summation in (2.14) has been replaced by integration. Comparing (2.17) with (2.2) we arrive at the desired relation

$$D = d - \alpha \tag{2.18}$$

which is a result widely used for the determination of D from the density correlations within a random fractal.

EXAMPLES

Next we give a few characteristic examples for the types of fractals mentioned in this section to illustrate the basic ideas discussed above. Because of the great variety of possible constructions the list is far from being complete, and those readers who are interested in more examples are advised to consult the book of Mandelbrot (1982).

Example 2.1. One of the simplest and best known fractals is the so called triadic Cantor set which is a finite size fractal consisting of disconnected parts embedded into one-dimensional space ($d = 1$). Its construction based on the subsequent division of intervals generated on the unit interval $[0, 1]$ is demonstrated in Fig. 2.3. First $[0, 1]$ is replaced by two intervals of length $1/3$. Next this rule is applied to the two newly created intervals, and the procedure is repeated *ad infinitum.* As a result we obtain a deterministic fractal and to calculate its fractal dimension we can use Eq. (2.10). Obviously, for the present example $n = 2$ and the reduction factor is $1/r = 1/3$. Therefore (2.10) gives for the dimension of the triadic Cantor set $D = \ln 2/\ln 3 = 0.6309...$ which is an irrational number less than 1. In general, Cantor sets with various n and r can be constructed. For example, keeping $n = 2$ and changing r between 2 and ∞ any fractal dimension $0 \leq D \leq 1$ can be produced. On the other hand, various Cantor sets with the same fractal dimension can be constructed as well. The two sets $n = 2,\ r = 4$ and $n = 3,\ r = 9$ have the same fractal dimension $D = 1/2$, but different overall appearance.

The intervals or gaps between the points belonging to the fractal set correspond to the empty regions mentioned in Section 2.2. They are distributed according to a power law which is another typical property of fractals. In particular, for a Cantor set of dimension D the number of gaps longer

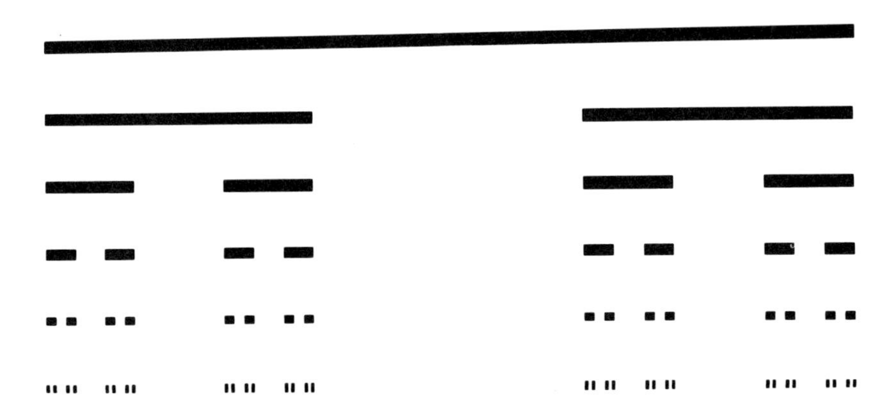

Figure 2.3. The triadic Cantor set shown in this figure is generated on the unit interval by replacing each of the intervals obtained at a given stage with two shorter ones.

than Δx_0 scales as $N(\Delta x > \Delta x_0) \sim \Delta x_0^{-D}$ (see rule g) in the previous section).

Example 2.2. One of the standard ways to construct a fractal surface is to replace the starting object with a single connected object of larger surface (made of reduced parts of the original one) and repeat this procedure using the reduced parts as originals. In two dimensions this method leads to a line (coastline) of infinite length with a fractal dimension larger than 1. Let us consider again the unit interval, and replace it with a curve consisting of 5 intervals of unit length as shown in Fig. 2.4. The fractal dimension is obtained from (2.10) and is equal to $D = \ln 5/\ln 3 \simeq 1.465$ which exactly coincides with D of the fractal shown in Fig. 2.1 and 2.2. In fact, the structure generated by this method is also analogous to that of Fig. 2.1.

Using a related procedure (Fig. 2.5) it is possible to define a single curve which can cover the unit square, i.e., it has a dimension equal to 2. For this so called Peano curve $D = \ln 9/\ln 3 = 2$, and in the limit of $k \to \infty$ it establishes a continuous correspondence between the straight line and the plane. The Peano curve is a very peculiar construction (it has infinitely fine details being arbitrarily close to each other, but does not have

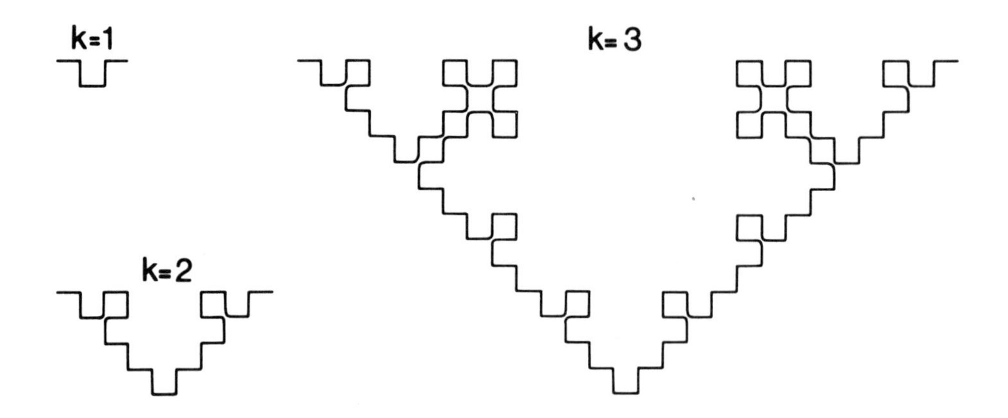

Figure 2.4. Construction of a growing fractal curve having the same fractal dimension as the objects shown in Figs. 2.1 and 2.2.

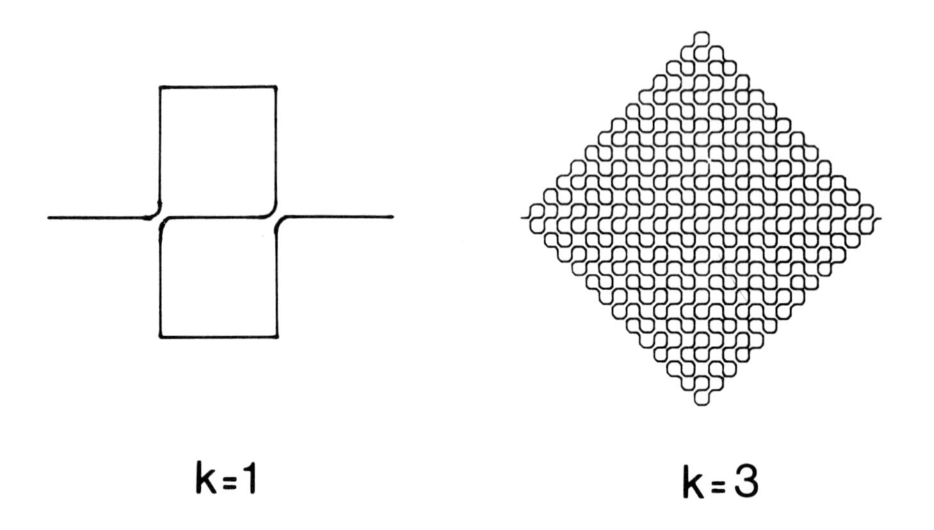

Figure 2.5. Application of Eq. (2.10) to the above displayed Peano curve gives $D = d = 2$ which means that this construction does not lead to a fractal according to the definition given in Section 2.2.

any intersections), however, it is not a fractal according to the definition given in Section 2.2. This is indicated by the absence of empty regions.

Example 2.3. The Sierpinski gasket shown in Fig. 2.6 is perhaps the most studied two-dimensional fractal structure since it can be regarded as a prototype of fractal lattices with an infinite hierarchy of loops. When constructing this fractal, three of the four triangles generated within the triangles obtained in the previous step are kept. Since the linear size of the triangles is halved in every iteration, the fractal dimension of the resulting object is $D = \ln 3/\ln 2 \simeq 1.585$.

Figure 2.6. The Sierpinski gasket shown in this figure has loops on all length scales.

Example 2.4. The iteration procedure described at the beginning of this section is not the only possibility to construct mathematical fractals. For example, the Julia sets are derived from the transformation

$$z' = f(z) = z^2 - \mu\,, \tag{2.19}$$

where z and μ are complex numbers. The set of z values (points in the plane) which is invariant under the transformation (2.19) can be called self-squared, and for a fixed value of μ this set is in most of the cases a fractal, sometimes with a very attractive appearance (Peitgen and Richter 1986). (More precisely, Julia sets do not contain the stable fixed points of (2.19), i.e, they represent those points which can be obtained by backward iteration of (2.19).) Figure 2.7 shows a few typical Julia sets.

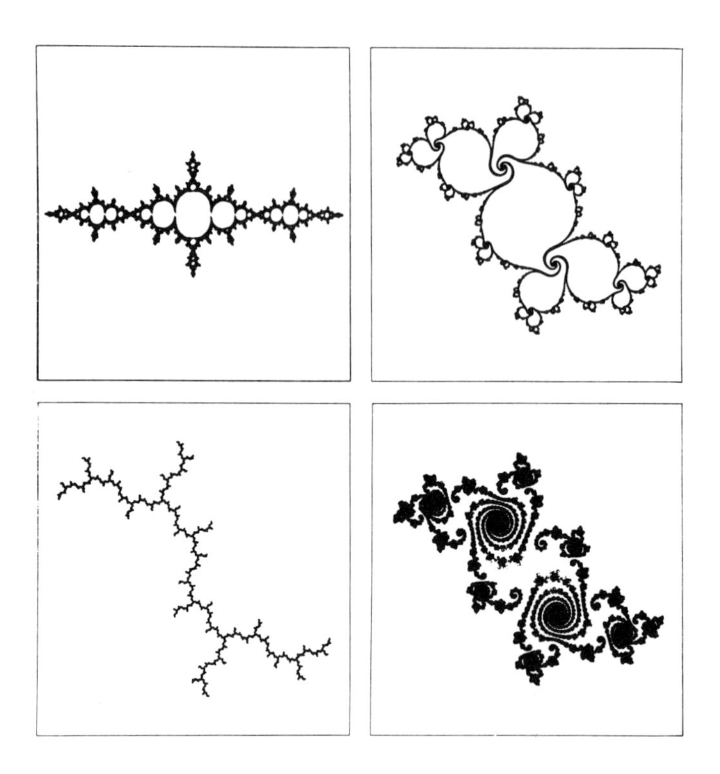

Figure 2.7. The rich variety of apparently self-similar Julia sets is well demonstrated by the above selected examples reproduced from Peitgen and Richter (1986).

Mandelbrot studied the convergence properties of the recursion

$$z_{k+1} = z_k^2 - \mu \tag{2.20}$$

corresponding to (2.19), as a function of μ. He found that the region of μ values for which the iterates of $z_0 = 0$ under (2.20) fail to converge to ∞ (the Mandelbrot set) is bounded by a fractal curve (Fig. 2.8). Moreover, marking the points in the μ plane with colours depending on the number of iterations k needed for $z_k > 2$ one obtains an extremely complex, beautiful picture with many self-similar structures in it (Peitgen and Richter 1986).

Figure 2.8. The region of μ values (Mandelbrot set) for which the iterates given by (2.20) remain finite for arbitrary k (Mandelbrot 1982).

Recursion relations are also used to construct fractal attractors characteristic of chaotic motion. As a simple example we briefly mention dynamical systems that period double on their way to chaos. At values $\lambda = \lambda_k$ the system gains a stable 2^k orbit. This series of period doublings accumulates at λ_∞, where the system follows a 2^∞ orbit. Such behaviour can be well represented by the following one-dimensional map

$$x' = 4\lambda x(1-x)\,, \tag{2.21}$$

with $\lambda_\infty \simeq 0.837005134$. Here x is real and (2.21) is a one-dimensional counterpart of (2.19) in the sense that starting with some initial x_0 the calculated x' values quickly converge to a fractal subset of the $[0,1]$ interval, i.e., this attractor is an invariant set corresponding to (2.21). However, the set invariant with regard to (2.19) is a repellor (not an attractor). It can be shown that the attractor forms approximately a non-uniform Cantor set (with $n = 2$ and $r_1 \neq r_2$). The fractal dimension of this set is $D \simeq 0.537$

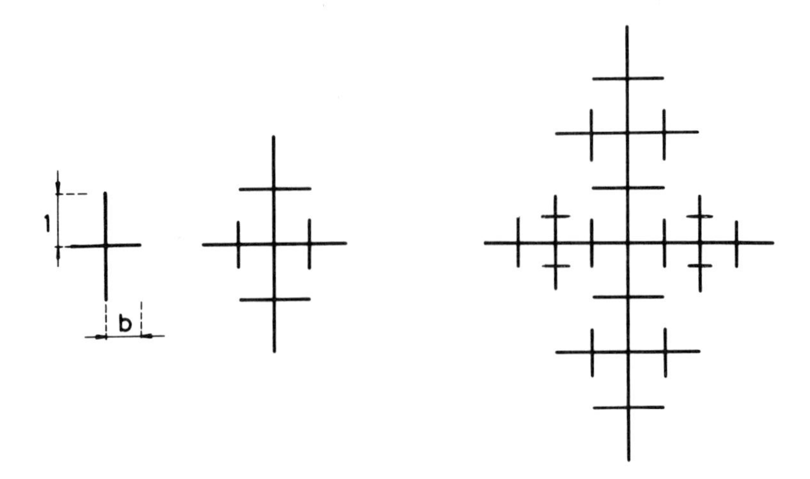

Figure 2.9. This non-uniform fractal grows by adding to the four principal tips of the $(n-1)$th configuration the structure itself without the lower main stem. This addition has to be done by applying appropriate rotation and shrinking to keep the ratio of the corresponding branches equal to $b < 1$.

(Hentschel and Procaccia 1983).

Example 2.5. The construction presented in Fig. 2.9 leads to a fractal which is both growing and non-uniform. To grow this fractal one adds to the four main tips of the already existing configuration a part of it in the following manner. The part to be added is the configuration itself minus one of the the main branches growing out from the centre vertically. Moreover, this part has to be rotated and reduced in an appropriate way (see Fig. 2.9) before attachment (reduction by a factor b is needed when the horizontal branches are updated).

To obtain the fractal dimension of this tree-like object we first note that reducing the configuration generated in the kth step by a factor 2^k one obtains a structure that would have been generated by the division technique with $1/r_1 = 1/r_2 = 1/2$ and $1/r_3 = 1/r_4 = b/2$ (i.e., by replacing the intervals in each step with four new ones, two of which are half as long, and two with size shrunk by a factor $b/2$). Now we use (2.13) and calculate D from the equation

$$2\left(\frac{1}{2}\right)^D + 2\left(\frac{b}{2}\right)^D = 1 \tag{2.22}$$

which can be solved numerically for any given b. For some of the b values the implicit equation (2.22) can be inverted, e.g., if $b = 1/2$, $D = 1 - \ln(\sqrt{3} - 1)/\ln 2 \simeq 1.45$.

Example 2.6. The random motion of a particle represents a particularly simple example of stochastic processes leading to growing fractal structures. A widely studied case is when the particle undergoes a random walk (Brownian or diffusional motion) making steps of length distributed according to a Gaussian in randomly selected directions. Such processes can be described in terms of the mean squared distance $R^2 = \langle R^2(t) \rangle$ made by the particles during a given time interval t. For random walks $R^2 \sim t$ independently of d (see Chapter 5) which means that the Brownian trajectory is a random fractal in spaces with $d > 2$. Indeed, measuring the volume of the trajectory by the total number of places visited by the particle making t steps, ($N(R) \sim t$), the above expression is equivalent to

$$N(R) \sim R^2 \tag{2.23}$$

and comparing (2.23) with (2.2) we conclude that for random walks $D = 2 < d$ if $d > 2$. In this case, rather unusually, the fractal dimension is an integer number. However, the fact that it is definitely smaller than the embedding dimension indicates that the object must be non-trivially scale invariant.

Brownian motion can be used to demonstrate random fractal curves in two dimensions as well. Consider the trajectory of a randomly walking particle on the plane. It separates the plane into two parts: an exterior which can be reached from a distant point without intersecting the trajectory and an interior (Fig. 2.10). The boundary of the interior part (the Brown hull) is a very complex curve resembling the coastlines mentioned earlier. According to the numerical results (Mandelbrot 1982) it is self-similar and has a dimension $D \simeq 4/3$ which is the same as that of self-avoiding random walks (see Section 5.4.2).

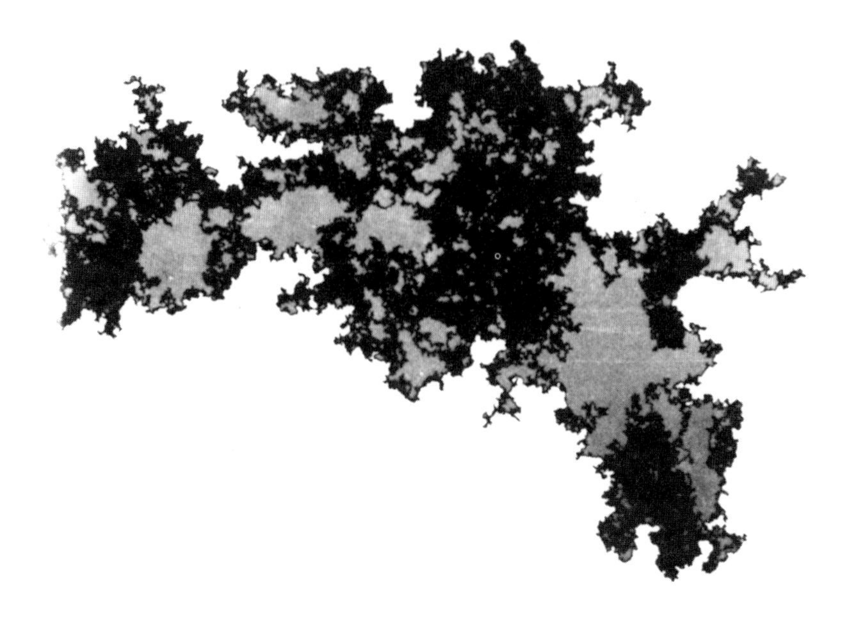

Figure 2.10. Example for a random coastline. This Brown hull represents the external perimeter of the trajectory of a looping random walk on the plane which is indicated by a darker line (Mandelbrot 1982).

Example 2.7. A straightforward generalization of the Brownian motion is called Levy flight which is another example of growing random fractals. As before, it is a sequence of jumps in random directions, but with a hyperbolic distribution of the jump distances. More precisely, all directions are chosen with the same probability, and the probability of making a jump longer than Δx_0 is $Pr(\Delta x > \Delta x_0) = \Delta x_0^{-D_L}$, except that $Pr(\Delta x > \Delta x_0) = 1$ when $\Delta x_0 < 1$. All jumps are independent, and the resulting trajectory has a fractal dimension $D = D_L$.

Figure 2.11 shows the positions of stopovers of a long Levy flight taking place on a plane. Most of the individual sites can not be seen for the given resolution. The displayed pattern is made of disconnected, but clustered positions, and was generated using $D_L = 1.26$ in the expression for the distribution of jump distances. The stochastic self-similarity of the configuration is manifested by the fact that enlarging a small part of the structure by about 100 times, the resulting configuration has the same general appearance

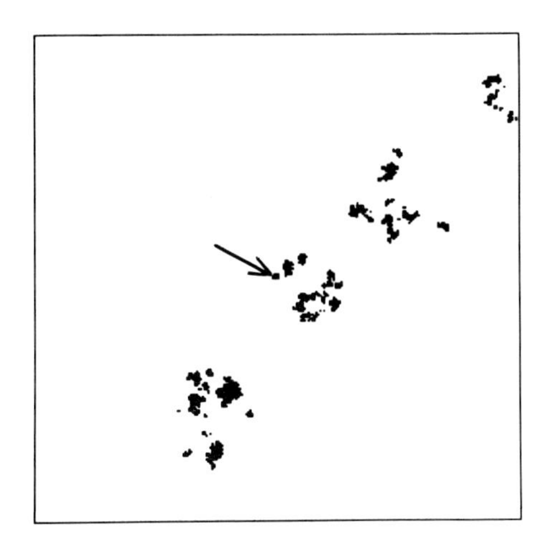

Figure 2.11. The shapes of disconnected clusters corresponding to the stopovers of a long Levy flight on the plane. The stochastic self-similarity of the clusters is demonstrated by blowing up small parts of the configurations. The picture on the right side is an approximately 100 times enlarged image of a tiny region in the left configuration indicated by the arrow (Mandelbrot 1982).

as the original one.

2.3.2. Self-affine fractals

Self-similarity of an object is equivalent to the invariance of its geometrical properties under isotropic rescaling of lengths. In many physically relevant cases the structure of the objects is such that it is invariant under dilation transformation only if the lengths are rescaled by direction dependent factors. These anisotropic fractals are called self-affine (Mandelbrot 1982, 1985 and 1986).

Single-valued, nowhere-differentiable functions represent a simple and typical form in which self-affine fractals appear. If such a function $F(x)$ has the property

$$F(x) \simeq b^{-H} F(bx) \tag{2.24}$$

it is self-affine, where $H > 0$ is some exponent. (2.24) expresses the fact that the function is invariant under the following rescaling: shrinking along the x axis by a factor $1/b$, followed by rescaling of values of the function (measured in a direction perpendicular to the direction in which the argumentum is changed) by a different factor equal to b^{-H}.

In other words, by shrinking the function using the appropriate direction-dependent factors, it is rescaled onto itself. For some deterministic self-affine functions this can be done exactly, while for random functions the above considerations are valid in a stochastic sense (expressed by using the sign $\simeq$).

Of course, there are self-affine fractals different from single-valued functions and at the end of this section, among others, a couple of examples will be given for such structures as well. However, the most typical physical process producing self-affine structures is the marginally stable growth of interfaces (see Chapter 7) leading to surfaces which can be well approximated by single-valued functions.

As we shall see self-affine fractals do not have a unique fractal dimension of the kind defined in Section 2.2. Instead, their global behaviour is characterized by an integer dimension smaller than the embedding dimension, while the local properties can be described using a local fractal dimension. To show this we shall concentrate on functions of a single scalar variable. Such a function is, for example, the plot of the distances measured from the origin, $X(t)$, of a Brownian particle diffusing in one dimension as a function of time t. It is obvious that a fractional Brown plot with $\langle X_H^2(t)\rangle \sim t^{2H}$ stochastically satisfies (2.24) by $F(t) = X(t)$.

Let us first construct a deterministic self-affine model, in order to have an object we can treat exactly (Mandelbrot 1985). The plot corresponding to this model, $M_H(t)$, is defined as a regular version of the above mentioned Brown plot. We assume that H is of the form

$$H = \ln b_2 / \ln b_1 \,, \tag{2.25}$$

where b_1 and b_2 are integers and $b_1 - b_2 > 0$. The idea is that the function $X_H(t)$, whose increments are Gaussian over all δt with a standard deviation $(\delta t)^H$, is replaced by a function $M_H(t)$ whose increments over suitable δt–s are binomial with the same mean equal to 0 and the same standard deviation. This means that we require

$$M_H(pb_1^{-k}) - M_H[(p+1)b_1^{-k}] = \pm(b_2)^{-k} = \pm(\delta t)^H \tag{2.26}$$

for all k and p, where (2.25) was used to get the last equality for $\delta t = b_1^{-k}$.

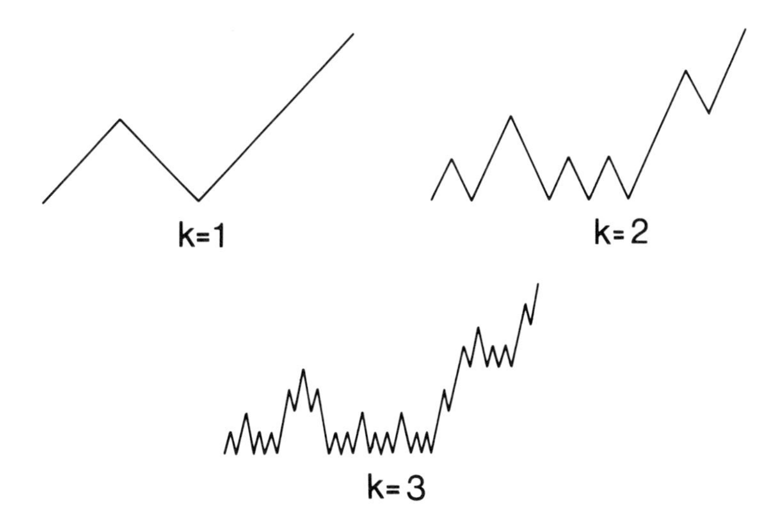

Figure 2.12. Deterministic model for a self-affine function defined on the unit interval. The single-valued character of the function is preserved by an appropriate distortion of the z-shaped generator ($k = 1$) of the structure.

An actual construction of such a bounded self-affine function on the unit interval is demonstrated in Fig. 2.12. In this example $b_1 = 4$ and $b_2 = 2$. The object is generated by a recursive procedure by replacing the intervals of the previous configuration with the generator having the form of an asymmetric letter z made of four intervals. However, the replacement this time should be done in a manner different from the earlier practice. Here every interval is regarded as a diagonal of a rectangle becoming increasingly elongated during the iteration. The basis of the rectangle is divided into four equal parts and the z-shaped generator replaces the diagonal in such a way that its turnovers are always at analogous positions (at the first quarter and the middle of the basis). In the kth stage we obtain $M_H^{(k)}(t)$, and the function becomes self-affine in the $k \to \infty$ limit.

Now we can apply an exact argument to determine the structure's local dimension using definition (2.5) which with $l = 1/b$ reads as $D = \lim_{b\to\infty} \ln N(b)/\ln b$ where $N(b)$ is the number of discs or boxes of linear size $1/b$ needed to cover the object. Let us cover $M_H(t)$ from $t = 0$ to $t = 1$ with boxes of size $1/b = b_1^{-k}$. For this purpose one needs $b_1^k = b$ bins or columns of boxes. The height of these columns (the amount by which $M_H(t)$ changes in an interval of length b_1^{-k}) is approximately b_2^{-k}, because of (2.26). Consequently, to cover $M_H(t)$ in a bin one needs (b_2^{-k}/b_1^{-k}) boxes. The number of boxes needed along the whole unit interval is

$$N(b) \sim b_1^k \left(\frac{b_2^{-k}}{b_1^{-k}}\right) = (b_1^2 b_2^{-1})^k\,. \tag{2.27}$$

From (2.25) we have $b_2 = b_1^H$, hence $N(b) \sim b^{2-H}$ which according to (2.5) leads to

$$D_B = 2 - H, \tag{2.28}$$

where D_B denotes the *local or box dimension* of the plot of $M_H(t)$.

To show the validity of the statement that the global dimension of an unbounded self-affine function is equal to 1 we slightly modify the construction of the previously introduced deterministic example. This new version has to be defined for $t \gg 1$, but have the same scaling properties on the unit interval as $M_H(t)$. We construct the kth stage of the new function $\tilde{M}_H^{(k)}(t)$ from $M_H^{(k)}(t)$ by expanding it along the t axis and multiplying its values by an appropriate factor

$$\tilde{M}_H^{(k)}(t) \sim \left(\frac{b_1}{b_2}\right)^{kH} M_H^{(k)}\left[\left(\frac{b_1}{b_2}\right)^{-k} t\right].$$

On the unit interval $\tilde{M}_H(t)$ behaves as $M_H(t)$ because of the self-affine property (2.24), therefore, it has the same local fractal dimension $2-H$ (2.28). The values of this new function are now defined up to $t \to \infty$ as $k \to \infty$, moreover, its largest value in this limit diverges as t^H with $H < 1$ (slower than t). This leads to the conclusion that the *global dimension*, D_G, determined for the structure in the limit of large t using boxes of size $1/b \gg 1$ is equal to 1. In this case "observing the function from a large distance" (measuring with large boxes) it looks like a ragged line nearly merged into the abscissa.

We have seen that by rescaling a bounded self-affine function, or in other words, changing the units in which the distances are measured, an extra dimension, called global dimension could be found. This raises the question of choosing the appropriate units to measure F and t. To see both the local and global behaviours one should make a choice for the unit of both F and t. Then a quantity t_c can be defined through

$$|F(t+t_c) - F(t)| \sim |t_c|. \qquad (2.29)$$

This t_c can be called the *crossover scale* for the given process. The most essential fact about t_c is that it depends on the units which happen to be selected for F and t, therefore, the position of the crossover is in general not intrinsic.

An *important consequence* of the above statement is that for self-affine structures with a lower or upper length scale, changing the units may lead to losing the possibility of detecting the local fractal or the global trivial scaling. Indeed, if the units we chose are such that t_c becomes the same order as the lower cutoff length, a local fractal dimension can not be observed. This is the case, e.g., for the record of a one-dimensional random walk on a lattice if the same unit is used for the increments of F and the time t, since then the lower cutoff and the crossover scale coincide. Similar arguments are valid for the detectability of a trivial global dimension.

Before the discussion of additional examples it should be noted that the most typical self-affine structures (and the ones we consider) are diagonally self-affine. This means that they are invariant under a transformation whose invariant sets include any collection of straight lines parallel to the coordinate axes. A diagonal affine transformation is specified by giving a fixed point of coordinates φ_m $(0 < m < d-1)$ and an array of reduction ratios r_m, and considering the map

$$x_m \rightarrow \varphi_m + r_m(x_m - \varphi_m)\,. \tag{2.30}$$

The ratios r_m must not be equal, because then the transformation would be isotropic. The most general kind of self-affinity would imply invariance with regard to a transformation whose matrix has off diagonal elements as well.

EXAMPLES

Example 2.8. It is possible to treat the random function $X_H(t)$ in a manner analogous to the approach which was used for the description of the deterministic function $M_H(t)$, however, the arguments in this case involve heuristic approximations. We recall that $X_H(t)$ denotes the distance of a particle from the origin, randomly walking on a straight line, as a function of the time t. The distances are measured from the starting point, the direction of the jumps is chosen randomly (but not necessarily independently), and it is assumed that the mean squared distance scales with time according to

$$\langle X_H^2(t)\rangle \sim t^{2H}\,. \tag{2.31}$$

A random walk of this kind with $0 < H \neq 1/2 < 1$ is called fractional Brownian motion in one dimension. It is well known that for the ordinary Brownian motion, when the jumps are independent and their distances have a Gaussian distribution, $H = 1/2$ (i.e., $D_B = 2-H = 1.5$) and $X_{1/2}(t)$ satisfies (2.24) stochastically. Similarly, $X_H(t)$ and $b^{-H}X_H(bt)$ can be shown to be identical in distribution.

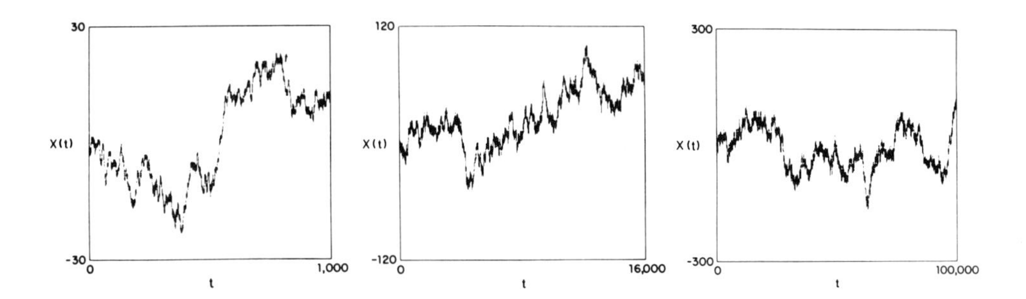

Figure 2.13. These plots of $X_{1/2}(t)$ were obtained by rescaling of Brownian plots of various lengths. For each of the three plots the vertical scale is proportional to the square root of the horizontal scale (Meakin 1986).

Figure 2.13 visualizes the statistical self-affinity of the Brownian plot with $H = 1/2$. Parts of originally different horizontal extension are scaled onto the same interval with a simultaneous rescaling of the heights by a factor which is equal to the square root of the factor used to shrink the horizontal size. The plots obtained are very similar as far as their appearance is concerned.

Since $0 < H < 1$, it follows from (2.31) that the global dimension corresponding to the behaviour for $t_c \gg 1$ is 1, because $X_H(t)/t \to 0$ as $t \to \infty$. The local dimension can be obtained from considerations similar to those used for the deterministic case. During a time interval δt, $|max\ [X_H(t)] - min\ [X_H(t)]|$ is of the order of $(\delta t)^H$. Covering the part of $X_H(t)$ on the interval δt by squares of side δt requires on the order of $(\delta t)^H/\delta t = \delta t^{H-1}$ squares. Therefore, covering $X_H(t)$ on the interval $[0,1]$ requires

$$N(\delta t) \sim \frac{\delta t^{H-1}}{\delta t} = \delta t^{-(2-H)} \tag{2.32}$$

which according to the definition (2.5) leads to $D_B = 2 - H$ just as for the deterministic construction (see Eq. 2.28).

The points at which $X_H(t) = 0$ form the zero set of the fractional Brownian motion. It is a random Cantor set of fractal dimension $D = 1 - H$ (Mandelbrot 1982). A fractional Brownian motion with $0 < H < 1/2$ is antipersistent which means that the walker tends to turn back to the point it came from. Alternatively, in the case $H > 1/2$ the increments have positive correlation with the direction of the previous jump. To see this we set $X_H(0) = 0$ and define the past increment as $-X_H(-t)$ and the future increment as $X_H(t)$. Then

$$\begin{aligned}\langle -X_H(-t)X_H(t)\rangle &= 2^{-1}\{\langle [X_H(t) - X_H(-t)]^2\rangle - 2\langle [X_H(t)]^2\rangle\} = \\ &= 2^{-1}(2t)^{2H} - t^{2H} .\end{aligned} \tag{2.33}$$

Dividing by $\langle X_H^2(t)\rangle = t^{2H}$, one obtains the correlation of increments which is independent of t; it is equal to $2^{2H-1}-1$ vanishing, as expected, for $H = 1/2$. Calculating the Fourier spectrum of a fractional Brown function one finds that the coefficients of the series, $A(f)$, are independent Gaussian random variables and their absolute value scales with the frequency f according to a power law

$$|A(f)| \sim f^{-H-\frac{1}{2}} . \tag{2.34}$$

Finally, as an example for random self-affine functions defined in higher dimensional spaces we mention the Brownian relief or Brown plane-to-line function shown in Fig. 2.14. Its vertical cross sections represent plots of one dimensional random walks $(X_H(t))$. It is not a trivial task to define and construct a fractional Brownian surface with a given H. There is, however, a relatively simple procedure generating a surface in $d = 3$ with $H = 1/2$. A horizontal plateau is broken along a straight line chosen at random and one of them shifted vertically. The difference between the levels of the two sides of the resulting precipice is also chosen randomly from a set of lengths distributed according to a Gaussian. Then we repeat the same and follow the kth stage by dividing all heights by $\sqrt{k}$. Generating surfaces with an arbritrary H requires other methods, e.g., involving the construction of a set of random Fourier coefficients with a distribution $f^{-H-3/2}$ and reconstructing the surface from its components or random addition algorithms (Voss 1985).

Figure 2.14. A Brownian surface having a local fractal dimension close to 2.4 (Mandelbrot 1982).

Example 2.9 In 1872 Weierstrass constructed a function which is continuous everywhere, but differentiable nowhere. Mandelbrot proposed a simple extension of this function which turned out to have no characteristic length scale. Let us consider the Fourier series

$$C(t) = \sum_{n=-\infty}^{\infty} \frac{1 - \cos(b^n t)}{b^{(2-D)n}}, \tag{2.35}$$

which is the real part of the more general Weierstrass-Mandelbrot function (Mandelbrot 1982). In the range of parameter values

$$1 < D < 2, \qquad b > 1 \tag{2.36}$$

$C(t)$ is continuous but the series defining $dC(t)/dt$ diverges everywhere. The frequencies b^n form a "Weierstrass spectrum", spanning the range from zero to infinity in a geometrical progression; this is the sense in which $C(t)$ possesses no scale. The self-affinity of $C(t)$ for $b > 0$ can be easily shown by a formal replacement of n by $n+1$ in (2.35) leading to the scaling relation $C(t) = b^{-(2-D)}C(bt)$ which is equivalent to the definition (2.24) of self-affine functions. This means that the graph of $C(t)$ on the interval $t_0 \leq t \leq bt_0$ can be obtained by magnifying the graph in the range $t_0/b \leq t \leq t_0$ with factors b and b^{2-D} in horizontal and vertical directions, respectively.

It can be argued and supported by numerical investigations that the local fractal dimension of $C(t)$ is equal to the parameter D. Consequently, for $D = 1.5$ it may also be regarded as a deterministic model for Brownian motion.

Example 2.10. So far we have discussed single-valued self-affine functions, because they seem to be more relevant from the point of view of applications than other possible self-affine structures. The two examples given below represent other types of self-affine objects, constructed in a spirit related to that used to generate some of the self-similar fractals.

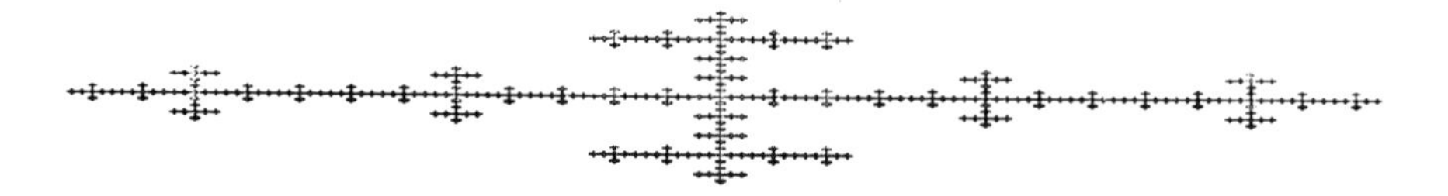

Figure 2.15. This growing self-affine fractal is generated by a procedure analogous to that used for constructing Fig. 2.1a, except that in the present case the seed configuration is not isotropic (Jullien and Botet 1987).

Figure 2.15 shows a growing structure which is a generalization of the fractal displayed in Fig. 2.1a. The rules of construction are analogous, but in the present case the seed configuration is anisotropic (Jullien and Botet 1987). As before, the self-affine structure is produced in the $k \to \infty$ limit, and the kth configuration is obtained by replacing the 7 subunits of the $k-1$th configuration with the whole structure generated in the $k-1$th step. Obviously, the global dimension of the resulting object will be equal to 1, since the width of the structure grows with k as 3^k, while its length as 5^k. This self-affine fractal has a lower cutoff length which is the size of the particles it is made of, therefore, *it has no local fractal dimension.*

The last example of this section is constructed by dividing the unit square into anisotropic subunits which serve as seeds for further divisions. Figure 2.16 demonstrates the actual procedure analogous to that used for generating the fractal shown in Fig. 2.1b. For this example the scales are chosen in such a way that the crossover scale is the same as the side of the unit square, and because of this the trivial global scaling is not manifested (although it could be seen by applying the construction "backward", i.e., growing the square). On the other hand, the structure has a local fractal dimension which can be calculated, for example, by covering the structure with squares of size corresponding to the shorter side of the elongated rectangles generated at each step of the construction. One can also use rule d) of Section 2.2. Let us assume that in the kth step the sides of the rectangles are $l_1^{(k)} = l_1^k$ and $l_2^{(k)} = l_2^k$, where $l_1 < l_2$. Since the fractal shown in Fig. 2.16 can be generated by multiplying two Cantor sets of dimensions $D_1 = \ln 2/\ln(1/l_1)$ and $D_2 = \ln 2/\ln(1/l_2)$, respectively, we obtain for the local fractal dimension of the resulting structure $D_L = \ln 2/\ln(1/l_1) + \ln 2/\ln(1/l_2)$.

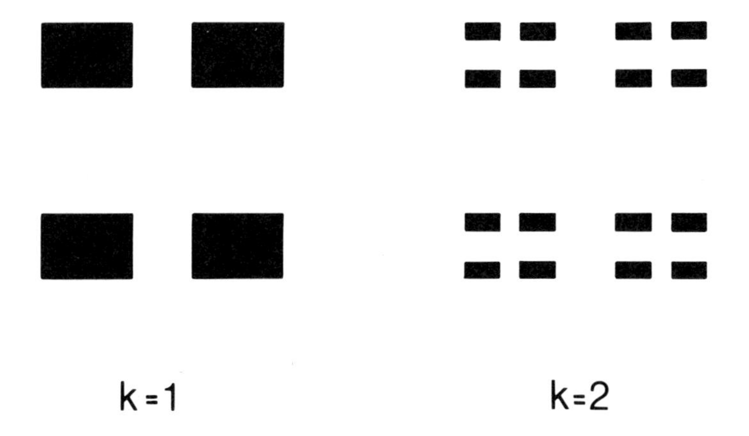

Figure 2.16. Generating a disconnected self-affine fractal embedded into two dimensions using elongated rectangles instead of squares during its construction.

2.3.3. Fat fractals

The most important feature of structures discussed in the previous sections was that they had a fractal dimension strictly smaller than the embedding dimension d. There are, however, structures for which $D = d$, but still exhibit a fractal behaviour (Mandelbrot 1982) in the following sense. When one calculates the volume $V(l)$ of such fat fractals using balls of decreasing size l, it *converges to a finite value algebraically* with a noninteger exponent. This is in contrast to ordinary or thin fractals, where $V(l) \to 0$ if $l \to 0$.

In general, the resolution dependent volume of an object can be written in the form

$$V(l) = V(0) + f(l)\,, \tag{2.37}$$

where $V(0) = V_0$ is the volume in the limit $l \to 0$. For thin (ordinary) fractals $V(0)$=0, and $f(l) \sim l^{d-D}$ with $D < d$. For fat fractals $V(0) > 0$, but $f(l)$ — as in the case of the thin ones — follows a power law with an exponent which can be regarded as a quantity characterizing the scaling properties of the structure. This fact can be expressed in the form (Farmer 1986)

$$V(l) \simeq V(0) + Al^{\beta} \tag{2.38}$$

where A is a constant and β is an exponent quantifying fractal properties. β can be calculated from

$$\beta = \lim_{l \to 0} \frac{\ln[N(l)l^d] - V(0)}{\ln l}\,, \tag{2.39}$$

where $N(l)$ is the number of d-dimensional balls needed to cover the structure. By definition $\beta > 0$, and it is equal to ∞ for non-fractal sets.

It is important to note that fat fractals are in general not self-similar objects. They have more in common with the Peano curve, since these structures are typically made of parts with dimension smaller than the embedding dimension d, while the whole object has a finite measure in d (lines with positive area, surfaces with positive volume, etc.). However, fat fractals are more inhomogeneous than the Peano-type objects, since for the latter $V(l)$ converges exponentially to its limiting value.

The example to be discussed below is constructed by generalizing the procedure leading to Cantor sets. However, there are many physical systems in which fat fractals are expected to occur. It has been shown that fat fractals can be associated with chaotic parameter values beyond the period-doubling transition to chaos, chaotic orbits of Hamiltonian systems or ballistic aggregation clusters. In addition, such biological objects as bronchia in the lung or coral-colonies are most likely to have the structure of fat fractals.

EXAMPLES

Example 2.11. As an illustration of fat fractals, consider the following modified Cantor set. In the original version (Example 2.1) first the central third of an interval is deleted, then the central third of the remaining intervals and so on *ad infinitum*. To "fatten" this thin fractal delete instead the central $\frac{1}{3}$, then $\frac{1}{9}$, then $\frac{1}{27}$, etc., of each remaining interval. The resulting set is topologically equivalent to the classical Cantor set, but the holes decrease in size sufficiently fast so that the limiting set has nonzero Lebesque measure and a dimension equal to 1.

The exponent β can be easily computed for the more general case of cutting out intervals of length $h_k(c) = c^k$ at stage k, where $0 < c < 1$ is a parameter (Umberger *et al* 1986). For a given c the total length (Lebesque measure) of the remaining set is larger than zero and is given by $0 < L_\infty = \lim_{n\to\infty} L_n < 1$, where

$$L_n = N_n \epsilon_n = 2^n \epsilon_n \,. \tag{2.40}$$

In the above expression the length of the covering intervals is chosen to be equal to the length of a single segment after the nth iteration is completed

$$\epsilon_n = \frac{1}{2^n} \prod_{k=1}^{n} (1 - c^k) . \tag{2.41}$$

To calculate β for this set we use (2.39)

$$\begin{aligned} \beta &= \lim_{n\to\infty} \frac{\ln(L_n - L_\infty)}{\ln \epsilon_n} \\ &= \lim_{n\to\infty} \frac{\ln[1 - \prod\limits_{k=n+1}^{\infty} (1 - c^k)]}{n \ln(1/2)} , \end{aligned} \tag{2.42}$$

where the terms $\prod_k^n (1-c^k) = o(n)$ are not written out. The above limit can be evaluated by using the identity $\ln\left[\prod(1-c^k)\right] = \sum \ln(1-c^k)$ to give

$$\beta = \frac{\ln(1/c)}{\ln 2} . \tag{2.43}$$

This example can be extended to arbitrary embedding dimension (Mandelbrot 1982). Suppose that we cut out a piece of volume v_1 from the unit hypercube of dimension d in such a way that the resulting structure is made of the 2^d hypercubes remaining at the corners of the starting object. Next, from each of these cubes a similar piece of relative volume v_2 is cut out, and this process is repeated infinitely many times. The total volume remaining after the kth iteration is

$$V_k = (1 - v_1)(1 - v_2)...(1 - v_k) = \prod_0^k (1 - v_k) . \tag{2.44}$$

V_k decreases as $k \to \infty$ to a limiting value V. For v_k fixed one has $V = 0$, however, if $\sum_0^\infty v_k < \infty$, the limiting volume is positive, $\prod_0^\infty (1 - v_k) > 0$. Since these objects are inhomogeneous in a rather specific way, measuring their volume with balls of size l one finds that it converges to $V(0) = V$ according to (2.38).

Example 2.12. To construct a fat fractal having a structure more typical for growth phenomena than Cantor sets, one can modify the method discussed in Example 2.5. In its original form the tree-like object shown in

Fig. 2.9 is a thin fractal with a fractal dimension depending on b, which is the length ratio of the first horizontal and vertical branches. It is obvious from the construction as well as from (2.22) that $b = 1$ results in a two-dimensional, "homogeneously fat" object for which $f(l)$ converges to zero exponentially fast. In order to generate an inhomogeneous fat fractal one should select a sequence of b_k values such that $\sum_0^\infty (1 - b_k) < \infty$.

Chapter 3

FRACTAL MEASURES

In the previous chapter such complex geometrical structures were discussed which could be interpreted in terms of a single fractal dimension. The present chapter is mainly concerned with the development of a formalism for the description of the situation when a singular distribution is defined on a fractal. As we shall see, the structure of fractals plays an essential role in the physical processes they are involved in and, as a result, one needs infinitely many dimension-type exponents to characterize these distributions as was first recognized by Mandelbrot (1974). This idea was later developed further by others, including (Hentschel and Procaccia 1983, Benzi *et al* 1984, Frisch and Parisi 1984, Halsey *et al* 1986).

It is typical for a large class of physical phenomena that the behaviour of a system is determined by the spatial distribution of a scalar quantity, e.g., concentration, electric potential, probability, etc. For simpler geometries this distribution function and its derivatives are relatively smooth, and they usually contain only a few (or none) singularities, where the word singular corresponds to a local power law behaviour of the function. (In other words, we call a function singular in the region surrounding point $\vec{x}$ if its local integral diverges or vanishes with a non-integer exponent when the integration size goes to zero). In the case of fractals the situation is quite different: a physical process involving a fractal may lead to a spatial distribution of the relevant quantities which possesses infinitely many singularities.

As an example, consider an isolated, charged object. If this object has sharp tips, the electric field around these tips becomes very large in accord with the behaviour of the solution of the Laplace equation for the potential. In the case of charging the branching fractals produced in the $k \to \infty$ limit of constructions shown in Fig. 2.1 or 2.5 one has infinite number of tips and corresponding singularities of the electric field. Moreover, tips being at different positions, in general have different local environments (configuration of the object in the region surrounding the given tip) which affect the strength of singularity associated with that position. There are further examples, some of which will be discussed later. At this point we only mention that additional physical phenomena leading to distributions with infinite number of singularities include, among many others, electric transport in percolation networks, viscous fingering, diffusion-limited aggregation, turbulence, chaotic motion, etc.

The interest in the properties of fractal measures has grown considerably in the last couple of years. A few reviews of the related results have been published very recently (Meakin 1987, Paladin and Vulpiani 1988, Tél 1988). When introducing certain concepts in the present chapter we follow the introductory paper by Tél (1988).

3.1. MULTIFRACTALITY

The above discussed time independent distributions defined on a fractal substrate are called *fractal measures*. In general, a fractal measure possesses an infinite number of singularities of infinitely many types. The term *"multifractality"* expresses the fact that points corresponding to a given type of singularity typically form a fractal subset whose dimension depends on the type of singularity. It is perhaps difficult to imagine how such an extremely complex distribution can appear as a result of simple physical processes. Let us, therefore, use a deterministic construction to demonstrate the mechanisms by which fractal measures can be generated.

We shall assume that the measure denoted by $\mu(\vec{x})$ is normalized so that its total amount on the fractal is equal to 1. Then, the fractal measure can be regarded as a probability distribution. Consider now a *deterministic recursive (multiplicative) process* generating a non-uniform fractal with varying weights or probabilities attributed to each rescaled part in a manner demonstrated in Fig. 3.1. In the first step the starting object (square) is replaced by its three smaller copies with corresponding reduction factors $1/r_1$, $1/r_2$ and $1/r_3$. In addition, the three newly created objects are given a weight factor (probability) denoted by P_1, P_2 and P_3. In the next step ($k = 2$) the same procedure is repeated for each of the squares, treating them as starting objects. To obtain a fractal measure this process has to be continued till $k \to \infty$ (Hentschel and Procaccia 1983).

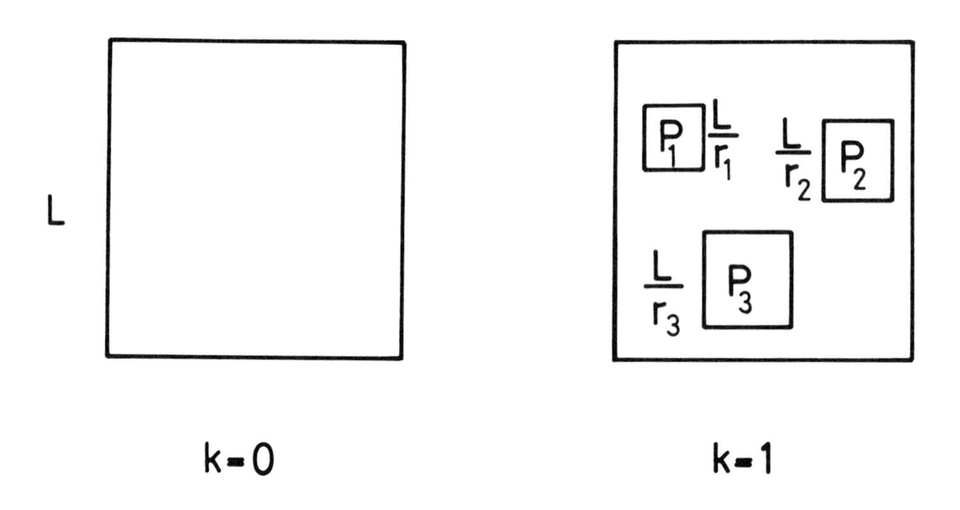

Figure 3.1. Construction of a fractal measure defined on a non-uniform fractal support embedded into two dimensions. The multifractal is obtained after infinitely many recursions.

It can be well seen from Fig. 3.1 that as the recursion advances a very complex situation emerges both concerning the size and the weight distribution of the objects. For example, if $P_1 > P_2$ and and $k \gg 1$, a square which appears as a result of going through reductions each time by the factor $1/r_1$, will have a much larger weight ($P_1^k \gg P_2^k$) than that being reduced each time by $1/r_2$.

In order to describe a fractal measure in a more quantitative way we imagine that a d dimensional hypercubic lattice with a lattice constant l is put on the fractal, and denote by p_i the stationary probability associated with the ith box of volume l^d, where

$$p_i = \int_{i\text{th box}} d\mu(\vec{x}) \tag{3.1}$$

with $\sum_i p_i = 1$. First we are interested in the behaviour of p_i as a function of the box size measured in units of the linear size L of the structure. As before this dimensionless unit will be denoted by $\epsilon = l/L$, and scaling of various quantities in the limit $\epsilon \to 0$ (corresponding to either $l \to 0$ or in the growing case to $L \to \infty$) will be studied. Trivially, for a homogeneous structure with a uniform distribution (density) $p_i(\epsilon) \sim \epsilon^d$. In the case of a uniform fractal ($r_i \equiv r$) of dimension D with a uniform distribution on it $p_i(\epsilon) \sim \epsilon^D$, since for a distribution with a uniform density p_i corresponds to the volume of the support in the ith box. (The term support is used for the fractal on which the measure is defined.)

In the more complex situation, when a non-uniform fractal with a distribution having infinitely many singularities is considered, we are led to assume a general form

$$p(\epsilon) \sim \epsilon^\alpha \,, \tag{3.2}$$

where $\epsilon \ll 1$ and α can take on a range of values depending on the given region of the measure. The non-integer exponent α corresponds to the strength of the local singularity of the measure and is sometimes also called crowding index or Hölder exponent. Although α depends on the actual position on the fractal, there are usually many boxes with the same index α. In general, the number of such boxes scales with ϵ as

$$N_\alpha(\epsilon) \sim \epsilon^{-f(\alpha)} \,, \tag{3.3}$$

where, in view of (2.4), $f(\alpha)$ is the fractal dimension of the subset of boxes characterized by the exponent α. The exponent α can take on values from the interval $[\alpha_\infty, \alpha_{-\infty}]$, and $f(\alpha)$ is usually a single humped function (see Example 3.1 in Section 3.3) with a maximum

$$\max_\alpha f(\alpha) = D\,. \tag{3.4}$$

The $f(\alpha)$ spectrum of an ordinary uniform fractal is a single point on the $f - \alpha$ plane.

Thus, a typical fractal measure is assumed to be made of interwoven sets of singularities of strength α, each characterized by its own fractal dimension $f(\alpha)$. This fact is the reason for the name multifractal frequently used for such fractal measures. In order to determine $f(\alpha)$ for a given distribution, it is useful to introduce a few quantities which are more directly related to the observable properties of the measure. Then, relations among these quantities and $f(\alpha)$ make it possible to obtain a complete description of a fractal measure (for a recent conceptual basis see Appendix C).

3.2. RELATIONS AMONG THE EXPONENTS

An important quantity which can be determined from the weights p_i is the sum over all boxes of the qth power of box probabilities

$$\chi_q(\epsilon) \equiv \sum_i p_i^q \tag{3.5}$$

for $-\infty < q < \infty$. This definition of χ is closely related to the Rényi entropies (Rényi 1970, Hentschel and Procaccia 1983). For $q = 0$ (3.5) gives $N(\epsilon)$, the number of boxes of size ϵ needed to cover the fractal support (the region, where $p_i \neq 0$). Therefore,

$$\chi_0(\epsilon) = N(\epsilon) \sim \epsilon^{-D}\,, \tag{3.6}$$

where D is the dimension of the support. Since the distribution is normalized, $\chi_1(\epsilon) = 1$.

Because of the complexity of multifractal distributions the scaling of $\chi_q(\epsilon)$ for $\epsilon \to 0$ generally depends on q in a non-trivial way

$$\chi_q(\epsilon) \sim \epsilon^{(q-1)D_q}\,, \tag{3.7}$$

where D_q is the so-called *order q generalized dimension.* It should be noted, however, that in spite of their name the D_q-s are not fractal dimensions, as we shall see later. The factor $(q-1)$ was chosen to satisfy the relation $\chi_1(\epsilon) = 1$ automatically. Correspondingly, for all q we have $D_q > 0$. It can be shown that the D_q values monotonically decrease with growing q. From the comparison of (3.6) with (3.7) it follows that $D_0 = D$. In the simple case of uniform fractals with a uniform distribution all D_q-s are equal to D. The typical behaviour of D_q as a function of q can be seen from Example 3.1 in the next section.

The distribution of p_i-s is extremely inhomogeneous concerning both their values and the number of boxes with the same p_i. As a result, when $\epsilon \to 0$ the dominant contribution to the sum (3.5) comes from a subset of all possible boxes. This subset forms a fractal with a fractal dimension f_q depending on the actual value of q, thus

$$N_q(\epsilon) \sim \epsilon^{-f_q}\,, \tag{3.8}$$

where $N_q(\epsilon)$ is the number of boxes giving the essential contribution in (3.5). In addition, all of these boxes have the same $p_i = p_q$. We shall denote the singularity strength for boxes with probability p_q by α_q, i.e.,

$$p_q \sim \epsilon^{\alpha_q} \tag{3.9}$$

for $\epsilon \to 0$. The f_q and α_q spectra defined by (3.8) and (3.9) provide an alternative description of a fractal measure with regard to (3.2) and (3.3).

What is the origin of the fact that a given q selects a fractal subset of dimension f_q with a corresponding crowding index α_q? When $\epsilon \to 0$ the contribution of a box depends very sensitively both on q and α, in particular, increasing $q > 1$ or α a bit, results in a dramatic decrease of the contribution coming from the same box. On the other hand, in the same limit the number of boxes having a little different probability than the given box, is very much different. The main contribution comes from boxes for which the product $N(\alpha, \epsilon)p_i^q$ is maximal, where $N(\alpha, \epsilon)$ is the number of boxes with α. For all other values of $\alpha \neq \alpha_q$ the contribution of the corresponding boxes is negligible.

According to (3.6) and (3.8) f_0 is equal to the dimension of the supporting fractal. It can be easily shown that the dimension of a fractal is always larger than the dimension of any of its subsets (see rule e) in Section 2.2), therefore, $f_0 > f_q$ for $q \neq 0$. f_q as a function of growing q first increases up to D then monotonically decreases. Since for increasing q the dominant contribution comes from boxes with larger probabilities p_q, α_q is a monotonically decreasing function. For a simple fractal $f_q = \alpha_q = D$.

There is a relation between D_q and the spectra f_q and α_q which can be easily obtained taking into account that

$$\chi_q(\epsilon) \simeq N_q(\epsilon) p_q^q \,, \tag{3.10}$$

since the essential contribution to $\chi_q(\epsilon)$ comes from boxes with p_q. Inserting (3.7), (3.8) and (3.9) into (3.10) we get

$$(q-1)D_q = q\alpha_q - f_q \,. \tag{3.11}$$

The above expression is consistent with the earlier observation that $D = D_0 = f_0$. Moreover, since f is finite we have $D_{\pm\infty} = \alpha_{\pm\infty}$.

Now it is possible to relate the generalized dimensions to the $f(\alpha)$ spectrum. First we express $\chi_q(\epsilon)$ for $\epsilon \to 0$ through $f(\alpha)$ using (3.2), (3.3) and (3.5)

$$\chi_q(\epsilon) \sim \int_{\alpha_\infty}^{\alpha_{-\infty}} \epsilon^{q\alpha' - f(\alpha')} d\alpha' \,, \tag{3.12}$$

where α is a quasi-continuous variable. For $\epsilon \ll 1$ the integral will be dominated by the value of α which minimizes the exponent. This leads to the condition

$$\frac{df(\alpha)}{d\alpha}|_{\alpha_q} = q \,, \tag{3.13}$$

where α_q is the value of α for which $q\alpha - f(\alpha)$ is minimal. We see that (3.12) is consistent with (3.11) if, in addition to (3.13),

$$f_q = f(\alpha_q) \,. \tag{3.14}$$

Thus, knowing $f(\alpha)$ we can find α_q from (3.13) and then f_q from (3.14) (Halsey *et al* 1986).

Alternatively, if D_q is given, one can calculate α_q from the relation

$$\alpha_q = \frac{d}{dq}[(q-1)D_q] \tag{3.15}$$

which can be obtained from (3.11) using (3.13) and (3.14). If we know α_q and D_q, f_q can be determined from (3.11) and finally, $f(\alpha)$ from (3.14). Therefore, the spectra D_q and $f(\alpha)$ represent equivalent descriptions of multifractals, since they are Legendre transforms of each other (Halsey *et al* 1986).

For the evaluation of experimental data the following procedure is usually applied. Using an appropriate normalization of the observable quantities, the set of p_i values is determined. Then the generalized dimensions are obtained from

$$D_q = \lim_{\epsilon \to 0} \left[\frac{1}{q-1} \frac{\ln \sum_i p_i^q}{\ln \epsilon} \right] \tag{3.16}$$

with the help of the corresponding log-log plot. Finally, the resulting plot of D_q versus q is numerically derived to obtain $f(\alpha)$.

Another alternative to present experimental results is based on the expressions (3.2) and (3.3) defining α and $f(\alpha)$ through the set p_i. For a fractal measure the plots $\ln[N(p_i, \epsilon)]/\ln(1/\epsilon)$ versus $\ln p_i/\ln(1/\epsilon)$ for various $\epsilon \ll 1$ should fall onto the same universal curve which is the $f(\alpha)$ spectrum of the multifractal (Meakin 1987). In this approach corrections proportional to $1/\ln \epsilon$ are expected to make the evaluation of data less effective.

Finally, we make a few remarks concerning the order $q = 1$ generalized dimension, for which we have from equation (3.11) and its derivative taken at $q = 1$ the following relation

$$D_1 = \alpha_1 = f_1 \,. \tag{3.17}$$

The above expression will be used in Example 3.1 to explain some of the properties possessed by the plots of $f(\alpha)$, f_q and α_q. Furthermore, inserting (3.5) into (3.7) and taking the $q \to 1$ limit (using l'Hospital's rule) one gets

$$-\sum_i p_i \ln p_i \sim D_1 \ln(1/\epsilon) \,. \tag{3.18}$$

The left-hand side of (3.18) is a familiar expression from information theory and corresponds to the amount of information associated with the distribution of p_i values. Therefore, according to (3.18) D_1 describes the scaling of information as $\epsilon \to 0$. This is why D_1 is called *information dimension*. A distribution with $D_1 < D$ is necessarily a fractal measure (Farmer 1982).

3.3. FRACTAL MEASURES CONSTRUCTED BY RECURSION

The formalism described in the previous section assumes the knowledge of p_i, the set of box probabilities. For a real system or computer models p_i has to be determined experimentally or numerically, but in the case of *deterministic fractal measures* constructed by an exact recursive procedure the box probability distribution can be obtained analytically, just like the fractal dimension for ordinary deterministic fractals (Section 2.3.1).

To construct a multifractal distribution we generalize the procedure used in Chapter 2 to produce non-uniform fractals, to the case when the weight or measure corresponding to a newly generated part is also changed by a given factor. Thus, we consider the following construction: in the first step ($k = 1$) the ith part obtained from the seed object by reduction using a rescaling factor $(1/r_j) < 1$ will be given a probability P_j, with $\sum_j^n P_j = 1$, where n is the number of pieces the starting configuration is substituted with. At the next stage ($k = 2$) each part is divided into n parts with probabilities reduced further by P_j and size rescaled by $1/r_j$. This procedure is illustrated for $n = 3$ in Fig. 3.1. The multifractal is obtained in the $k \to \infty$ limit.

As a result of the above construction the fractal structure (support) on which the measure is defined can be divided into n parts each being a rescaled version of the whole support by a factor $1/r_j$. The total measure associated with the jth such part is P_j. Therefore,

$$\chi_{q,j}(\epsilon) = \sum_i p_{j,i}^q = P_j^q \chi_q(\epsilon r_j)\,, \tag{3.19}$$

where $\chi_{q,j}(\epsilon)$ is the quantity defined by (3.5) evaluated for the jth part and $p_{j,i}$ is the probability of the ith box in the jth part. For the whole system

$$\chi_q(\epsilon) = \sum_{j=1}^n \chi_{q,j}(\epsilon)\,. \tag{3.20}$$

Using (3.7) and (3.19) in (3.20) we get (Hentschel and Procaccia 1983, Halsey *et al* 1986)

$$\sum_{j=1}^n P_j^q r_j^{(q-1)D_q} = 1 \tag{3.21}$$

which is an implicit equation for the generalized dimensions D_q.

Depending on the particular choice for q or the P_j and r_j values in (3.21) various special cases can be recovered. As expected, for $q = 0$ (3.21) provides the fractal dimension of the support, since in this limit it becomes identical with (2.13). On the other hand, for general q, but identical rescaling factors $r_j \equiv r$ equation (3.21) can be solved for D_q

$$D_q = \frac{1}{q-1} \frac{\ln\left(\sum\limits_{j=1}^{n} P_j^q\right)}{\ln(1/r)} . \tag{3.22}$$

Multifractal properties are trivially lost, if all P_j-s and r_j-s are equal. Furthermore, distributing the measure on the support with a constant density, i.e., choosing

$$P_j = \frac{r_j^{-d}}{\sum\limits_{j=1}^{n} (1/r_j)} \tag{3.23}$$

results in a non-trivial D_q spectrum which will be discussed in the next section.

EXAMPLES

Example 3.1. First we consider a multifractal distribution which, in spite of its simplicity, possesses the relevant features of more general fractal measures (Farmer 1982). The $f(\alpha)$ spectrum of this construction can be treated by exact calculations (Tél 1988), in addition to the self-similarity considerations leading to (3.21). The measure will be defined on the unit interval instead of a fractal support, but this fact does not change its characteristic scaling properties.

Thus, consider the unit interval divided into three equal parts of length (1/3), with the corresponding weights or probabilities P_1, P_1 and P_2, where $P_2 = 1 - 2P_1$. We shall assume that the probability of the middle interval is larger than that of the two others having equal weights, i.e., $P_2 > P_1$.

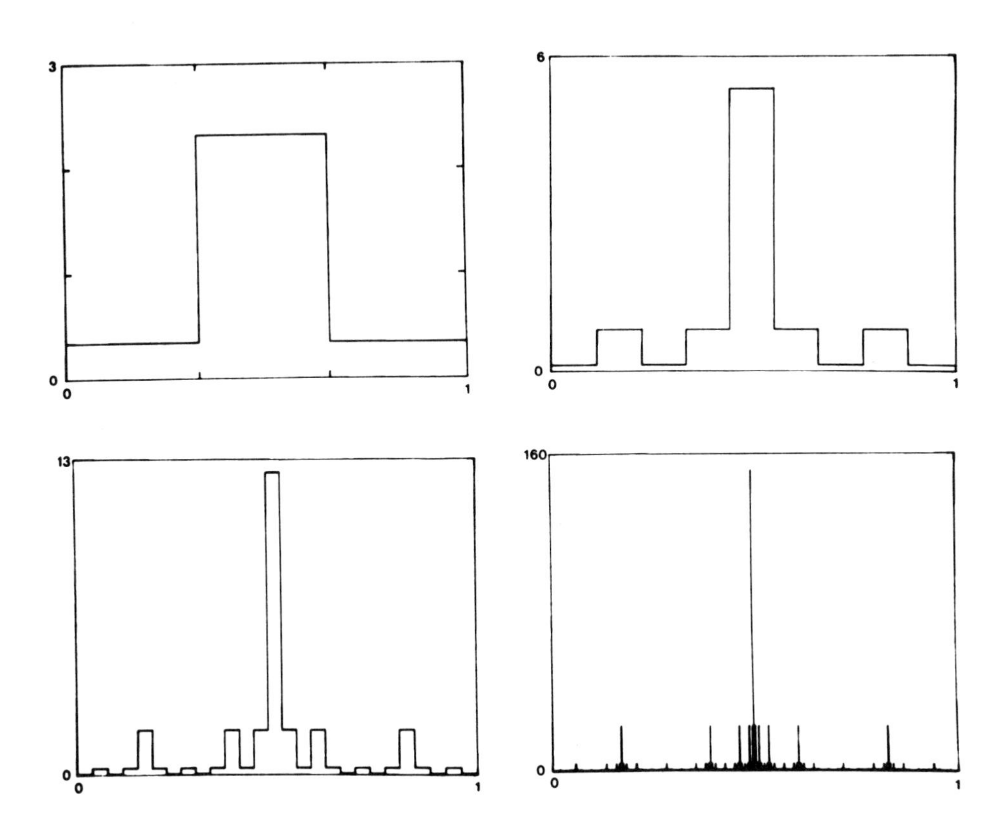

Figure 3.2. The first steps of constructing a fractal measure on the unit interval (Farmer 1982).

In the next step ($k = 2$) each of the three intervals is further divided and the probability is redistributed within the 9 new intervals according to the proportions used in the first step. This construction, which corresponds to the special case $n = 3$ and $1/r_j \equiv 1/3$, is illustrated in Fig. 3.2. In a few more steps the distribution becomes so inhomogeneous that its structure becomes visible only on a logarithmic scale. The density distribution in the limit $k \to \infty$ turns into a single-valued, everywhere discontinuous function.

Application of (3.22) yields an explicit expression for D_q through the probabilities P_1 and P_2

$$D_q = \frac{1}{(q-1)\ln(1/3)} \ln(2P_1^q + P_2^q)\,. \tag{3.24}$$

From the above equation α_q can be obtained using (3.15)

$$\alpha_q = \frac{1}{\ln(1/3)} \frac{2P_1^q \ln P_1 + P_2^q \ln P_2}{2P_1^q + P_2^q}\,. \tag{3.25}$$

Similarly, we can determine f_q substituting (3.24) and (3.25) into (3.11)

$$f_q = \frac{1}{\ln(1/3)} \left[\frac{2P_1^q \ln P_1^q + P_2^q \ln P_2^q}{2P_1^q + P_2^q} - \ln(2P_1^q + P_2^q) \right]. \tag{3.26}$$

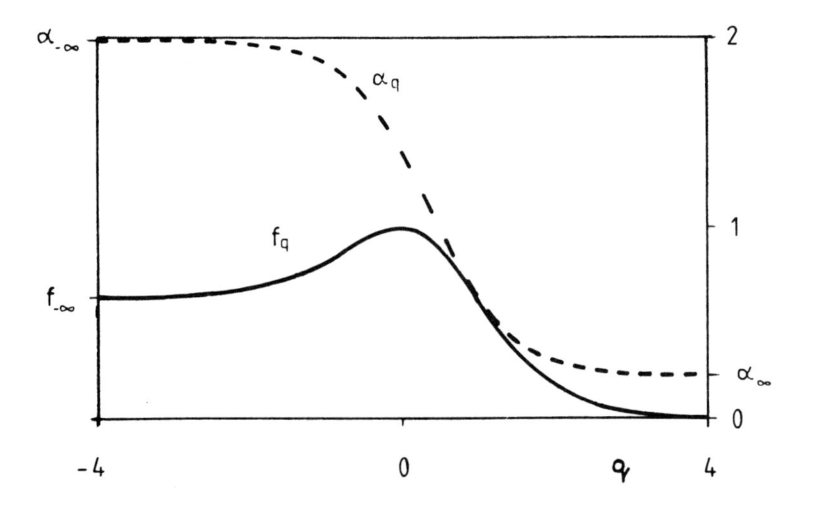

Figure 3.3. The q-dependence of the fractal dimension f_q and the exponent α_q for the multifractal shown in Fig. 3.2 (Tél 1988).

The dependence of f_q and α_q on q is shown in Fig. 3.3. Knowing f_q and α_q we can determine $f(\alpha)$ from (3.14). The result is displayed in Fig. 3.4.

To demonstrate how the singularity strength α_q with the corresponding fractal dimension f_q is selected for a given q we determine the distribution of p_i-s explicitly (without using (3.22)). Let us cover the unit interval

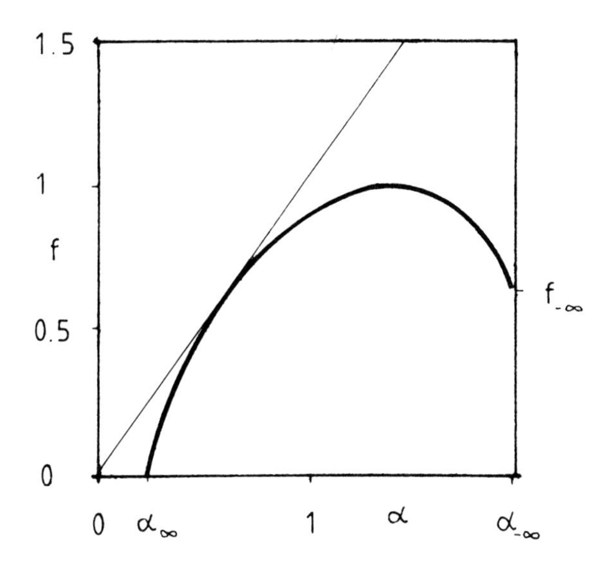

Figure 3.4. The $f(\alpha)$ spectrum of fractal dimensionalities for the multifractal of Fig. 3.2. The straight line corresponds to $f = \alpha$ (Tél 1988).

with one-dimensional boxes of size $l = \epsilon = (1/3)k$ and assume that $k \gg 1$. According to the construction of the present example the possible values of the box probabilities are the following

$$p_m = P_1^m P_2^{k-m}\,, \tag{3.27}$$

where m is an integer number and $0 \le m \le k$. The number of boxes with probability p_m is given by

$$N_m = \binom{k}{m} 2^m\,. \tag{3.28}$$

Next we investigate which set of the boxes gives a finite contribution to the total probability in the limit $k \to \infty$. For this purpose we calculate $\ln N_m p_m$ using Stirling's formula $\ln n! \simeq n(\ln n - 1)$ and get

$$\ln N_m p_m = k\ln k - m\ln m - (k-m)\ln(k-m) + m\ln 2 + m\ln P_1 + (k-m)\ln P_2\,. \tag{3.29}$$

The maximum of (3.29) is at

$$m_1 = 2kP_1\,, \tag{3.30}$$

and the value of $\ln N_{m_1} p_{m_1}$ is equal to 0 within the accuracy of Stirling's formula. This means that the contribution of boxes with m_1 is approximately equal to 1, therefore these boxes contain nearly all of the total measure. As concerning boxes with other values of m, their contribution is negligible, because either their number or the measure they carry is too small.

To find the fractal dimension of the structure made of boxes with p_{m_1} we first express $\ln N_{m_1}$ using (3.28) and (3.30)

$$\ln N_{m_1} = -k(2P_1\ln P_1 + P_2\ln P_2)\,. \tag{3.31}$$

Then, from (3.31) and $\epsilon = (1/3)^k$ it follows that

$$N_{m_1}(\epsilon) = \epsilon^{-f_1} \tag{3.32}$$

with

$$f_1 = \frac{1}{\ln(1/3)}(2P_1\ln P_1 + \ln P_2)\,. \tag{3.33}$$

Therefore, boxes forming a fractal of dimension f_1 give the dominant contribution to the stationary distribution.

The following step is to study the qth power of the box probabilities with $-\infty < q < \infty$. Their total amount is $\sum_m N_m p_m^q = (2P_1^q + P_2^q)^k$. As before, the dominant contribution is given by boxes with a characteristic probability, p_{m_q}, depending on q. Simple algebra analogous to the above derivation leads to (Tél 1988)

$$m_q = \frac{2kP_1^q}{2P_1^q + P_2^q}\,.$$

Writing N_{m_q} and p_{m_q} in the form $N_{m_q}(\epsilon) = \epsilon^{-f_q}$ and $p_{m_q}(\epsilon) = \epsilon^{\alpha_q}$ it is easy to show that these expressions hold with f_q and α_q identical with those obtained by the similarity argument (see Eqs. (3.25) and (3.26)).

Example 3.2. In order to discuss an example which has direct connection with physical systems let us consider a hierarchical network of resistors shown in Fig. 3.5. At the kth step of the iteration procedure each bond of length $l = 1$ is replaced with a configuration having a linear size equal to $r = 3$ and made of $n = 6$ bonds of unit length. This growing fractal in the $k \to \infty$ limit has a fractal dimension $D = \ln 6/\ln 3$. To complete the model, we assume that the bonds have the same resistivity and the voltage between the two terminal bonds is equal to $V_0 = 1$. Then, in the unit cell there will be two bonds with voltage drop $V_1 = 1/3$ and four others with $V_2 = 1/6$. The sum of all voltage drops is not equal to 1, therefore, the distribution is not normalized. (It can be normalized by simply choosing $V_0 = 3/4$.)

The multifractal properties can be deduced from (Arcangelis *et al* 1985, 1986)

$$\sum_i v_i^q \sim L^{-(q-1)D_q}\,, \tag{3.34}$$

where v_i is the voltage drop across the ith bond and $L = (l/\epsilon) = 3^k$ is the linear size of the network. Because of the hierarchical structure of the model the sum of the v_i^q-s after completing the kth recursion is

$$\sum_m N_m v_m^q = (2V_1^q + 4V_2^q)^k\,. \tag{3.35}$$

In the above expression

$$N_m = 2^{k+m}\binom{k}{m} \tag{3.36}$$

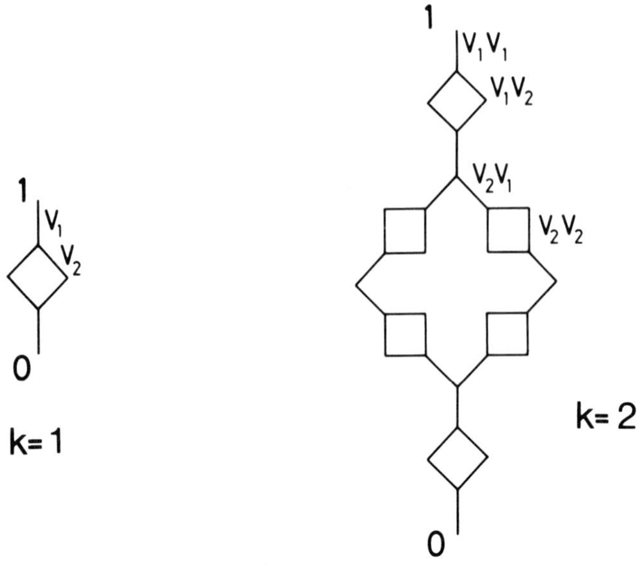

Figure 3.5. Fractal model for hierarchical networks of resistors. The distribution of voltage drops exhibits multifractal scaling.

is the number of bonds with the voltage drop

$$v_m = V_1^{k-m} V_2^m \tag{3.37}$$

with $0 \le m \le k$. Substituting (3.35) into (3.34) and using the actual voltage values we get

$$D_q = \frac{1}{q-1} \frac{\ln[2\left(\frac{1}{3}\right)^q + 4\left(\frac{1}{6}\right)^q]}{\ln(1/3)} \tag{3.38}$$

showing that the moments of the distribution of voltage drops can be described by an infinite number of generalized dimensions. Finally, we remark that the distribution (3.36) as a function of m is a simple binomial modified by a factor 2^m. Since the voltage drops depend on m approximately exponentially, we conclude that the distribution $N(v_m)$ has a rather special shape which is close to the log-binomial. Finally, it should be noted that using an appropriately modified definition of the exponents corresponding to the moments of the voltage distribution it is possible to obtain an infinite hierarchy of exponents which is independent of the actual fractal dimension of the network (i.e., which reflects only its topological properties).

3.4. GEOMETRICAL MULTIFRACTALITY

The results discussed in the previous two sections were obtained for the general case when an inhomogeneous measure was defined on the support. The special case of a distribution with constant density on a non-uniform fractal, however, deserves particular attention. In this case it is the geometry of the fractal only which is described by the formalism. Since this book concentrates on the structure of growing fractals rather than on their physical properties, in the following the application of the theoretical approach developed above will be discussed for structures with no singular density defined on them.

The fact that non-uniform fractals with a uniform distribution can also be described in terms of generalized dimensions has already been indicated by expression (3.21). This equation valid for a recursive fractal leads to a D_q spectrum even if the weight factors P_j are defined in such a way that the measure of a newly created part becomes the same as its volume, i.e., the probability distribution is uniform on the support. In the case of Example 3.1 this can be achieved by choosing $P_j = (1/r_j)/\sum_j(1/r_j)$ which together with (3.21) results in the non-trivial dependence of D_q on q.

Let us now consider the problem for the general case of growing fractals. We assume that the structures grow on a lattice and are built up by identical particles. The actual linear size and mass of the cluster will be denoted by L and M, respectively. The structure is to be covered by boxes of size l such that

$$a \ll l \ll L\,, \tag{3.39}$$

where a is the lattice constant. (In some cases the condition $a/l \to 0$ has to be satisfied.) One can then determine the mass M_i of the ith non-empty box. The mass index or singularity exponent α of this box is defined by

$$M_i \sim M(l/L)^{\alpha} \tag{3.40}$$

for $l/L \ll 1$. Boxes with the same mass index α form a fractal subset of dimension $f(\alpha)$. Their number $N(\alpha)$ is, therefore, related to $f(\alpha)$ via

$$N(\alpha) \sim (l/L)^{-f(\alpha)}\,. \tag{3.41}$$

If there exists a set of different mass indices the growing structure will be called a *geometrical (or mass) multifractal* (Tél and Vicsek 1987), since the measure generating the spectrum is a uniform mass distribution (Lebesgue measure), and thus $f(\alpha)$ characterizes the geometry of the system directly.

To determine the generalized dimensions one can use the scaling relation (3.7) which in the present notation has the form

$$\chi_q(M, l, L) \equiv \sum_i M_i^q \sim M^q (l/L)^{(q-1)D_q}\,. \tag{3.42}$$

As for general multifractals, knowing D_q we can determine $f(\alpha)$ using expressions (3.11) and (3.13).

Before discussing the examples one should make a few important comments. i) Non-uniform recursive fractals are multifractals in a geometrical sense. ii) Since geometrical multifractality as defined above is a consequence of local density fluctuations, it should be most pronounced in inhomogeneous growth processes. iii) A necessary condition for observing this phenomenon is the existence of three well separated length scales (see (3.39)) which may require the linear size L to be close to the largest cluster size ever produced in numerical simulations.

Finally, one can simply check that in the case of the examples to be discussed below the values obtained for the fractal dimension using two different methods, *do not coincide*. First, one can apply the so-called sandbox method which is equivalent to (2.4), i.e., to calculating the number of particles $M(L)$ within boxes of increasing linear size L centred at the same point.

Applying this method at each step of the construction of a non-uniform fractal (2.9) gives a fractal dimension different from that obtained from equation (2.13) derived for the fractal dimension from self-similarity arguments. In fact, the dimension obtained from the sandbox method is equal to $D_{-\infty}$. The contradiction is resolved by the unusual behaviour of $\ln M(L)$ versus $\ln L$ as a function of the position chosen for the centre of the boxes. Because of the extreme inhomogeneity of the structure the quantity

$$\ln M(L)/\ln L \tag{3.43}$$

which for large L usually is equal to the fractal dimension does not converge to a unique limiting value as $L \to \infty$. The right fractal dimension can be obtained by averaging over the position of sandbox centres.

EXAMPLES

Example 3.3. The construction we shall investigate is a growing version of the non-uniform Cantor set (Fig. 3.6). This growth process embedded into one dimension enables us to determine how the multifractal behaviour emerges as the system grows (Tél and Vicsek 1987). The first unit of the process consists of three particles in the first, third and fourth sites. At the next stage the twice enlarged copy of the "seed" configuration is added between the 9th and 16th sites (leaving out four sites equal to the length of the first configuration) and this procedure is repeated with the new configuration playing the role of the seed. Figure 3.6 shows the objects obtained in the first three steps of the construction.

After k steps the linear size is 4^k, it is, therefore, convenient to use $l = 2^k$ as the box size. It can be easily observed that the number of non-empty boxes is then F_k, a Fibonacci number defined by $F_o = 1$, $F_1 = 2$, $F_k = F_{k-1} + F_{k-2}$. Let $M_i^{(k)}$, $i = 1, ..., F_k$ denote the mass of the ith nonempty box at the kth step of construction (i=1 corresponds to the leftmost box). The distribution $M_i^{(k)}$ is related to that of two previous generations for $k > 2$ by

Figure 3.6. The first few steps in the construction of the growing non-uniform Cantor set (Tél and Vicsek 1987).

$$M_i^{(k)} = \begin{cases} 3M_i^{(k-2)} & i = 1, ...F_{k-2}, \\ 2M_{F_{k-2}+i}^{(k-1)} & i = 1, ...F_{k-1}, \end{cases} \tag{3.44}$$

($M = 3^k$) as can be checked directly. The "initial values" for this recursion are $M_i^{(1)} = 1, 2$, $M_i^{(2)} = 3, 2, 4$. At this point it is already possible to calculate a few characteristic exponents for the system. Since the number of nonempty boxes is F_k and $F_{k+1}/F_k \approx w$ for large k, where $w = (\sqrt{5}-1)/2$ is the golden mean, we find from (3.42) $D_o = \ln(w)/\ln(2)$ for the fractal dimension of the complete set. The densest box is the rightmost one with mass 2^k, therefore, the smallest α is obtained from (3.40) as $\alpha_{min} = D_\infty = \ln(2/3)/\ln(1/2) = 0.585...$. The most rarified nonempty interval is the first or the second one. For its mass we find $3^{(k-1)/2}$ for an odd k and $2 \times 3^{k/2-1}$ for even k. Consequently, the largest α is $\alpha_{max} = D_{-\infty} = \ln(1/3)/\ln(1/4) = 0.792...$.

The picture simplifies further by observing that there are only $k+1$ different mass values in the set $M_i^{(k)}$. Denoting these quantities by $\tilde{M}_j^{(k)}$, $j = 1, ..., k+1$ one obtains for $k > 1$

$$\tilde{M}_j^{(k)} = \begin{cases} 3^{[k/2]} & j = 1 \\ 2\tilde{M}_{j-1}^{(k-1)} & j = 2, ..., k+1 \end{cases} \tag{3.45}$$

with $\tilde{M}_j^{(1)} = 1, 2$ where [ ] denotes the integer part. Let the numbers $N_j^{(k)}$ denote how many times the value $\tilde{M}_j^{(k)}$ appears in $M_j^{(k)}$. They are found to follow a two step recursion

$$N_j^{(k)} = N_{j-1}^{(k-1)} + N_j^{(k-2)}, \qquad 1 < j < k \tag{3.46}$$

and $N_j^{(k)} = N_k^{(k)} = N_{k+1}^{(k)} = 1$. Now it is possible to derive an equation for the $f(\alpha)$ spectrum using (3.40), (3.41), (3.45) and (3.46). The solution is

$$f(\alpha) = \left[(\alpha - \alpha_{min}) \ln\left(\frac{\alpha - \alpha_{min}}{1-\alpha}\right) + 2(\alpha_{max} - \alpha) \ln\left(\frac{2(\alpha_{max} - \alpha)}{1-\alpha}\right)\right] \times \frac{1}{2\ln(1/2)(\alpha_{max} - \alpha_{min})}. \tag{3.47}$$

Finally, one can obtain from (3.11) and (3.47) an expression for the D_q spectrum.

The knowledge of recursions (3.45) and (3.46) makes it possible to calculate χ_q numerically for any finite stage, k, of the construction. Comparing two subsequent stages of the growth process we define a k *dependent* $D_q(k)$ by

$$\frac{\chi_{q,k+1}/M_{k+1}^q}{\chi_{q,k}/M_k^q} = \left(\frac{l_{k+1}/L_{k+1}}{l_k/L_k}\right)^{(q-1)D_q(k)}, \tag{3.48}$$

where the subscript denotes quantities at the k-th stage of the growth. For large k $D_q(k) \to D_q$ as can be seen from (3.42). The results plotted in Fig. 3.7 show a rapid convergence as a function of k towards D_q, but note that the linear size in this model grows as 4^k.

It is also possible to calculate numerically the function

$$f_j^{(k)} = -\ln(N_j^{(k)})/\ln(l_k/L_k) \tag{3.49}$$

versus $\alpha_j^{(k)} = \ln(\tilde{M}_j^{(k)}/M_k)/\ln(l_k/L_k)$ for *increasing values of* k. These quantities converge to $f(\alpha)$ and α as was shown above. The convergence is, however, slow since the constant factors not written out explicitly in Eqs. (3.40) and (3.41) lead to contributions proportional to k^{-1}. The result for $k = 100$ together with the exact solution is displayed in Fig. 3.8.

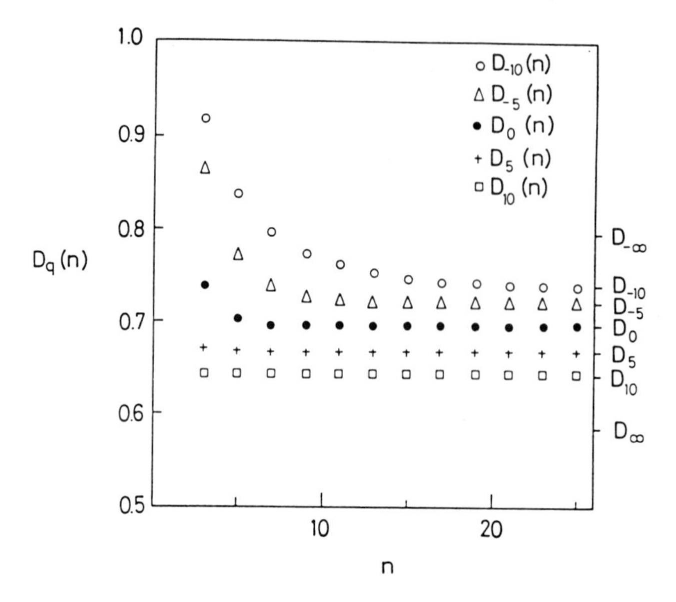

Figure 3.7. Convergence towards the generalized dimensions as a function of the actual size of the growing multifractal (Tél and Vicsek 1987).

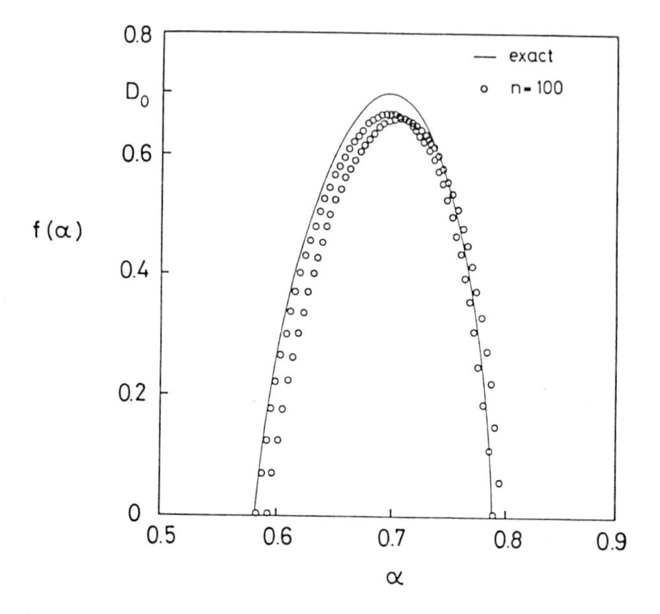

Figure 3.8. The $f(\alpha)$ spectrum obtained from (3.47) is shown by the full curve. The open circles represent the function $f_j^{(k)}$ versus $\alpha_j^{(k)}$ for $k = 100$. The special doubled internal structure of the latter plot is a consequence of the regular geometry of the model (it would be destroyed for a random configuration) (Tél and Vicsek 1987).

Example 3.4. To demonstrate geometrical multifractality for a growing fractal embedded into two dimensions we generalize the construction shown in Fig. 2.1a. The recursion procedure leading to a non-uniform fractal cluster is explained in Fig. 3.9. In the kth stage the twice enlarged version of the configuration corresponding to the $k-1$th stage of growth is added to the four corners of the cluster obtained for $k-1$. Although the large homogeneous plaquettes at stage k have an increasing size 2^{k-1} their size relative to that of the cluster is $(2/5)^{k-1}$, approaches zero as $k \to \infty$.

Figure 3.9. Growing geometrical multifractal embedded into two dimensions. In the kth step the twice enlarged copy of the previous stage is added to the four corners of the configuration.

We shall use (3.21) to get an implicit equation for the generalized dimensions. First note that at each step the configuration is replaced by five new parts, one of which has the same size as the previous configuration and the other four are twice enlarged versions of it. Simultaneously, the linear size of the cluster becomes 5 times larger. Let us reduce the growing structure at every step by a factor 1/5 so that (3.21) could be applied directly with rescaling factors to $1/r_1 = 1/5$ and $1/r_2 = ... = 1/r_5 = 2/5$. Furthermore, the part with r_1 has a mass which is $P_1 = 1/17$ times smaller than that of the whole structure. Analogously, $P_2 = ... = P_5 = 4/17$. Therefore, we have

$$\sum_{j=1}^{5} r_j^{(q-1)D_q} P_j^q = \left(\frac{1}{17}\right)^q 5^{(q-1)D_q} + 4\left(\frac{4}{17}\right)^q \left(\frac{5}{2}\right)^{(q-1)D_q} = 1\,. \qquad (3.50)$$

The above equation can be solved for general q numerically, while in the limits $q \to \pm\infty$ it provides the explicit expressions $D_\infty = \ln \frac{17}{4} / \ln \frac{5}{2} \simeq 1.579$ and $D_{-\infty} = \ln 17 / \ln 5 \simeq 1.760$. For $-\infty < q < \infty$ the corresponding D_q-s are between the above limiting values. Note that the sandbox method (i.e., calculating the number of particles within boxes of size 5^k centred at the seed particle) would incorrectly predict $D = D_{-\infty}$.

Chapter 4

METHODS FOR DETERMINING FRACTAL DIMENSIONS

When one tries to determine the fractal dimension of growing structures in practice, it usually turns out that the direct application of definitions for D given in the previous two chapters is ineffective or can not be accomplished. Instead, one is led to measure or calculate quantities which can be shown to be related to the fractal dimension of the objects.

Three main approaches are used for the determination of these quantities: experimental, computer and theoretical. *Experiments* represent a standard way of examining phenomena in every field of physics and they have been playing an important role in the development of research concerning fractal growth as well. The situation is less typical in the case of the other two approaches. Since the physics of fractal growth lacks a unified theoretical description, most of the investigations prompted by theoretical motivations are based on *computer simulation.* The only *theoretical* principle which seems to be applicable to a relatively wide range of growth processes is renormalization which will be discussed in the last section following a discussion of the experimental and numerical methods for determining D.

4.1. MEASURING FRACTAL DIMENSIONS IN EXPERIMENTS

A number of experimental techniques have been used to measure the fractal dimension of scale invariant structures grown in various experiments. The most widely applied methods can be divided into the following categories: (a) digital image processing of two-dimensional pictures, (b) scattering experiments, (c) covering the structures with monolayers, and (d) direct measurement of dimension-dependent physical properties.

(a) *Digitizing the image* of a fractal object is a standard way of obtaining quantitative data about geometrical shapes. The information is picked up by a scanner or an ordinary video camera and transmitted into the memory of a computer (typically a PC). The data are stored in the form of a two-dimensional array of pixels whose non-zero (equal to zero) elements correspond to regions occupied (not occupied) by the image. Once they are in the computer, the data can be evaluated using the methods described in the next section, where calculation of D for computer generated clusters is discussed.

The only principal question related to processing of pictures arises if two-dimensional images of objects embedded into three dimensions are considered. In Section 2.2 it has already been mentioned that the fractal dimension of the projection of an object onto a $(d-m)$-dimensional plane is the same as its original fractal dimension, if $D < d-m$. Unfortunately, there are only heuristic arguments supporting this assumption, and considerable deviations may occur from it, especially when D is only a bit smaller than $d-m$. In addition, if $D > d-m$ the method breaks down completely, since in this case the projection is simply a $(d-m)$-dimensional object.

(b) *Scattering experiments* represent a powerful method to measure the fractal dimension of microscopic structures (Teixeira 1986). Depending on the characteristic length scales associated with the object to be studied, light, X-ray or neutron scattering can be used to reveal fractal properties. There are a number of possibilities to carry out a scattering experiment. One can investigate i) the structure factor of a single fractal object, ii) scattering

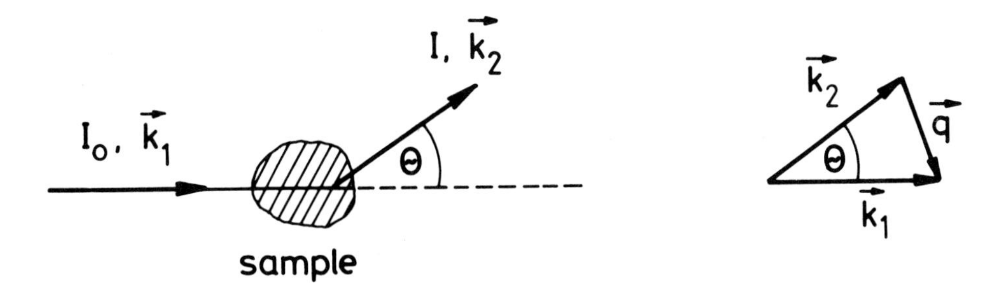

Figure 4.1. Schematic representation of the incident and scattered beams in a scattering experiment.

by many clusters growing in time, iii) the scattered beam from a fractal surface, etc.

In scattering experiments a beam of intensity I_0 is directed on the sample and the scattered intensity is measured as the function of the angle θ between the incident and the scattered beam. Let us denote the difference between the wave vectors corresponding to these beams denoted by $\vec{q} = \vec{k}_1 - \vec{k}_2$ (Fig. 4.1). In the case of small θ (small angle scattering) the main contribution to the scattered intensity comes from quasi-elastic processes with $|\vec{k}_1| = |\vec{k}_2| = k = 2\pi/\lambda$, where λ is the wavelength of the incident beam. Therefore, from Fig. 4.1

$$q = |\vec{q}| = 2k\sin(\theta/2)\,. \tag{4.1}$$

Most of the fractal structures studied experimentally are made of small spherical particles whose size exceeds the spatial resolution typical in small angle X-ray (SAXS) or neutron (SANS) scattering experiments. Thus, it is useful to identify a single scatterer with a corresponding form factor $P(q)$ and separate the scattered intensity into two factors

$$I(q) = \rho_0 P(q)[1 + S(q)]\,, \tag{4.2}$$

where ρ_0 is the average density in the sample, $S(q)$ is the interparticle structure factor and we assume that the particles are spherical having a radius r_0.

It can be shown that for $qr_0 \ll 1$ the form factor is approximately constant (Guinier regime), while for $qr_0 \gg 1$, $P(q) \sim q^{-4}$ which is called the Porod law.

According to the theory of scattering (see e.g., Squires 1978), the structure factor $S(q)$ is the Fourier transform of the density-density correlation function $c(r)$ defined by the expression (2.14). In a three-dimensional isotropic system this means that

$$S(q) = 4\pi \int_0^\infty c(r) r^2 \frac{\sin(qr)}{qr} dr \,. \tag{4.3}$$

To calculate the actual shape of $S(q)$ we recall that for fractals the density correlations decay with a power law depending on D in the form $c(r) \sim r^{D-d}$ (See Eqs. (2.16) and (2.18)). For a finite object of average radius R, $c(r)$ is expected to decrease very quickly to zero for $r > R$ which, for $d = 3$, can be taken into account by the assumption

$$c(r) \sim r^{D-3} f(r/R) \,, \tag{4.4}$$

where $f(x) \simeq$ constant for $x \ll 1$ and $f(x) \ll 1$ if $x \gg 1$. The cutoff function $f(x)$ is presumed to depend only on the ratio r/R because of the self-similar nature of the structure. Inserting (4.4) into (4.3) and changing the variable of integration $r = z/q$ we get

$$S(q) \sim q^{-D} \int_0^\infty z^{D-2} f(z/qR) \sin z dz \,. \tag{4.5}$$

This expression is expected to be valid in the range $qR \gg 1$ and $qr_0 \ll 1$, when the scattered beam probes the density correlations of particles within the object. Since in this case $f(x)$ is approximately constant up to large values of z, the integral in (4.5) only weakly depends on q and we can conclude that

$$I(q) \simeq S(q) \sim q^{-D} \qquad \text{for} \quad 1/R \ll q \ll 1/r_0\,, \tag{4.6}$$

since in this regime $P(q)$ is close to a constant. This is a result often used to estimate the fractal dimension of an experimental object. The statement that the integral in (4.5) is only weakly dependent on q can be supported by further calculations based on an assumption concerning the actual form of $f(x)$. Supposing that $f(r/R) \sim e^{-r/aR}$, where a is a constant, one can integrate (4.5) explicitly and arrive at (4.6) if $aqR \gg 1$.

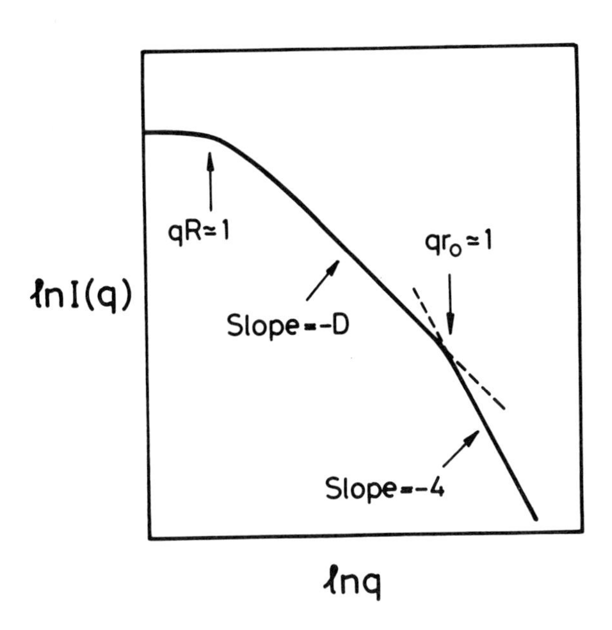

Figure 4.2. Schematic scattering curve showing the three main regimes which can be observed for an ensemble of fractal aggregates.

Thus, in scattering experiments one can distinguish three major regimes (Fig. 4.2):

i) $\quad qr_0 \gg 1 \gg qb$,

where b denotes the interatomic distance. In this case one probes the shape of the individual particles the structure is made of. This regime is characterized by a simple power law decay $I(q) \sim q^{-4}$ (Porod's law).

ii) $qR \gg 1 \gg qr_0$
This is the region of q values where (4.6) is expected to describe the spatial fluctuations of particles on a length scale smaller than the average radius of the object, R. In this case the fractal dimension can be determined from the slope of $\ln I(r)$ against $\ln q$.
iii) $1 \geq qR$
In this limit the fractal object behaves as a single particle from the point of view of small angle scattering. If a sufficiently dilute solution of clusters is present in the system, this regime allows the application of an independent method for the determination of D which will be briefly discussed below.

According to the standard theory of scattering (Squires 1978), in the region corresponding to case iii) the structure factor can be approximated by the expression

$$S(q) \sim \frac{\rho_0 M_w}{(1 + q^2 R_z^2/3 + ...)}, \tag{4.7}$$

where $M_w = S(0)$ is the weight average molecular weight of the clusters and R_z is a quantity proportional to the average radius of the the clusters (it is equal to the so called z average radius of gyration). One expects that there is a relation between M_w and R_z (Schaefer *et al* 1984)

$$M_w \sim R_z^D \tag{4.8}$$

analogous to (2.2). Then one can determine D measuring $S(q)$ in a diluted system of aggregates *growing in time*. Because of (4.7) the intercept of $S(q)$ with the $q = 0$ axis provides M_w, while $S(q)$ starts to bend downward at $qR_z \simeq 1$. D can be obtained by making a log-log plot of these quantities as a function of time.

The above analysis was concerned mainly with stochastic structures. It can be shown that light diffraction on deterministic fractals embedded into two dimensions results in a self-similar diffraction pattern (Allain and Cloitre 1986). In the related experiment a laser beam is directed onto the sample which is a structure obtained after a few steps of one of the deterministic

Figure 4.3. Optical diffraction pattern and the corresponding calculated structure factor of the fractal shown in Fig. 2.1a (k=5) (Allain and Cloitre 1986).

constructions discussed in Section 2.3.1. The diffraction pattern observed on a screen represents an optical Fourier transform of the object. The corresponding structure factor $S(p,q)$ can also be calculated and its value averaged over the frequency bands scales according to (4.6). In particular, the optical diffraction pattern of the fractal shown in Fig. 2.1 and its calculated structure factor were found to be self-similar. This is demonstrated by Fig. 4.3.

(c) To measure fractal dimensions by *covering the structure* with probe particles of varied radii (e.g., Pfeifer and Avnir 1983) is an obvious idea which is directly related to the definitions of D discussed in Chapter 2. In order to carry out an investigation of this sort one has to find materials which are well adsorbed by the surface of the objects. In addition, the difference $\epsilon_{max} - \epsilon_{min}$ between the smallest and largest radii of the molecules has to be large enough so that at least two or three decades could be covered by the method. The fractal dimension is then obtained from the relation $n(\epsilon) \sim \epsilon^{-D}$ equivalent to (2.4), where $n(\epsilon)$ moles/g is the number of adsorbent molecules forming a monolayer on the surface. This method is limited to measuring surface properties, since closed but empty regions inside a fractal object are not accessible to the molecules. Obviously, the monolayer technique would give $D = 1$ for the Sierpinski gasket (Section 2.3.1), provided we were confined to two dimensions, while the mathematical definition allows covering the fractal everywhere. On the other hand, it is expected to work well for open branching structures and surfaces.

In a simple variation of this method the size of the molecules is kept constant and R, the radius of the particles having a fractal surface, is increased. In this case $n(\epsilon) \sim R^{D-3}$, where it is assumed that the number of particles/g scales as R^{-3}, i.e., they are not volume fractals.

Several experiments are based on the determination of the *cumulative volume* $V(r > \epsilon)$ of empty regions or holes with a characteristic radius larger than ϵ. A typical measurement of this type is used to study the structure of porous media. For example, in porositometry, mercury is injected into the object with a given capillary pressure p. The non-wetting mercury can only enter pores with a radius larger than the radius of curvature inversely proportional to p. The pore size distribution which can be related to the fractal dimension is then obtained from the change of the volume of injected mercury as a function of the increased pressure.

To find the ϵ dependence of $V(r > \epsilon)$ one can use the following heuristic argument (Pfeifer 1986). Let us cover the fractal structure with a minimum number $N(\epsilon)$ of balls of radius ϵ. Then the volume covered by the balls is

$$V(\epsilon) \sim N(\epsilon)\epsilon^3 \sim \epsilon^{3-D} . \tag{4.9}$$

This is the volume which is not available for particles (invading fluid) having a larger radius (radius of curvature) than ϵ, since by definition we do not cover empty regions with radius larger than ϵ. Obviously, with decreasing ϵ, $V(\epsilon)$ decreases by an amount equal to the increase of $V(r > \epsilon)$, i.e., $-dV(\epsilon)/d\epsilon = dV(r > \epsilon)/d\epsilon$. From here

$$\frac{dV(r > \epsilon)}{d\epsilon} \sim -\epsilon^{2-D}, \tag{4.10}$$

where the left-hand side can be determined experimentally for various ϵ and the corresponding log-log plots allow the calculation of D. Expression (4.10) corresponds to a rule mentioned in Section 2.3.1 according to which the number of holes of radius larger than ϵ usually scales as ϵ^{-D}.

When one uses a wetting fluid to cover a fractal (this might be especially useful in the case of macroscopic objects) the situation is similar because the capillary forces lead to a characteristic curvature in such experiments as well (see Section 10.4). However, the geometry is inversed: Empty regions with a radius larger than that of the meniscus are not filled. Nevertheless, one expects that the volume of the wetting fluid surrounding the object depends on the radius of curvature ϵ according to the same law as above, namely, $V \sim \epsilon^{3-D}$ for a range of ϵ values (de Gennes 1985).

(d) Measurements of *physical properties* of fractal objects can also be used for the experimental determination of D. A number of methods have been suggested, most of them based on electrical properties including measurements of current, electromagnetic power dissipation and frequency dependence of the complex impedance of fractal interfaces. These methods typically provide an indirect estimate of D and have been used less extensively than the above discussed approaches.

4.2. EVALUATION OF NUMERICAL DATA

Throughout this section we assume that the information about the stochastic structures is stored in the form of d-dimensional arrays which correspond to the values of a function given at the nodes (or sites) of some underlying lattice. In the case of studying geometrical scaling only, the value of the function attributed to a point with given coordinates (the point being defined through the indexes of the array) is either 1 (the point belongs to the fractal) or 0 (the site is empty). When multifractal properties are investigated the site function takes on arbitrary values. In general, such discrete sets of numbers are obtained by two main methods: i) by digitizing pictures taken from objects produced in experiments, ii) by numerical procedures used for simulation of various growth phenomena.

In the case of random growth numerically generated data are typically produced by variations of the Monte Carlo method. In addition, exact enumeration techniques and numerical integration of the corresponding equations can also be used. In Section 4.1, we discussed a number of techniques one can use to get information from an experimentally grown structure. Analogously, there are many ways of determining the fractal dimension D from numerical data. Below we discuss how to measure D for a single object. To make the estimates more accurate one usually calculates the fractal dimension for *many clusters* and averages over the results.

Perhaps the simplest method is to use the definition of D as given in (2.2) and (2.3). In our case the unit length corresponds to the lattice constant, and the number of balls of unit volume $N(R)$ needed to cover the structure within a sphere of radius R is the same as the number of sites with a site function equal to 1 in the sphere. Since for growing fractals $N(R) \sim R^D$, plotting $\ln N(R)$ versus $\ln R$ results in a curve which has an asymptotic slope equal to D. (Strictly speaking $N(R) \sim R^D$ is valid only if there is an equivalence between the scaling observed by covering the structure with a lattice of boxes (box counting) and using boxes of increasing size centred on the same point (sandbox method). This equivalence exists only for uniform fractals with no multifractal spectrum of their "mass" distribution (see Section 3.4).)

Thus, the fractal dimension can be obtained by fitting a straight line to the asymptotic part of the $N(R)$ data, e.g., using the method of least squares. In practice one chooses a point belonging to the fractal (usually close to its centre of mass) and counts the number of sites belonging to the object within a sequence of spheres of growing radius. Instead of spheres one can also use boxes of linear size L. Figure 4.4 shows a schematic plot of this kind demonstrating the crossovers which take place when R becomes smaller than the lattice constant and R is larger than the size of the structure.

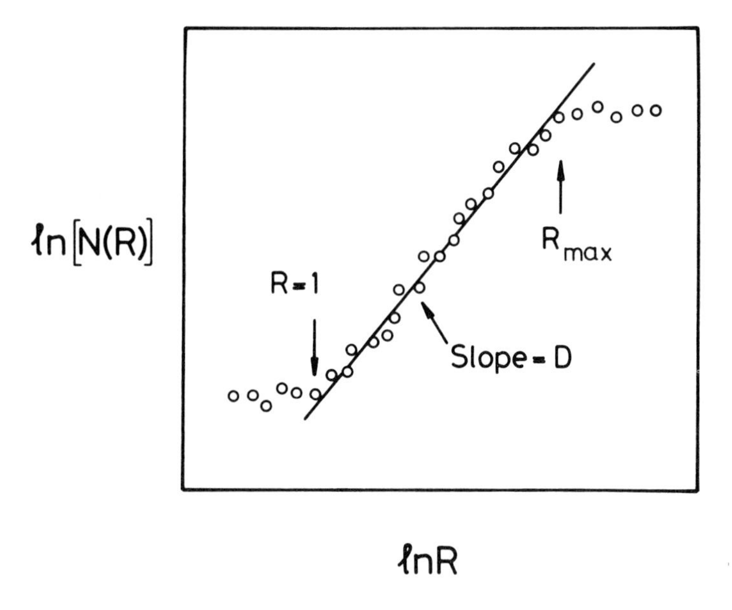

Figure 4.4. Schematic log-log plot of the numerically determined number of particles $N(R)$ belonging to a fractal and being within a sphere of radius R. If R is smaller than the particle size or larger than the linear size of the structure a trivial behaviour is observed. The fractal dimension is obtained by fitting a straight line to the data in the scaling region.

If the fractal object consists of small, nearly identical particles, one can think of $N(R)$ as the number of particles within a region of volume R^d, i.e., $N(R) \sim M(R)$, where $M(R)$ is the mass of the cluster of radius R. For convenience, in the following we shall frequently use the terminology "particle" for a lattice site which belongs to the fractal (is filled) and cluster for the objects made of connected particles.

A variation of the above method is generally used if the total number of particles within a cluster is recorded during the growth. Such a situation is common for example in Monte Carlo simulations, where the structure is typically grown by subsequent addition of particles to the object. In this approach one first calculates a quantity $R_g(N)$ called radius of gyration using the expression

$$R_g(N) = \left(\frac{1}{N} \sum_{i=1}^{N} r_i^2 \right)^{1/2} , \tag{4.11}$$

where r_i is the distance of the ith particle from the center of mass of the cluster and N is the total number of particles in the cluster at the given stage of the growth process. Then, it is assumed that

$$R_g(N) \sim N^{\nu} , \tag{4.12}$$

where $\nu = 1/D$. Therefore, $1/D$ can be obtained from the slope of the plot $\ln R_g$ as a function of $\ln N$. (4.12) corresponds to the assumptions that i) in the asymptotic regime R_g is linearly proportional to the total radius of the cluster, ii) corrections due to the boundary effects can be neglected and iii) the structure is not a geometrical multifractal.

The fractal dimension of random structures can be also estimated from their *density-density correlation function* $c(r)$. According to its definition (2.14), $c(r)d^dr$ is the probability of finding a particle in the volume d^dr being at a distance r from a given particle. As was discussed in Section 2.3.1 $c(r) \sim r^{D-d}$ and this expression allows the determination of D from the corresponding log-log plot. When calculating $c(r)$ the following procedure is followed. One chooses a particle within the cluster and counts the number of particles which are within a spherical shell of radius r and width δr, where typically $\delta r \simeq 0.1r$. Then the same calculation is repeated for other particles and the result is normalized taking into account the number of centres and the volume of the shells used. In order to avoid undesirable effects caused by anomalous contributions appearing at the edge of the cluster one should not choose particles as centres close to the boundary region.

Calculation of the correlation function is obviously closely related to the previously mentioned methods. Counting the number of particles in shells corresponds to determining the derivative of $N(r)$. The most advantageous feature of calculating D by determining $c(r)$ is provided by the fact that using this method one averages over many points within a single cluster which is expected to improve the statistics.

Sometimes there is a large, slowly decaying *correction* to the simple power law behaviour of $N(R)$ or $c(r)$. This correction may have various origins and forms. For example, if the growth takes place along a *surface*, the presence of the surface usually has an effect on the overall behaviour of the quantities used for determining D. To extract the information concerning D one assumes a special functional dependence of the correction. Then, instead of fitting a straight line to the $N(R)$ data one fits a curve of the following form

$$N(R) \simeq AR^D[1 + f(R)], \tag{4.13}$$

where A is a constant and $f(R)$ is a function which is typically chosen to be decaying as an exponential or a power law (or a combination of these).

Selecting the most appropriate variables when plotting the results is another effective way to obtain more accurate data. Many times there already exists a theoretical result or a good guess of other source for the value of the fractal dimension. In such cases it is the deviation from this value which can be a quantity of interest. A common procedure is to plot for example $\ln[N(R)/R^{D_g}]$ versus $\ln R$, where D_g is a guess for the fractal dimension. If the true D is approximately equal to D_g, the straight part of the plot is close to a horizontal line, and any deviation can be magnified.

In some cases it is not only the fractal dimension one has a hypothesis for, but the entire functional dependence of $N(R)$ or $c(r)$ on their variables. The finite size of the samples which are investigated necessarily leads to a cutoff in the behaviour of these quantities at R or r values comparable with the cluster size. Because of the self-similar nature of fractals the actual form of the cutoff also scales with the total number of particles in a cluster, and its scaling behaviour is characterized by the same fractal dimension as that of the radius. This fact can be expressed by the assumption that

$$c(r) \sim (r/R)^{D-d} f(r/R), \tag{4.14}$$

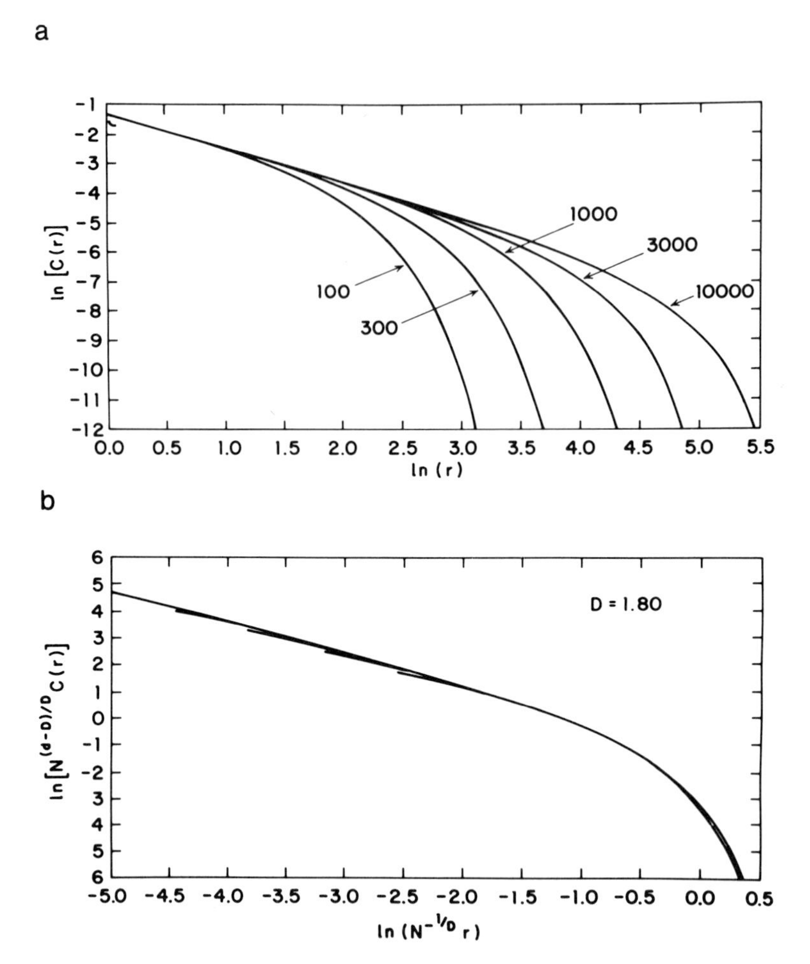

Figure 4.5. (a) Density-density correlation functions for three-dimensional off-lattice cluster-cluster aggregates (Section 8.1.1). (b) Using the assumption (4.14) the data can be scaled onto a single curve (Meakin 1987).

where $R \sim N^{1/D}$ is the radius of the cluster consisting of N particles and $f(x)$ is a cutoff function with $f(x) \simeq Constant$ for $x \ll 1$ and $f(x) \ll 1$ (exponentially small) for $x \gg 1$. According to the scaling assumption (4.14), if a structure is stochastically self-similar, the data obtained for $c(r)$ for various values of N should collapse onto the same universal curve $f(x)$, when $\ln[(rR)^{(d-D)}c(r)]$ is plotted against $\ln(r/R)$ using the correct value for D. A plot of this type is displayed in Fig. 4.5. The scaling shown in Fig. 4.5b both provides a check of self-similarity and leads to a more reliable estimate of the fractal dimension (Meakin 1987) than that obtained by attempting to fit a straight line to the plot of $\ln c(r)$ versus $\ln r$ over an intermediate range of length scales.

In case of a *fractal measure* defined on a growing structure, there is a weight or probability attributed to each particle. The generalized dimensions D_q for such objects can be obtained using a procedure analogous to the box counting method described at the beginning of this section. The only difference is that instead of simply counting the number of particles within a region of radius R, one calculates $\chi_q(R)$, the sum of the qth power of probabilities associated with the particles as a function of R. Then the generalized dimensions can be determined from the slopes of plots of $\ln \chi_q(R)$ versus $\ln R$, since according to (3.7) $\ln \chi_q(R) \sim (q-1)D_q \ln R$. The $f(\alpha)$ spectrum is obtained from D_q by Legendre transformation (see (3.11), (3.14) and (3.15)).

When discussing various evaluation techniques we assumed that the necessary data are already available. To obtain the arrays of coordinates corresponding to the positions of particles belonging to a fractal usually requires numerical procedures depending on the particular physical process which is to be investigated. There are, however, a few general remarks which should be considered when one simulates the growth of fractals in a computer.

For example, when applying the flexible Monte Carlo method one typically adds particles to the growing cluster according to some rules given by the model which is used to simulate the phenomenon. The true fractal behaviour is manifested only for $N \to \infty$ while, obviously, a computer simulation has limitations concerning the total number of particles N. In case of slow crossovers between different kinds of scaling behaviour – which is typical for many growth phenomena – the data have to be extrapolated with special care. Taking into account correction-to-scaling terms (4.13) or finite size effects (4.14) are examples for such analysis. Another possibility is to use periodic boundary conditions when it is possible, and in this way mimic an infinite system by a periodic sequence of finite subsystems.

Finally, in stochastic simulations randomness is introduced with the help of *random number generators*. Here again, one has to be careful, because it can be easily shown that the simplest methods fail to produce a long sequence of statistically uncorrelated random numbers. This problem can be avoided by "mixing" two random number generators (Stauffer 1986).

Exact enumeration techniques provide an alternative to Monte Carlo methods to generate data for the determination of fractal dimensions. Here the philosophy is quite different; instead of generating large clusters with stochastic deviations from the true average behaviour, one studies small clusters exactly and extracts results from careful extrapolation to the large system limit. Growth phenomena are not very suitable for such approaches because in most of the models the same configuration can be realized in many ways each contributing to the average scaling with a different weight. This fact usually makes the calculations prohibitively expensive.

In the case of modelling growing self-avoiding walks (Section 5.4.2) the situation is less complicated since there is a unique sequence leading to a given chain. For self-interacting growing walks each configuration of N steps (particles) has its own weight (probability) which is associated with it when calculating the average radius of the walks. Let us denote by $\langle R_e^2(N)\rangle$ the mean-square end-to-end distance and suppose that it is proportional to the mean-squared radius of a chain. To obtain D from the enumeration data one assumes the following scaling form (Djordjevic *et al* 1983)

$$\langle R_e^2(N)\rangle = AN^{2/D}(1 + BN^{-\Delta} + CN^{-1} + ...)\,, \tag{4.15}$$

where A, B and C are constants and Δ is a non-trivial correction-to-scaling exponent. (4.15) is expected to be a good approximation for $N \gg 1$. From the above expression one finds an estimate for the fractal dimension defined for clusters consisting of N particles

$$D(N) = 2\frac{\ln[(N+i)/N]}{\ln[\langle R_e^2(N+i)\rangle/\langle R_e^2(N)\rangle]} = D + \frac{\Delta B}{2}N^{-\Delta} + \frac{C}{2}N^{-1} + ... \tag{4.16}$$

Assuming that $\Delta > 1$ one finds D from the intercept of the plot of $D(N)$ against $1/N$ with the $D(N)$ axis at $1/N = 0$ (see Section 5.4.2). The value of the integer number i is usually chosen to be 1 or 2 depending on the type of lattice on which the growth takes place.

To complete the analysis it has to be shown that the correction-to-scaling exponent is indeed larger than unity. This can be done by plotting the quantity $\ln[D(N) - D]$ versus $\ln N$, where D is the asymptotic value as determined by the above extrapolation. If $\Delta > 1$, one gets a slope equal to -1; otherwise the slope is equal to $-\Delta$.

4.3. RENORMALIZATION GROUP

It has already been mentioned in the Introduction that there is a close relationship between fractals and critical phenomena. In the experiments on systems exhibiting second order phase transition a power law dependence of the relevant physical quantities was observed. The exponents characterizing the scaling of these quantities were found to have non-integer values just like the mass of a growing fractal scales with its radius according to an exponent D which is not an integer number. In fact, the analogy is deep, since it is self-similarity which is behind non-standard scaling in both cases. This was shown in the investigations of critical phenomena where the scale invariance of the systems at the critical point was demonstrated by both experimental and theoretical approaches.

The above scale invariance forms the basis of renormalization group theory which has been successfully applied to the description of continuous phase transition through the calculation of the critical exponents and the so-called phase diagrams revealing the relevance of the parameters effecting the transition. The idea of the Position-Space Renormalization Group (PSRG) approach is to renormalize a system with many degrees of freedom into a system having less degrees of freedom. The origin of many (infinite) degrees of freedom and scale invariance is the same: at the critical point the system possesses large fluctuations (regions belonging to one of the two phases) with no characteristic size.

In the course of application of PSRG a part of the system defined on a lattice is replaced with a cell containing smaller number of sites (Fig. 4.6). In order to account for scale invariance, however, during this transformation

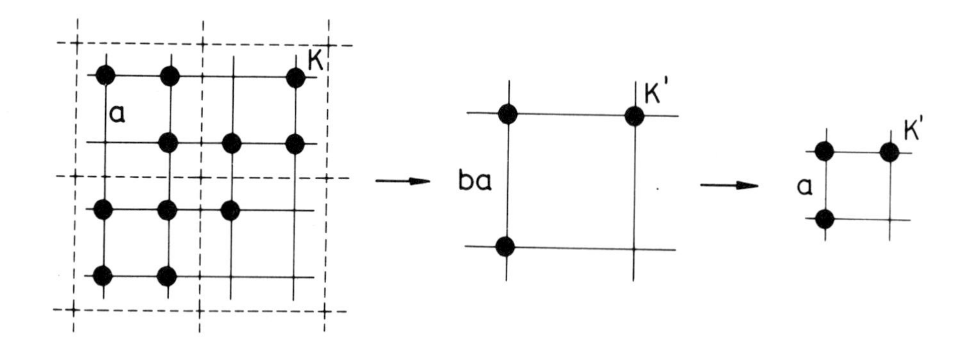

Figure 4.6. During this simple version of position-space renormalization four sites of the original system are replaced with a single new site having a renormalized fugacity K'. In this example the factor by which the linear size of the system is rescaled is equal to $b = 2$.

one has to change the weights associated with the filled sites so that the two systems (the original and the renormalized one) behave in the same way. The successive replacement of larger parts of the system with smaller ones is an *inverse analogue* of the recursive generation of deterministic fractals. At its critical point an infinite system is invariant under the renormalization procedure.

It is a quite obvious idea to try to apply this concept to growth phenomena leading to fractal structures. The renormalization scheme to be described below is analogous to the PSRG used for calculating critical exponents of equilibrium systems (Stanley *et al* 1982), and is based on the *generating function* (Nakanishi and Family 1985)

$$G(K) = \sum_{N=1}^{\infty} \sum_{i} P_{N,i} K^N , \tag{4.17}$$

where $P_{N,i}$ is the probability of creating the ith configuration consisting of N particles and K is a fugacity associated with each element of a cluster so that the total fugacity of a cluster is K^N. In the case of growth phenomena the total probability of generating clusters of size N is

$$\sum_i P_{N,i} = \sum_i \prod_{j=1}^{N} p_{N,i}(j) = 1\,, \tag{4.18}$$

since the creation of a cluster of N particles is a certain event (occurs with a probability equal to one) for all N, because of the ever growing character of the process. In the above expression $p_{N,i}(j)$ is the probability of adding the jth particle to the ith configuration of N particles. According to (4.18)

$$G(K) = \sum_{N=1}^{\infty} K^N\,. \tag{4.19}$$

This power series has a radius of convergence K_c=1.

Next we calculate R_g, the average radius of gyration of clusters using the generating function (4.19). Since the radius of gyration of a cluster consisting of N particles scales as $R_g(N) \sim N^\nu$ with $\nu = 1/D$ (see preceding section), we can determine the fractal dimension by finding a relation between $R_g(N)$ and N for a given growth process. The grand canonical average of $R_g(N)$ is

$$R_g = \frac{\sum_N N^\nu K^N}{\sum_N K^N} \sim (K_c - K)^{-\nu}\,. \tag{4.20}$$

To see this, one approximates the above sums by integrals (this can be done because the singular contribution to the sums come from the region $N \gg 1$). For the numerator one has

$$\begin{aligned}\int_0^\infty N^\nu K^N dN &\sim \int_0^\infty N^\nu e^{N\ln[1-(K_c-K)]} dN \\ &\sim (K_c - K)^{-\nu-1} \int_0^\infty z^\nu e^{-z} dz \sim (K_c - K)^{-\nu-1}\,,\end{aligned} \tag{4.21}$$

while similarly, the denominator in (4.20) diverges as $(K_c - K)^{-1}$. Therefore, if a quantity scales with growing N according to an exponent ν, the same quantity diverges with an exponent $-\nu$ when $K \to K_c = 1$. The calculation of D then reduces to the determination of ν for which the standard renormalization group method can be applied with the modifications discussed below.

As in usual critical phenomena the condition for renormalization is to *conserve a generalized form* of the generating function (4.17), while rescaling all lengths by a factor denoted by b. One should not consider the generating function (4.17) as a quantity to be conserved; this would always yield $K' = K$ and does not lead to any reasonable conclusion. Hypothetically exact renormalization would generate "further range interactions" which would require the consideration of many parameter generating functions and many-cell renormalization.

Instead, as a pragmatic simplification, one chooses a one-parameter formalism corresponding to a modified generating function based on the calculation of the most relevant contributions. This is achieved by equating the renormalized fugacity K' to the contribution of spanning configurations to $G(K)$ within a cell of size b^d

$$K' = \sum_N \sum_{i'} P_{N,i'} K^N \,, \tag{4.22}$$

where $P_{N,i'}$ is given in (4.18). Here the second summation is quite non-trivial, it is taken over all spanning configurations (labelled by i') consisting of N particles, where spanning must be defined appropriately for each problem (Nakanishi and Family 1985).

From the known renormalization transformation $K'(K)$, the fractal dimension can be calculated by the usual fixed-point analysis. Let us linearize $K'(K)$

$$K_c - K' \simeq \lambda(K_c - K) \tag{4.23}$$

around its fixed point K_c, where $K'(K_c) = K_c$ and $\lambda = dK'/dK|_{K_c}$. On the other hand, R_g in the system with K is proportional to $(K_c - K)^{-\nu}$, while in the renormalized one $R_g \sim (K_c - K')^{-\nu}$. Since the system with fugacity K'

is obtained by rescaling of the lattice units by a factor b, the condition that the radius of gyration should be invariant under the renormalization leads to

$$b(K_c - K')^{-\nu} \simeq (K_c - K)^{-\nu} . \tag{4.24}$$

Comparing (4.23) and (4.24) we find

$$D = \frac{1}{\nu} = \frac{\ln \lambda}{\ln b} , \tag{4.25}$$

where λ, the eingenvalue of the recursion relation (4.22) linearized around K_c can be calculated for small cells analytically, while for large cells it can be obtained using numerical methods.

To determine D in a small cell renormalization method one needs to calculate the sum of $P_{N,i}$-s corresponding to spanning configurations, where the definition of spanning depends on the particular model to be renormalized. This will be demonstrated on the example of true self-avoiding walks at the end of this section. In general, small cell renormalization does not lead to good estimates for the fractal dimension in the case of growth models.

Large-cell Monte Carlo (MC) renormalization (Stanley *et al* 1982, Nakanishi and Family 1985) represents an alternative way to improve the accuracy without including more than one parameter. In this method cells with large b are considered, and the actual configurations are generated by a computer. Using larger cells is expected to lead to more reliable results for a number of reasons. Within a cell the behaviour of the system can be well approximated, and going to larger b increases the size of region which is treated with a good accuracy. In addition, undesirable surface effects are gradually eliminated as $b \to \infty$.

The basic idea is that the sum of $P_{N,i'}$-s taken over i' is nothing else than the fraction of spanning configurations of size N among all configurations consisting of N particles. Therefore, if we generate configurations randomly (according to the rules of the given model), then the sum of $P_{N,i'}$-s correspond to the fraction of spanning configurations among all the configurations generated. In this approach first one determines D_b, the fractal dimension obtained using cells of linear size b, and then plots these values against $1/\ln b$ to obtain the extrapolated value corresponding to the presumably exact $b \to \infty$ limit.

The Monte Carlo renormalization method can not be used with satisfactory results if the specific properties of clusters are disadvantageous with regard to the spanning rule. In the case of growth processes it is quite common that the behaviour of the clusters' surface is qualitatively different from their global properties. This fact is likely to be the reason for the unusually slow convergence of large-cell MC results.

The results can be improved by modifying the spanning rule. According to this modification, a cluster is considered as spanning the cell if its radius of gyration (R_g) becomes equal to $b/2$. R_g represents an averaging over the shape of the cluster and it is expected that the effects caused by strong fluctuations in the surface structure can be eliminated by using R_g for characterizing the spatial extent of a cluster. A further improvement can be achieved by introducing a phenomenological parameter κ and assuming that κR_g should become equal to $b/2$. The "fixed point" κ^* of this optimization parameter is defined as the value for which the estimates of D for various b are the same. This method makes it possible to obtain accurate results using relatively small cells (see Section 6.1.3).

EXAMPLE

In this example we shall demonstrate how to apply small cell renormalization to cluster growth processes (Nakanishi and Family 1984), by calculating the fractal dimension of the so-called true self-avoiding walks (TSAW-s). This is a simple, but inherently kinetic growth model with non-trivial behaviour different from that of ordinary random walks in dimensions not larger than two. A TSAW is a random walk which attempts to avoid itself whenever it is possible. In its simplest version treated here (more details about TSAW-s will be given in Section 5.4.1) a true self-avoiding walk can cross itself only if there is no way to proceed without jumping into a site which already has been visited.

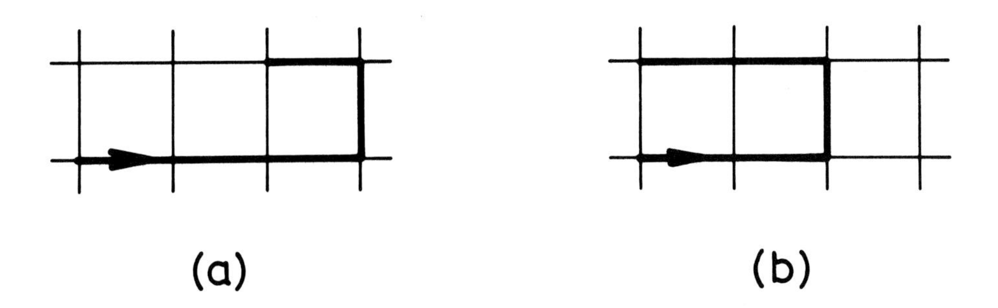

Figure 4.7. Two examples for true self-avoiding walks. Their weights are different, because the configuration 4.7b has two constrained steps, while 4.7a has only one (the last step).

By definition such a walk grows indefinitely and the statistics of a TSAW is controlled by its past history. This statement is illustrated by Fig. 4.7 showing two five steps walks. They are created with different probabilities $P_{5,1} = (\frac{1}{4})(\frac{1}{3})^3(\frac{1}{2})$ and $P_{5,2} = (\frac{1}{4})(\frac{1}{3})^2(\frac{1}{2})^2$, since the configuration Fig. 4.7b contains a constrained step at the end. Moreover, a walk has a different weight when traced in the opposite direction.

Application of the small cell renormalization approach on a 2×2 cell requires the calculation of the relative probability of spanning configurations. To make the spanning rule specific, we assume that all walks start at a corner site of the cell, and are restricted to stay within the cell until they exit via one of the external bonds. The probabilities $p_{N,i}(j)$ are calculated by counting at each step only those open bonds which are within the cell. Because of symmetry spanning in one direction is considered. Figure 4.8 shows the spanning configurations together with their weights and the renormalized cell. According to this figure the renormalization transformation (4.22) for TSAW on a 2×2 cells has the form

$$\frac{1}{2}K' = \frac{1}{4}K^2 + \frac{1}{6}K^3 + \frac{1}{12}K^4 . \qquad (4.26)$$

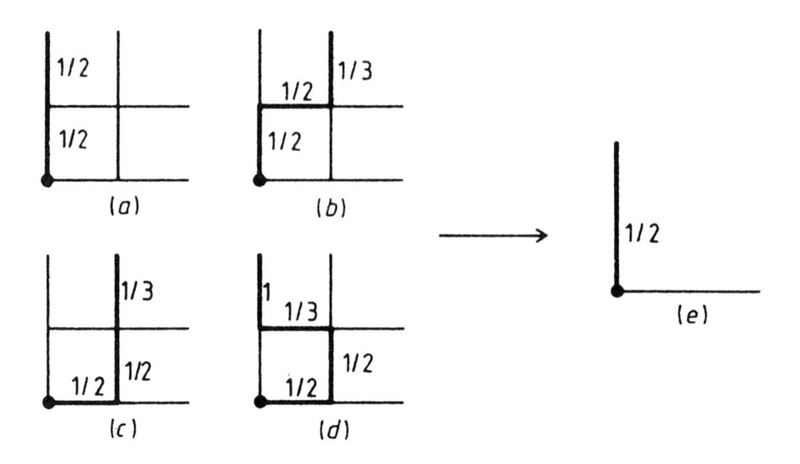

Figure 4.8. Spanning configurations of true self-avoiding walks in a 2×2 cell which are renormalized into one in (e). The probabilities corresponding to a given step are also indicated (Nakanishi and Family 1984).

From here $K_c = 1$ and $\lambda = dK'/dK|_{K_c} = \frac{8}{3}$. Using (4.25) yields for the fractal dimension $D = \ln(8/3)/\ln 2 \simeq 1.415$. The corresponding result for a 3×3 cell is somewhat larger, $D \simeq 1.419$. Calculations for the three-dimensional case can be carried out analogously. For the $2 \times 2 \times 2$ cell one gets

$$\begin{aligned} \frac{1}{3}K' = & \frac{1}{9}K^2 + \frac{1}{9}K^3 + \frac{11}{180}K^4 + \frac{29}{1080}K^5 \\ & + \frac{127}{8640}K^6 + \frac{79}{12960}K^7 + \frac{61}{25920}K^8 \end{aligned} \tag{4.27}$$

which leads to $D \simeq 1.70$.

This example demonstrates the advantages and problems associated with small cell renormalization of growth models. The numerical results obtained are rather poor since it can be shown that the fractal dimension of TSAW is 2 for $d \geq 2$. On the other hand, small cell renormalization can provide qualitative information about D using simple algebra, even for highly non-trivial models. This method is more useful for studying growth processes depending on a parameter. Then, application of a two-parameter version of

PSRG may reveal whether this parameter is relevant enough to change the fractal dimension. However, calculations of this sort published so far are based on approximations whose justification is not sufficiently satisfactory.

REFERENCES (PART I)

Allain, C. and Cloitre, M., 1986 *Phys. Rev.* **B33**, 3566

de Arcangelis, L., Redner, S. and Coniglio, A., 1985 *Phys. Rev.* **B31**, 4725

de Arcangelis, L., Redner, S. and Coniglio, A., 1986 *Phys. Rev.* **B34**, 4656

Bentley, W. A. and Humpreys, W. J., 1962 *Snow Crystals* (Dover, New York)

Besicovitch, A. S., 1935 *Mathematische Annalen* **110**, 321

Benzi, R., Paladin, G., Parisi, G. and Vulpiani, A., 1984 *J. Phys.* **A17**, 3521

Boccara, N. and Daoud, M., editors, 1985 *Physics of Finely Divided Matter* (Springer, New York)

Djordjevic, Z. V., Majid, I., Stanley, H. E. and dos Santos, 1983 *J. Phys.* **A16**, L519

Engelman, R. and Jaeger, Z., editors, 1986 *Fragmentation, Form and Flow in Fractured Media Ann. Isr. Phys. Soc*, **8**

Family F. and Landau, D. P., editors, 1984 *Kinetics of Aggregation and Gelation* (North-Holland, Amsterdam)

Farmer, J. D., 1982 *Z. Naturforsch.* **37a**, 1304

Farmer, J. D., Ott, E. and Yorke, J. A., 1983 *Physica* **7D**, 153

Farmer, J. D., 1986 in *Dimensions and Entropies in Chaotic Systems* edited by E. Mayer-Kress (Springer, Berlin) p. 54

Feder, J., 1988 *Fractals* (Plenum Press, New York)

Frisch, U. and Parisi, G., 1984 in *Turbulence and Predictability in Geophysical Fluid Dynamics and Climate Dynamics*, International School of Physics "Enrico Fermi", Course LXXXVIII, edited by M. Ghil, R. Benzi and G. Parisi (North-Holland, New York) p.84

de Gennes, P. G., 1985 in *Physics of Disordered Materials* edited by D. Adler, H. Fritzsche and S. R. Ovshinsky (Plenum Press, New York)

Halsey, T.C., Jensen, M. H., Kadanoff, L. P., Procaccia, I. and Shraiman, B. I., 1986 *Phys. Rev.* **A33**, 1141

Hausdorff, F., 1919 *Mathematische Annalen* **79**, 157

Hentschel, H. G. E. and Procaccia, I., 1983 *Physica* **8D**, 435

Herrmann, H., 1986 *Phys. Rep.* **136**, 154

Jullien, R. and Botet, R., 1987 *Aggregation and Fractal Aggregates* (World Scientific, Singapore)

Jullien, R., 1986 *Ann. Telecomm.* **41**, 343

Kolmogorov, A. N. and Tihomirov, V. M., 1959 *Uspekhi Math. Nauk* **14**, 3

Mandelbrot, B. B., 1974 *J. Fluid. Mech.* **62**, 331

Mandelbrot, B. B., 1975 *Les Objects Fractals: Forme, Hasard et Dimension* (Flammarion, Paris)

Mandelbrot, B. B., 1977 *Fractals: Form, Chance and Dimension* (Freeman, San Francisco)

Mandelbrot, B. B., 1980 *Annals New York Acad. Sci.* **357** 249

Mandelbrot, B. B., 1982 *The Fractal Geometry of Nature* (Freeman, San Francisco)

Mandelbrot, B. B., 1985 *Physica Scripta* **32**, 257

Mandelbrot, B. B., 1986 in *Fractals in Physics* edited by L. Pietronero and E. Tosatti (Elsevier, Amsterdam) p. 3

Mandelbrot, B. B., 1988 *Fractals and Multifractals: Noise, Turbulence and Galaxies* (Springer, New York)

Meakin, P., 1987a in *Phase Transitions and Critical Phenomena* edited by C. Domb and J. L. Lebowitz, Vol. 12. (Academic Press, New York)

Meakin, P., 1987b *CRC Crit. Rev.* **13**, 143

Nakanishi, H. and Family, F., 1984 *J. Phys.* **A17**, 427

Nakanishi, H. and Family, F., 1985 *Phys. Rev.* **A32**, 3606

Paladin, G. and Vulpiani, A., 1987 *Phys. Rep.* **156**, 147

Peitgen, H. O. and Richter, P. H., *The Beauty of Fractals* (Springer, Berlin)

Pfeifer, P. and Avnir, D., 1983 *J. Chem. Phys.* **79**, 3558

Pfeifer, P., 1986 in *Fractals in Physics* edited by L. Pietronero and E. Tosatti (Elsevier, Amsterdam) p. 47

Pietronero P. and Tossati, E., editors, 1986 *Fractals in Physics* (North-Holland, Amsterdam)

Pynn, R. and Skjeltorp, A., editors, 1986 *Scaling Phenomena in Disordered Systems* NATO ASI Series B133 (Plenum Press, New York)

Rényi, A., 1970 *Probability Theory* (North-Holland, Amsterdam)

Sander, L. M., 1986 *Nature* **322**, 789

Schaefer, D. W., Martin, J. E., Wiltzius, P. and Cannel, D. S., 1984 in *Kinetics of Aggregation and Gelation* edited by F. Family and D. Landau (North-Holland, Amsterdam) p. 71

Shlesinger, M., editor, 1984 *Proceedings of a Symposium on Fractals in Physical Sciences* NBS Gaithersburg, J. Stat. Phys. **36** No. 8/6

Squires, G. L., 1978 *Introduction to the Theory of Thermal Neutron Scattering* (Cambridge Univ. Press, London)

Stanley, H. E., 1977 *J. Phys.* **A10**, L211

Stanley, H. E., Reynolds, P. J., Redner, S. and Family, F., 1982 in *Real-Space Renormalization* edited by T. Burkhardt and J. M. J. van Leeuwen (Springer, New York)

Stanley, H. E. and Ostrowsky, N., editors, 1986 *On Growth and Form* NATO ASI Series E100 (Martinus Nijhoff, Dordrecht)

Stauffer, D., 1986 in *On Growth and Form* edited by H. E. Stanley and N. Ostrowsky (Martinus Nijhoff, Dordrecht) p. 79

Tél, T. and Vicsek, T., 1987 *J. Phys.* **A20**, L835

Tél, T., 1988 to appear in *Z. Naturforschung*

Teixeira, J., 1986 in *On Growth and Form* edited by H. E. Stanley and N. Ostrowsky (Martinus Nijhoff, Dordrecht) p. 145

Umberger, D. K., Mayer-Kress, G. and Jen, E., 1986 in *Dimensions and Entropies in Chaotic Systems* edited by G. Mayer-Kress (Springer, Berlin) p. 42

Vicsek, T., 1983 *J. Phys.* **A16**, L647

Voss, R. F., 1985 in *Fundamental Algorithms in Computer Graphics* edited by E. A. Earnshaw (Springer, Berlin) p. 805

Weierstass, K., 1895 *Mathematische Werke* (Mayer and Müller, Berlin)

Witten, T. A. and Cates, M. E., 1986 *Science* **232**, 1067

Part II

CLUSTER GROWTH MODELS

Chapter 5

LOCAL GROWTH MODELS

In Part II models based on growing structures made of identical particles will be treated. While in Chapter 2 mostly artificial examples were discussed, here we shall concentrate on more realistic models which are constructed in order to reflect the essential features of specific growth phenomena occurring in nature. Various models allowing exact or numerical treatment have been playing an important role in the studies of growth. Because of the complexity of the phenomena it is usually a difficult task to decide which of the factors affecting the growth plays a relevant role in determining the structure of the growing object. In a real system the number of such factors can be relatively large, and this number is decreased to a few by appropriate model systems. Thus, the investigation of these models provides a possibility to detect the most relevant factors, and demonstrate their effects in the absence of any disturbance.

Structures consisting of connected particles are usually called *clusters* or *aggregates*. In most of the cases the growth will be assumed to take place on a *lattice* for computational convenience, and two particles are regarded as connected if they occupy nearest neighbour sites of the lattice. However, for studying universality and related questions, *off-lattice* or further neighbour versions of clustering processes can also be investigated. A lattice site with a particle assigned to it is called occupied or filled. An important additional feature included into the majority of models to be described is stochasticity which is typical for growth phenomena.

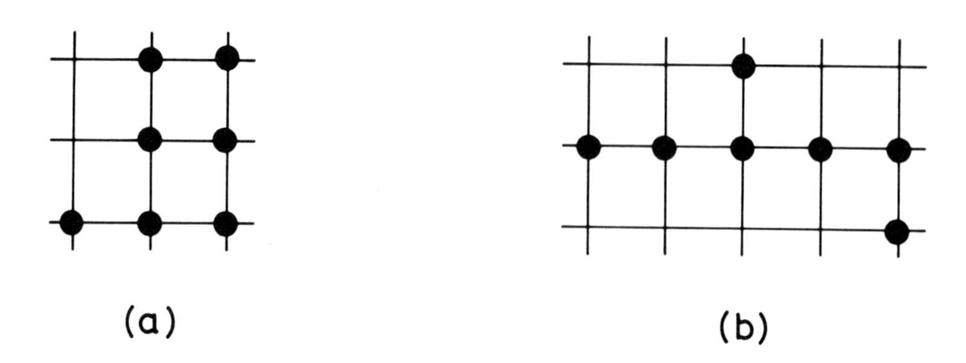

Figure 5.1. Two possible configurations (clusters) consisting of the same number of particles (black sites). The statistical weight of a given cluster depends on its geometry. For example, the probability associated with configuration (a) is larger for a growth process which preferably produces compact clusters.

In general, a stochastic cluster growth model may lead to all possible configurations which can be formed from a given number of particles. What makes these models differ from each other is the weight or probability $P_{N,i}$ associated with a given configuration i consisting of N units (Fig. 5.1). $P_{N,i}$ can be different for the same configuration even in the same model, because generally it depends on the sequence, according to which the individual particles are added to the cluster (Herrmann 1986). The distribution of $P_{N,i}$ as a function of i is uniquely related to the particular model investigated, and determines the value of the quantity

$$S_N(q) = -\frac{1}{N(q-1)} \ln \sum_i P_{N,i}^q \tag{5.1}$$

analogous to the order q Rényi information. On the basis of (5.1) it is possible to define (Vicsek *et al* 1986)

$$S = \lim_{q \to 1} \lim_{N \to \infty} S_N(q) = - \lim_{N \to \infty} \frac{1}{N} \sum_i P_{N,i} \ln P_{N,i} \,. \qquad (5.2)$$

corresponding to the configurational entropy of a given cluster growth model.

In the following we shall distinguish two main types of cluster growth processes, depending on the global character of the rule which is used in the course of adding a particle (or a cluster of particles) to the growing cluster. *The rule will be called local if it depends only on the immediate environment* of the position where the new particle is to be added. In other words, when deciding whether to add a particle at site $\vec{x}$ only the status (filled or not) of the nearest or next nearest neighbours of this site is taken into account. On the contrary, *in non-local models the structure of the whole cluster* can affect the probability of adding a site at a given position.

5.1. SPREADING PERCOLATION

In this section we shall consider a model which represents perhaps the simplest growth process leading to a branching fractal structure (Leath 1976, Alexandrowitz 1980). The process starts with a single seed particle placed onto a site of a lattice. Its neighbouring sites are considered live in the sense that they potentially may become occupied in the future. Next, one of these live sites is chosen randomly and (i) filled with a particle with a probability p or (ii) killed for the rest of time with a probability $1 - p$. Occupation of a site with probability p is realized by generating a random number r (with a uniform distribution on $(0, 1])$ and filling the site if $r < p$. The filled site becomes part of the growing cluster and its new neighbours become living sites. A large cluster is grown by repeating the same procedure many times. In a variation of this model at each time step all of the living or *growth sites* are considered for occupation, instead of one at a time, therefore, the cluster grows by adding shells to it. In addition, one can replace the seed particle with a hyperplane of seed particles.

The above process is relevant to a number of spreading phenomena, including epidemics, chemical reactions, flame propagation, etc. For exam-

Figure 5.2. Result of a typical run of growing a percolation cluster along a line for $p = p_c$. The cluster is generated on a triangular lattice by adding to it all of the growth sites at each time step. The growth sites are denoted by heavy dots. (Grassberger 1985).

ple, using the language of epidemic, the live sites are susceptible to infection, the killed sites are immune, while the occupied sites correspond to infected individuals (Grassberger 1983). An epidemic will spread over the whole population if there is always at least one live site. Figure 5.2 shows a configuration of filled sites generated on a triangular lattice using a straight interval as the seed configuration. It seems to have a complex structure with holes and fjords having no characteristic size that is typical for fractal clusters.

Note, that although the random numbers r are generated during the growth, the same configurations are obtained as if we had assigned to all of the sites of the lattice a random number previously, then defined equilibrium percolation clusters as connected objects consisting of sites for which $r < p$ and started the process afterwards. Therefore, the above model is equivalent to a simple type of growth on a static percolation cluster and as the available sites of a given configuration are filled we gradually recover an equilibrium percolation cluster. The growth stops when all sites belonging to the cluster containing the seed particle are filled. Starting the spreading algorithm many times with different initial conditions one can reproduce ordinary percolation clusters with a size distribution corresponding to p.

Equilibrium percolation is a widely used model for describing various properties of inhomogeneous media (Stauffer 1985). Here we only recall those results of percolation theory which are related to the fractal nature (Stanley 1977) of spreading percolation clusters. In particular, when p is increased, at $p = p_c$ a transition takes place which is manifested in the appearance of a connected infinite cluster having a density for $0 < p - p_c \ll 1$

$$P(p) \sim (p - p_c)^{\beta}, \tag{5.3}$$

where $\beta > 0$ is the critical exponent of the percolation probability, $P(p)$ and p_c is called percolation threshold. In the following we consider the properties of the infinite cluster. The correlation length diverges at p_c according to

$$\xi \sim |p - p_c|^{-\nu_p} \tag{5.4}$$

where ξ corresponds to the radius at which the power law decay of the pair correlation function $c(r)$ (see (2.14)) crosses over into a constant behaviour. As before, because of scaling we can assume that for $\xi \gg a$ and $r \gg a$ (a is the lattice spacing) the only relevant length is ξ, and correspondingly (Kapitulnik *et al* 1985),

$$c(r) \sim P(p) f(r/\xi), \tag{5.5}$$

where the scaling function $f(x)$ approaches a constant for $x \to \infty$. For $x \ll 1$ we expect that $c(r)$ is independent of ξ which can be satisfied only if $f(x) \sim x^{-\beta/\nu_p}$ for $x \ll 1$. From here and using (5.3) and (5.4) we find $c(r) \sim r^{-\beta/\nu_p}$ for $r \ll \xi$. Then, in analogy with the arguments leading to (2.18) one obtains for the fractal dimension of the infinite cluster at p_c ($\xi \to \infty$)

$$D = d - \frac{\beta}{\nu_p} \,. \tag{5.6}$$

The exponents β and ν_p are known exactly for $d = 2$ and can be calculated by numerical or theoretical methods for higher dimensions. For example, in two dimensions $\nu_p = \frac{4}{3}$ and $\beta = \frac{5}{36}$ gives for the fractal dimension of the infinite cluster at p_c $D = \frac{91}{48} \simeq 1.896$.

As mentioned above the spreading percolation process reproduces a static percolation cluster, when it terminates. However, if the growth on the given finite cluster is not completed yet, the structure of the growing and equilibrium percolation clusters is not exactly the same. According to the simulations the fractal dimension determined during the growth is not affected by the algorithm, but the same is not true for the overall shape of the clusters (Family *et al* 1985). This can be shown by calculating the radius of gyration tensor and determining the ratio, $\langle A_N \rangle$, of the principal radii of gyration. The results presented in Fig. 5.3 demonstrate that spreading percolation clusters are in the asymptotic limit elongated, but to a smaller extent than the static ones.

The truly *dynamic behaviour* of growing percolation clusters can be interpreted in terms of the total number and distribution of *growth sites* which are here defined as filled sites having live neighbours. Let us first consider spreading at $p = p_c$ occupying one site at each time step. The total number of growth sites N_G scales with the number of particles in the cluster N as (Family and Vicsek 1985, Herrmann and Stanley 1985)

$$N_G \sim N^{\delta} \,, \tag{5.7}$$

where N plays the role of time. Since the radius of the region over which the growth sites are scattered increases as $N^{1/D}$, we conclude that the set of growth sites forms a fractal of dimension $D_G = \delta D$. In the case of filling shells (considering all live sites simultaneously for occupation) at a given time step t, one has a different law (Alexandrowitz 1980)

$$N_G(t) \sim t^{d_s - 1}, \tag{5.8}$$

where d_s is called the spreading or chemical dimension. Comparing (5.7) and (5.8) one gets

$$\delta = \frac{d_s - 1}{d_s}. \tag{5.9}$$

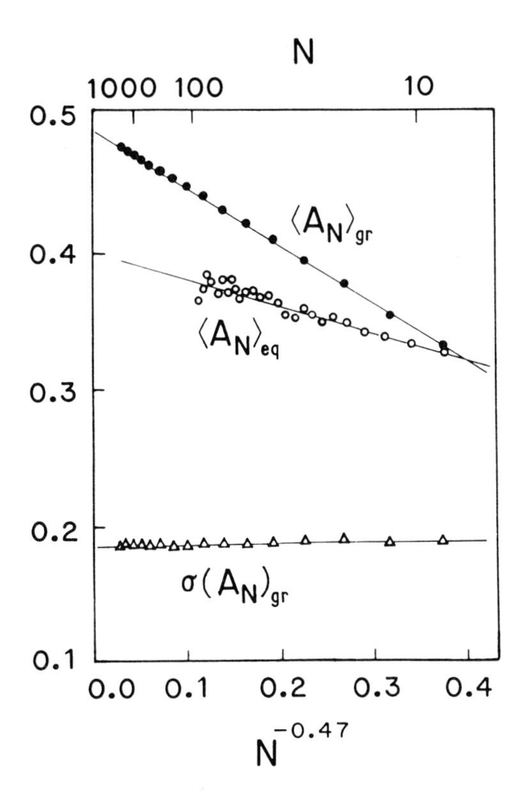

Figure 5.3. The average anisotropy of equilibrium ($\langle A_N \rangle_{eq}$) and growing ($\langle A_N \rangle_{gr}$) percolation clusters as a function of $N^{-\theta}$, where $\theta = 0.47$ is the correction-to-scaling exponent. The fluctuation in the anisotropy of growing percolation clusters, $\sigma(A_N)_{gr}$, is also shown (Family *et al* 1985).

Introducing the exponent $\nu_\parallel$ through the relation $R(t) \sim t^{\nu_p/\nu_\parallel}$, where $R(t)$ is the average radius of the infected region, it is possible to derive an additional relation (Grassberger 1983, 1985) involving the exponents already defined

$$d_s = \frac{d\nu_p - \beta}{\nu_\parallel} . \tag{5.10}$$

Both (5.9) and (5.10) are supported by numerical simulations (Herrmann and Stanley 1985, Grassberger 1985). For $d = 2$ the following results were obtained: $\delta \simeq 0.402$, $d_s \simeq 1.675$ and $\nu_\parallel \simeq 1.509$.

A more detailed description of the dynamics of spreading is provided by the radial distribution of growth sites $N_g(r, N)$, where $N_g(r, N)\Delta r$ is the number of growth sites located from the seed at a distance between r and $r + \Delta r$ and N is the actual number of already occupied sites. Scaling considerations suggest, and related simulations support that (Herrmann and Stanley 1985)

$$N_g(r, N) \sim N^{\delta - 1/D} f\left(\frac{r}{N^{1/D}}\right) . \tag{5.11}$$

This scaling form expresses the fact that the distribution is determined by only one relevant length scale proportional to $R \sim N^{1/D}$. The prefactor $N^{\delta-1/D}$ is included to satisfy $\int N_g(r, N)dr \sim N^\delta$ (see Eq. (5.7)). Here and in the following the scaling behaviour of such integrals is determined by performing a change of variable $z = r/N^{1/D}$.

Similarly, one can determine the function $P(r, N)$ which is the probability of choosing a growth site being at a distance r from the seed after N particles have been added to the cluster. Again, the only relevant parameter is N and (Bunde *et al* 1985a)

$$P(r, N) \sim N^{-1/D} g\left(\frac{r}{N^{1/D}}\right) , \tag{5.12}$$

where the scaling function $g(x)$ typically is close to a Gaussian. Scaling laws like (5.11) or (5.12) can be checked numerically by collecting data for various r values and examining whether the results fall onto a universal curve when plotting the appropriately rescaled variables.

The problem of spreading percolation has attracted considerable interest recently. It was analysed by field theoretical formalism (Cardy and

Grassberger 1985), modified by i) taking into account revival of the dead sites (Ohtsuki and Keyes 1986) and by ii) introducing various rules for choosing a growth site before making a decision about filling one (e.g., Bunde *et al* 1985b).

5.2. INVASION PERCOLATION

Invasion percolation was introduced in order to simulate the *displacement of one fluid by another in porous media* under the condition that the capillary forces dominate the motion of the interface (Lenormand and Bories 1980, Chandler *et al* 1982). Many porous media may be represented conveniently as a network of pores joined by narrower connecting throats. Consider the process of a non-wetting fluid, say oil, being displaced from such a medium by a wetting fluid, say water, at a constant but very small flow rate. In this limit the capillary forces dominate which are the strongest at the narrowest places in the medium. The interface moves quickly through a throat but gets stuck in the larger pores. This motion can be represented as a series of discrete jumps in which at each time step the water displaces oil from the smallest available pore.

As the water behind the interface advances, it may entirely surround regions filled with oil. Since oil is incompressible, one must take into account in a model that water can not invade finite, isolated regions of "residual oil".

The model for computer simulation of the above process is defined on a lattice. i) A random number drawn from the uniform distribution on the unit interval is assigned to each site of a cell of linear size L. ii) As in the growing percolation model the process starts with a seed particle or a surface and goes on by subsequent occupation of one of the perimeter sites (empty sites which are nearest neighbours of the cluster). iii) However, the perimeter site to be occupied is not selected randomly, but the one with the smallest random number r (corresponding to the smallest capillary force) is occupied.

In this version of the invasion percolation model (which simulates the interface motion, if one of the fluids is infinitely compressible) the process does not stop until the finite cell is filled in completely, since we do not have a temperature-like parameter analogous to the occupation probability p of ordinary percolation. On the other hand, as a well defined configuration one can study the structure of the region filled in by the invader fluid at the point in time when the invader first percolates, i.e., first forms a connected path between the two opposite edges of the cell. According to the simulations (Wilkinson and Willemsen 1983) the number of sites occupied by the invader at this moment can be expressed as $N \sim L^D$, where $D \simeq 1.89$ in two and $D \simeq 2.52$ in three dimensions. These values for the fractal dimension of the invasion percolation clusters are in good agreement with those obtained for the ordinary percolation clusters, showing the similarity between the static properties of the two models.

To take into account the incompressibility of the fluids one needs an additional rule: Once a region filled by "defender" sites has been surrounded by the invading fluid none of the sites belonging to this "trapped" area is available for occupation by the invading fluid. In this case the fractal dimension of the cluster which is made of the invaded sites at the moment the invider first percolates the two dimensional cell was found to be $D \simeq 1.82$ (Chandler et al 1982). This value is different from $D \simeq 1.89$ and indicates that invasion percolation with trapping and ordinary percolation belong to distinct universality classes.

One way to quantify the difference between static and invasion percolation is to investigate the cumulative acceptance profile $a_N(r)$ defined by (Wilkinson and Willemsen 1983)

$$a_N(r) = \frac{\langle \text{no of random numbers in } [r, r+dr] \text{ accepted into cluster} \rangle_N}{\langle \text{no of random numbers in } [r, r+dr] \text{ considered} \rangle_N} . \tag{5.13}$$

It can be shown that asymptotically

$$a_\infty = \begin{cases} 1 & \text{if } r < p_c, \\ 0 & \text{if } r > p_c . \end{cases} \tag{5.14}$$

This is the same as the acceptance profile of static percolation, which is valid in the present case only in the large N limit. Furthermore, for invasion percolation with trapping (5.14) *breaks down* and $a_N(r)$ does not approach a step function.

There is an additional threshold in the trapping version of invasion percolation in $d > 2$. It takes place when the largest, originally spanning region of invader free (defender) sites breaks into finite isolated clusters. At this point further deviations from the ordinary percolation behaviour can be observed. The finite but large clusters of defender sites can be considered as fractals. Their fractal dimension in $d = 3$ was found to be $D_{def} \simeq 2.13$ which is different from both D of the invaded region and D of the ordinary percolation clusters.

In an analogy with static percolation one can define the defender cluster size distribution n_s as the normalized number of clusters consisting of s defender particles. The simulation results support the scaling assumption (Willemsen 1984)

$$n_s \sim n_0(L)s^{-\tau} \tag{5.15}$$

with $\tau \simeq 2.07$ in three and $\tau < 2$ in two dimensions. This τ does not satisfy the relation $\tau = 1 + d/D$ known from static percolation. Instead, in $d = 3$

$$\tau = 1 + \frac{d'}{D_{def}} \tag{5.16}$$

holds with $d' \simeq 2.24$. An important consequence of (5.16) is that $n_0(L)$ should scale with L. This is even more explicitly manifested in two dimensions, because of $\tau < 2$.

At first sight the latter result is surprising since $\tau > 2$ is a condition which can be derived for ordinary percolation from mass conservation (the sum $\sum_s sn_s$ must converge as $s \to \infty$). However, the set of isolated defender regions at the threshold can be associated with the so called volatile fractals (Herrmann and Stanley 1984) for which $\tau < 2$ has been shown to be the consistent value (Vicsek and Family 1984, Herrmann and Stanley 1984). In the present case mass conservation with $\tau < 2$ is provided by the L dependence of the prefactor $n_0(L)$ which should decay with growing L compensating the divergence of the above sum.

This behaviour can be interpreted by the following mechanism. As one goes to a larger cell, some of the smaller isolated clusters may turn out to be connected through parts which were outside of the original cell, and form larger clusters. Consequently, when L increases, there is a net decrease in the density of clusters of a given size. In general, the condition for the scaling of $n_0(L)$ is $d' < d$.

5.3. KINETIC GELATION

The term gelation is generally used for a transition when some of the finite clusters in the system join to form a single large cluster spanning the whole sample. This phenomenon is accompanied by relatively drastic changes in the physical properties of the system, e.g., its shear viscosity sharply increases. The clusters can be made of many kinds of particles ranging from molecules to red blood cells.

In the following we shall use the language of sol-gel transition which takes place in a liquid mixture of sol molecules and neutral solvent molecules under specific conditions. Originally the sol molecules are separated and are called monomers. With time the monomers form chemical bonds with each other producing dimers, trimers and so on. During this irreversible process the mean cluster size gradually increases, and at a given moment, t_g, called gelation time, the linear size of the largest molecule becomes equal to that of the system. This single spanning cluster is called gel and its weight grows further after t_g. Its mass represents only a small percentage of the mass of all sol and solvent molecules which are typically trapped in the holes formed by the gel, thus the sample behaves as a soft but elasctic material (gelatine). The gelation process is largely influenced by the maximum number of chemical bonds f (functionality) associated with a monomer, for example, molecules with $f = 2$ do not contribute to the sol-gel transition.

Depending on the mechanism by which two molecules can form a chemical bond various models can be used to describe gelation. In polycondensation a bond appears suddenly as a result of either i) an external excitation or ii) the collision of molecules. In the first case the motion of the molecules can be neglected and static percolation can be used to characterize the phenomenon. In the other limit, when it is exclusively the diffusion of clusters which determines the bond formation rate, diffusion-limited cluster-cluster aggregation (to be discussed in Chapter 8) is a good candidate for an appropriate model.

The kinetic gelation model was introduced to simulate another type of gelation called *addition polymerization*. In this process unsaturated electrons jump from one molecule to the other, meanwhile assisting in producing bonds. As an important deviation from static percolation in this case the bonds are strongly correlated in space. Generally, one assumes that the mobility of sol molecules is much smaller than that of the electrons.

Let us consider the following *lattice model* of the above process (Manneville and de Seze 1981). i) Monomers are placed randomly at the sites of a cell of linear size L and periodic boundary conditions are imposed. Empty sites are solvents with functionality $f = 0$ and have a concentration $1-\sum_i c_i$, where c_i is the concentration of monomers of functionality $f = i$. ii) Initiators (radicals) are randomly added to sites with a monomer already sitting on them. The concentration of initiators is usually chosen to satisfy the condition $c_I = c_I(t = 0) \ll 1$. iii) At each growth step one of the radicals is randomly selected. Next, the radical attempts a jump in a random direction along one of the bonds leading out from the site. Such a transfer is prohibited if the site at the other end of the bond has zero functionality. The time is increased by an amount $\delta t = \left[c_I(t)L^d\right]^{-1}$ even if the attempt fails. iv) After a successful jump the functionality of the sites connected by the given bond is decreased by one and the bond becomes occupied. v) If a radical happens to be at the new site, the two radicals annihilate, thus $c_I(t)$ is slowly decreasing in time. A radical can also get trapped in a site which has no nearest neighbour sites of functionality larger than zero. In a finite cell all of the radicals become trapped within a finite time.

This extensive list of rules can be programmed relatively easily and the resulting configurations are evaluated in terms of quantities analogous to those used in static percolation. A cluster is defined as a set of sites which can be connected through occupied bonds. The actual status of the gelation process is characterized by the cluster size distribution function $n_s(t) = L^{-d}N_s(t)$, where $N_s(t)$ is the number of clusters consisting of s monomers at time t. The concentration of occupied bonds, p, increases monotonically with t and can be used to monitor the development of gelation with a sol-gel transition taking place at p_c.

Large scale Monte Carlo simulations of three-dimensional kinetic gelation (Herrmann *et al* 1982) have shown that the critical exponents β and ν_p defined in (5.3) and (5.4) within the errors are the same for this process as for ordinary percolation. Consequently, the fractal dimension $D = d - \beta/\nu_p$ of the spanning gel molecule also coincides with the fractal dimension of the infinite static percolation cluster at p_c.

This universal behaviour of the two models seems to *break down* when one studies the quantity A which is the *ratio of the critical amplitudes* of the second moment of n_s

$$\chi_{q=2} = \sum_{s<\infty} s^2 n_s \sim \begin{cases} a^+(p-p_c)^{-\gamma}, & \text{if } p \to p_c^+ \\ a^-(p_c-p)^{-\gamma}, & \text{if } p \to p_c^- . \end{cases} \tag{5.17}$$

Here γ is another critical exponent and a^+ and a^- are the amplitudes, so that $A = a^-/a^+$. The cluster size distribution in systems exhibiting critical behaviour generally scales with s and $p - p_c$ as (Stauffer 1985)

$$n_s \sim s^{-\tau} f(|p-p_c|s^\sigma)\,, \tag{5.18}$$

where τ and σ are further critical exponents connected to the previously introduced exponents through scaling relations, and $f(x)$ is a scaling function. In the theory of equilibrium critical phenomena it is assumed that $f(x) = BF(Cx)$, where B and C are constants which may depend on the properties of the system investigated, but $F(x)$ is supposed to be universal, i.e., the same for all systems belonging to a relatively broad class. This assumption together with (5.17) and (5.18) leads to the conclusion that A, the ratio of a^+ and a^-, has to be universal. According to the numerical simulations of three-dimensional percolation $A \simeq 10$.

A careful analysis of the amplitude ratio A for kinetic gelation in $d = 3$ (Herrmann *et al* 1982) showed that it is generally considerably smaller than 10. In fact, it decreases with the initial concentration of radicals c_I, and in the limit $c_I \to 0$ tends to become equal to 1. This means that $F(x)$ has to be different from that describing the scaling in ordinary percolation, therefore, the two models *belong to different universality classes.* An interesting observation was recently made concerning A in a system with regularly distributed initiators, where A was found to be approximately equal to 10 (Herrmann 1986). This result suggest that the randomness of the positions of initiators is essential, in apparent analogy with the problem of diffusion in a medium with randomly or regularly distributed traps and (continuum) percolation of overlapping spheres (Balberg 1987).

An additional sign of non-universality of kinetic gelation is manifested in the behaviour of n_s. Although the scaling assumption (5.18) for the cluster size distribution function is known to describe n_s adequately in many different systems, it is not consistent with the data obtained for kinetic gelation. Instead, it shows oscillations of period $s^* \simeq p/c_I$, and scales as (Chhabra *et al* 1984)

$$n_s \sim \tilde{s}^{-\tau} e^{-b|p-p_c|\tilde{s}^{\sigma}} f(\tilde{s}) e^{c(p-p_c)}\,, \tag{5.19}$$

where b and c are constants and $\tilde{s} = s/s^*$. Figure 5.4 shows the scaling function $f(x)$ determined from numerical simulations and using (5.19). Note, that if $f(x)$ tends to a constant value for large x, the last two terms represent only a correction to the original scaling form. This, however, does not seem to be the case, since for $c_I \to 0$ the oscillations are not found to be damped entirely.

Returning to the structure of the gel molecule at p_c we recall that it is a fractal having a dimension equal (or close) to D of the infinite cluster in static percolation at p_c. In addition to its dimension a fractal can be characterized by other quantities related to its internal structure. For example, in Chapter 3 it has been shown that a singular distribution on a fractal defines an infinite number of subsets all having a fractal dimension smaller or equal to D. One

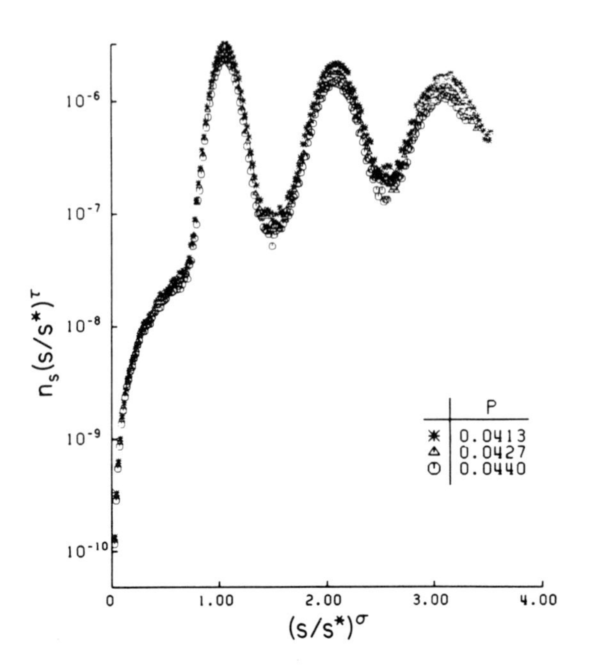

Figure 5.4. Scaling plot of the data for the cluster-size distribution in three-dimensional kinetic gelation. The data are for $L = 60$ and $c_I = 0.0003$ (Chhabra *et al* 1984).

of the simplest subsets of a fractal is its "*backbone*" which can be introduced through the voltage distribution on a cluster (see Example 3.2). Consider now a large cluster of resistors with two electrodes – one at each of its opposite edges. Then the backbone is defined as the set of bonds (resistors) with a voltage drop different from zero. It is easy to see that $D_{BB} = D_0$, where D_{BB} is the dimension of the backbone and D_0 is the generalized dimension corresponding to the zeroth moment of the voltage distribution. According to an equivalent definition of backbone, it is a set of bonds (sites) which can be connected staying on the fractal to both electrodes via at least two routes having no overlaps. In short, routes leading to deadends are thrown away and only relevant loops contribute to the backbone (Fig. 5.5).

The fractal dimension of the backbone of a cluster is an independent exponent characterizing the "loopness" of the structure. For example, in three-dimensional static percolation $D_{BB} \simeq 1.74 < D \simeq 2.5$. D_{BB} has

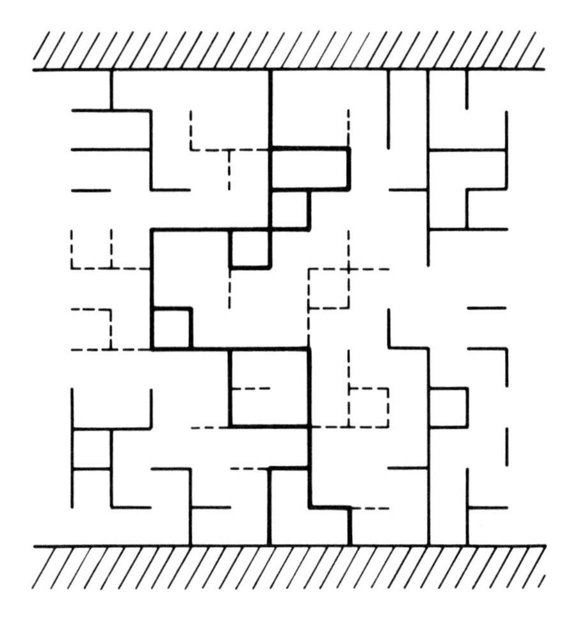

Figure 5.5. Schematic picture of the backbone of a percolation cluster. The bonds belonging to the backbone are drawn with heavy lines, while the bonds leading to deadends are denoted by dashed lines.

also been determined for kinetic gelation in $d = 3$. It was found to be approximately equal to 2.22 (Chhabra *et al* 1985), a value considerably larger than D_{BB} for percolation. This can be understood; the main contribution to the mass of the clusters comes from loops formed by chains generated by randomly walking radicals.

To take into account experimental realizations many variations of the kinetic gelation model have been considered, including simulations with poison sites disabling radicals (Pandey and Stauffer 1983) and mobile solvent molecules (Bansil *et al* 1984).

5.4. RANDOM WALKS

Investigation of various types of random walks represents a particularily effective approach to the description of systems consisting of growing, nonbranching objects (de Gennes 1979). The most important example is *poly-*

merization, where monomer molecules of functionality 2 (being able to form two chemical bonds) are joined together leading to a long chain called linear polymer molecule. In most of the cases walks on a lattice are considered, the lattice sites corresponding to a monomer. The set of sites visited by the walker is regarded as a chain of molecules or particles (in general, multiple occupation of a site is allowed). For simplicity we shall use the term random walk for this trail.

In an actual realization, a random walk starts from an initial position and proceeds by making jumps of one lattice unit in a direction which is selected randomly from the directions allowed by the given model. For example, when strictly self-avoiding walks are considered, there is no allowed direction if all of the nearest neighbours of the current position of the walking particle are occupied, and the walk must terminate. In this section we shall concentrate on walks which are truly growing, i.e., are never trapped by themselves. Assuming that the particle makes one step in a unit interval, the number of steps in a walk N is equal to the duration of the walk, t.

The most fundamental quantity used for characterization of random walks is their *mean squared end-to-end distance* $\langle R_e^2(t)\rangle = R_0^2(t)$, where $R_e(t)$ is the distance between the walking particle and the initial position at time t. It measures the square of the average spatial extent of a walk, $R_0(t)$, and its time dependence will allow us to make conclusions about the fractal nature of the process. In particular, if

$$R_0(t) \sim t^\nu \tag{5.20}$$

the quantity $D = 1/\nu$ will be identified with the fractal dimension of the chain.

5.4.1. Self-intersecting random walks

The simplest version of random walks is a sequence of jumps in random directions, where there is no constraint on the probability of jumping to any of the neighbouring sites. This process can also serve as a lattice model for diffusion, and we shall also refer to unrestricted random walks as trails of diffusing particles. As was briefly discussed in Example 2.6, the trail of a diffusing particle has fractal properties. To see this let us denote the end-to-end vector at time t by $\vec{R}_e(t) = \sum_{i=1}^{t} \vec{a}_i$, where $\vec{a}_i$ is a vector of length a equal to the lattice spacing and it has z possible orientation with z being the coordination number. Then

$$\langle R_e^2(t)\rangle = \sum_{i,j}\langle \vec{a}_i\vec{a}_j\rangle = a^2 t + 2\sum_{i>j}\langle \vec{a}_i\vec{a}_j\rangle \sim t\,, \tag{5.21}$$

since due to the independence of the orientations of $\vec{a}_i$-s the cross products vanish. Accordingly, the characteristic linear size of a walk is $R_0(t) \sim t^{1/2}$. For a walk starting at $t = 0$ it is also straightforward to show that the probability of finding the walker at a distance R from the origin is asymptotically $P(R,t) \sim e^{-R^2/Dt}$, where $D = a^2/\Delta t$ with $\Delta t = 1$ in our case.

Let us consider the set of sites which are visited by a walk of duration t at time intervals $t \gg \Delta t' \gg 1$. The typical distance of these sites is $l \sim (\Delta t')^{1/2}$, since the walks which take place during each time interval $\Delta t'$ are themselves independent random walks. Consequently, the whole walk can be covered by $N(l)$ boxes of linear size l, where

$$N(l) = \frac{t}{\Delta t'} \sim t l^{-2}\,. \tag{5.22}$$

This expression corresponds to the definition of fractal dimension with $D = 2$. In obtaining (5.22) we implicitly assumed that the random walks of size l do not overlap and we need a separate box to cover each of them. This is obviously not true in one dimension which is indicated by the fact that $D > d$ in this case. The $D = d = 2$ case is marginal, while in $d > 2$ the fraction of overlaps becomes negligible and a true self-similar object is generated by random walks. Note, however, that one can consider the number of steps t made within a region of radius R_0, instead of the number of distinct sites visited by the walker. Then regarding t as the number of particles within this region the scaling of mass $M(R_0) \sim R_0^2$ has the same form as (5.22) even in $d < 2$. Figure 2.10 shows a long random walk on the plane together with its external perimeter.

The situation does not change qualitatively if the jump direction depends merely on a finite number of previously completed steps. Let us suppose that only the last $\Delta t'$ steps affect the probability to jump in a given direction. In the next step this probability has to be strictly independent of the history preceding the last $\Delta t'$ steps. Then one can define a walk which is made of sites visited at times separated by a time interval larger than $\Delta t'$ and the previously considered model is recovered. Therefore, finite time memory has no effect on the asymptotic behaviour.

The *overall shape* of unrestricted random walks is asymptotically highly *anisotropic*. Numerical simulation shows that the ratios of the squared principal radii of gyration of three dimensional walks are the following (Solc 1973)

$$\langle R_1^2 \rangle : \langle R_2^2 \rangle : \langle R_3^2 \rangle = 11.8 \ : \ 2.69 \ : \ 1.00\,. \tag{5.23}$$

There is an exact result for the quantity

$$A_d = \frac{\sum\limits_{i>j}^{d} \langle (R_i^2 - R_j^2)^2 \rangle}{(d-1)\langle \left(\sum\limits_{i=1}^{d} R_i^2 \right)^2 \rangle}\,, \tag{5.24}$$

measuring the asphericity of a walk in d dimensions. Here R_i^2 is the square of the ith principal radius of a walk (the ith eigenvalue of the corresponding radius of gyration tensor). According to the analytical calculations (Rudnick and Gaspary 1986)

$$A_d = \frac{2(d+2)}{(5d+4)}\,. \tag{5.25}$$

There are several ways to modify the above discussed simplest case. If the direction of each step is correlated by the direction of the previous one, the situation changes qualitatively. In this case arbitrarily distant points also become correlated, and the mean square distance diverges according to an exponent $R_0^2(t) \sim t^{2H}$, where H and the correlations are related (see Example

2.8). The fractal dimension corresponding to the set of points visited by a walker for a given H is

$$D = \frac{1}{H} \tag{5.26}$$

a result following from an argument analogous to that leading to (5.22).

Random walks, therefore, may have a highly non-trivial structure exhibiting fractal scaling. However, an unrestricted walk does not represent a real polymer chain appropriately, since it does not account for the repulsion between two molecules which are close together. In other words, a random walk can cross itself, while two monomers can not occupy the same position in space (excluded volume effect). The simplest modification to avoid this problem is to prohibit intersections: whenever a random walk would cross itself it is removed from the statistics. According to their definition *self-avoiding random walks* (SAW-s) do not grow indefinitely. They rather reproduce the equilibrium statistics of linear polymers in a good solvent. Since we are more interested in non-equilibrium, truly growing phenomena we shall consider next a never stopping version of SAW.

The following rules define a model called *true self-avoiding walks* (TSAW) (Amit *et al* 1983) already discussed in part in Example 4.1. The walker jumps at each time step to one of the neighbouring sites with a probability depending on the number of times the new site has been visited in the past. In particular, the probability p_{ij} of jumping from site i to the neighbouring site j is in the model equal to

$$p_{ij} = \frac{e^{-g_{ij}n_j}}{\sum\limits_{k=1}^{z} e^{-g_{ik}n_k}}, \tag{5.27}$$

where g_{ij} is a parameter controlling the degree of inhibition associated with the particular bond ij, n_k is the number of times the site k has been visited before and z is the lattice coordination number. The strongest inhibition is provided by the limit when all $g_{ij} \to \infty$, but TSAW can cross itself even in this limit. This occurs when there is no previously unvisited neighbour, and is made possible by the normalization included in (5.27).

The upper critical dimension and the fractal dimension $D = 1/\nu$ of TSAW have been investigated by various approaches. According to heuristic arguments (Pietronero 1983, Obukhov and Peliti 1983) one first assumes that the root mean square end-to-end distance $R_0(t)$ is the only relevant length measuring the size of a self-similar TSAW. In this case we expect the density $\rho(r,t)$ of points visited in a walk of duration t to have the form

$$\rho(r,t) \sim tR_0^{-d} f\left(\frac{r}{R_0}\right) . \tag{5.28}$$

Furthermore, assuming as usual that $R_0(t) \sim t^{\nu}$, the increase of $R_0(t)$ due to the prolongation of a walk for additional Δt steps is given by

$$\Delta R_0 = R_0(t + \Delta t) - R_0(t) \sim t^{\nu-1}\Delta t . \tag{5.29}$$

This increase is primarily due to the repulsion effects forcing the walk to expand. The magnitude of this effect is expected to be proportional to the gradient of $\rho(r,t)$ calculated at a distance from the origin approximately equal to R_0. From (5.28) we have

$$\frac{d\rho(r,t)}{dr}|_{r\sim R_0} \sim tR_0^{-d-1} f(1) \sim t^{1-(d+1)\nu} . \tag{5.30}$$

Comparison of (5.29) and (5.30) leads to

$$D = 1/\nu = \frac{d+2}{2} . \tag{5.31}$$

The above result does not hold above $d = 2$, because in this case the repulsion effects become negligible compared with the outward expansion of the walk due to ordinary diffusion. Therefore, allowing self-intersection even by taking into account long living memory does not change the upper critical dimension. However, in $d = 1$ TSAW exhibits different scaling behaviour from ordinary random walks and has a mass scaling exponent close to that predicted by (5.31).

Finally, there is a model called growing self-avoiding trail (GSAT) which seems to lead to a fractal dimension different from $D = 2$ in $d = 2$, although it allows for self intersection (Lyklema 1985). This is a growing self avoiding walk during which the condition of avoiding sites is changed so as not to allow the walker to pass through the same bond twice. Such a process never terminates except at the origin, because on a lattice with an even coordination number there are either at least two free bonds leading out from a site or none. The origin is a special point, where the number of free bonds is always odd. Extensive Monte Carlo simulations indicate that $D \simeq 1.87$ in two dimensions and the unrestricted random walk value $D = 2$ is recovered only in $d = 3$. Therefore, GSAT has an upper critical dimension $d = 3$ at which logarithmic corrections of the form

$$R_0(t) \sim At(\ln t)^{\alpha} \tag{5.32}$$

are expected to modify the asymptotic behaviour. The simulations confirmed the above expression.

5.4.2. Growing self-avoiding walks

In this section strictly self-avoiding walks will be considered, i.e., the walk will not be allowed to have intersections. However, in contrast to the equilibrium version of SAW, the walk is designed in such a way that it tries to avoid itself whenever it is possible. There are two possibilities: depending on the particular model a walk can be trapped by itself or it is bound to grow indefinitely.

The interest in growing self-avoiding walks is motivated by a well known transition in the conformation of polymer chains as a function of temperature T (de Gennes 1979). At high temperatures the molecules behave as ordinary SAW with a fractal dimension $D = \frac{4}{3}$ in $d = 2$. Lowering T results in a collapse of the chains at a temperature T_θ due to an attractive two-body interaction. In this tricritical θ point the fractal dimension of chains is $D \simeq 1.75$ ($d = 2$). At $T = 0$ one has a completely space filling SAW with a trivial dimension. As we shall see below, a never terminating self-avoiding walk is a good candidate to describe the geometry of long molecules at the θ point.

The *growing self-avoiding walk* (GSAW) can be defined as a simple modification of TSAW with $g \to \infty$ (Majid *et al* 1984, Lyklema and Kremer 1984). In the case of GSAW the walker chooses an unoccupied neighbouring site randomly, with a probability $p = 1/n$, where n is the number of free neighbours. If $n = 0$, the walk terminates; this property represents the main difference from TSAW. The scaling of $R_0(t)$ with the number of steps t in principle could be studied by exact enumeration of the possible walk configurations (Section 4.3). When calculating the average end-to end distance each configuration enters the sum by its own weight $P_{t,i} = \prod_j p_{t,i}(j)$ which is the product of the probabilities $p_{t,i}(j)$ of adding the jth new step to the ith configuration of length t (for ordinary SAW all $P_{t,i}$-s are equal). Knowing $R_0(t)$ one can use expression (4.20) to estimate the fractal dimension. However, it turns out that in the case of GSAW the typical cluster size available in the exact enumeration approach ($t \simeq 20$ steps) is not enough to see the true behaviour.

To see the scaling of R_0 one should carry out high precision Monte Carlo simulations up to sizes $t \simeq 200$ (Kremer and Lyklema 1985a), and analyse the data by the application of expression (4.16) which normally can be used only for exact enumerations. From this approach the effective exponent $\nu(t) = 1/D(t)$ can be estimated for chains of length $t \leq 200$. Figure 5.6 shows the results obtained for the square lattice. An unusually slow crossover seems to take place from the small to the large cluster behaviour, and the extrapolated value $D(t \to \infty) \simeq 1.5$ coincides with the best results for the fractal dimension of ordinary SAW. This means that the growing and the original version of self-avoiding walks belong to the same universality class; the kinetic rule does not change the asymptotic scaling. The observation of a *slow crossover*, however, is a qualitatively new result of these investigations and as we shall see such behaviour is *manifested in many other growth phenomena*.

The goal of constructing a non-trapping SAW was achieved by introducing the model called *indefinitely growing self-avoiding walk* (IGSAW)

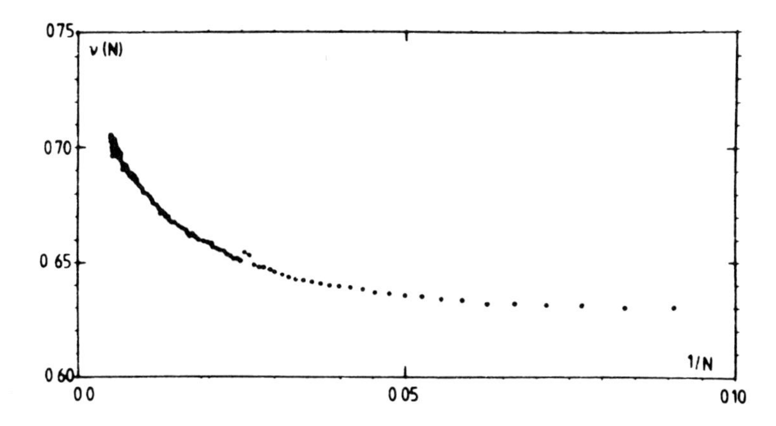

Figure 5.6. Extrapolation of the effective exponent $\nu(t) = 1/D(t)$ obtained from high precision Monte Carlo simulations of growing self-avoiding walks up to a chain length of $t = 200$ steps (Kremer and Lyklema 1985a).

(Kremer and Lyklema 1985b). This walk is analogous to GSAW, except that the walker recognizes possible traps (cages from which he can not return) in advance, and avoids them. To generate such a walk in two dimensions one needs local information on the walk only. This is demonstrated in Fig. 5.7, where a short IGSAW is shown on the square lattice. It is clear from this figure that when the walker approaches his previous path, in order to avoid trapping, he can use the knowledge of the local configuration and a quantity, called the winding number, attributed to the already visited sites. This number is equal to $n_l - n_r$, where n_l and n_r denote respectively the number of left and right turns made previously by the walker. In addition, Fig. 5.7 demonstrates the irreversibility of such walks which can be checked by simply calculating the corresponding jump probabilities for a reverse walk along the same path. In order to recognize whether a new possible step leads into a trap it is enough to consider all sites which form the smallest closed path in the forward direction. This means that for the triangular lattice only nearest neighbour (nn) sites, on the square lattice next to nn sites, while on the honeycomb lattice all next to next nn sites have to be examined together with the corresponding winding numbers.

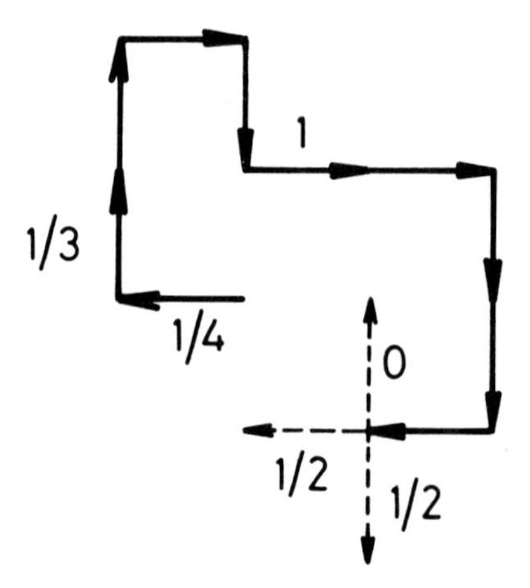

Figure 5.7. Example for a short IGSAW on the square lattice. To recognize whether the next step would lead to a trap it is enough to know the local configuration and the direction of the old part.

An early numerical study based on exact enumeration of the configurations gives for the fractal dimension of IGSAW $D \simeq 1.75$ (Kremer and Lyklema 1985b) indicating that IGSAW has different scaling behaviour from the previously considered models. Actually, the value

$$D = 1.75 = \frac{7}{4} \tag{5.33}$$

can be shown to be exact by establishing a correspondence between IGSAW and walks along the external perimeter or "hull" of an infinite percolation cluster (Weinrib and Trugman 1985). Then (5.33) follows from an exact result for the fractal dimension of the latter process (Saleur and Duplantier 1987).

First we show that the ring-forming version of IGSAW on a honeycomb lattice reproduces the perimeter of critical percolation clusters (Weinrib and Trugman 1985). The rules of this version of the original model are somewhat different: the origin is an allowed site, and a walk is considered to become trapped if it enters a region from which there is no path to the origin. However, one has good reasons to expect that the fractal dimension of large walks is the same in both versions. Honeycomb lattice is considered because for this lattice there is a one-to-one correspondence between the walk and the external perimeter of the site percolation clusters on a triangular lattice at its percolation threshold. (The equivalence is exact only for the ring-forming version.)

Consider the site percolation problem on the triangular lattice which is dual to the honeycomb lattice. The perimeter of a cluster can be defined as the bonds on the dual lattice separating filled from empty sites. Let us imagine that we build up the edge of a percolation cluster by deciding the occupation of a site determining the perimeter as we proceed. At each step one makes a decision whether a given site is occupied (with probability $p_c = \frac{1}{2}$) or empty (with probability $1 - p_c = \frac{1}{2}$). This choice determines whether the perimeter turns left or right at that step. Moreover, as demonstrated in Fig. 5.8, when the perimeter approaches itself the condition of self-avoidance is automatically satisfied, since the occupation of the corresponding sites has already been determined. The perimeter separates occupied from empty sites, thus it can not cross itself or enter a trap.

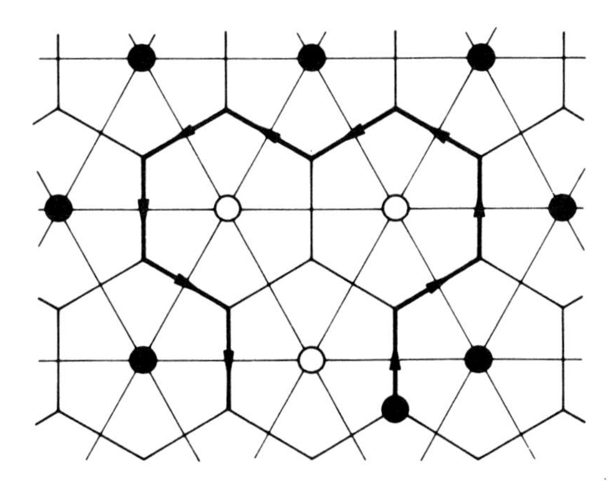

Figure 5.8. Example for a short IGSAW on the hexagonal lattice. It follows the perimeter of a site percolation cluster on the dual triangular lattice.

In this way IGSAW on a honeycomb lattice traces out the *hull of critical percolation clusters.* In this sense it is equivalent to an analogous walk designed with the purpose of generating the external perimeter of percolation clusters (Ziff *et al* 1984, Ziff 1986). On the basis of universality valid for the critical behaviour of percolation the same statement can be extended to other types of two-dimensional lattices. A long perimeter walk on the square lattice is shown in Fig. 5.9 to demonstrate the structure of such walks.

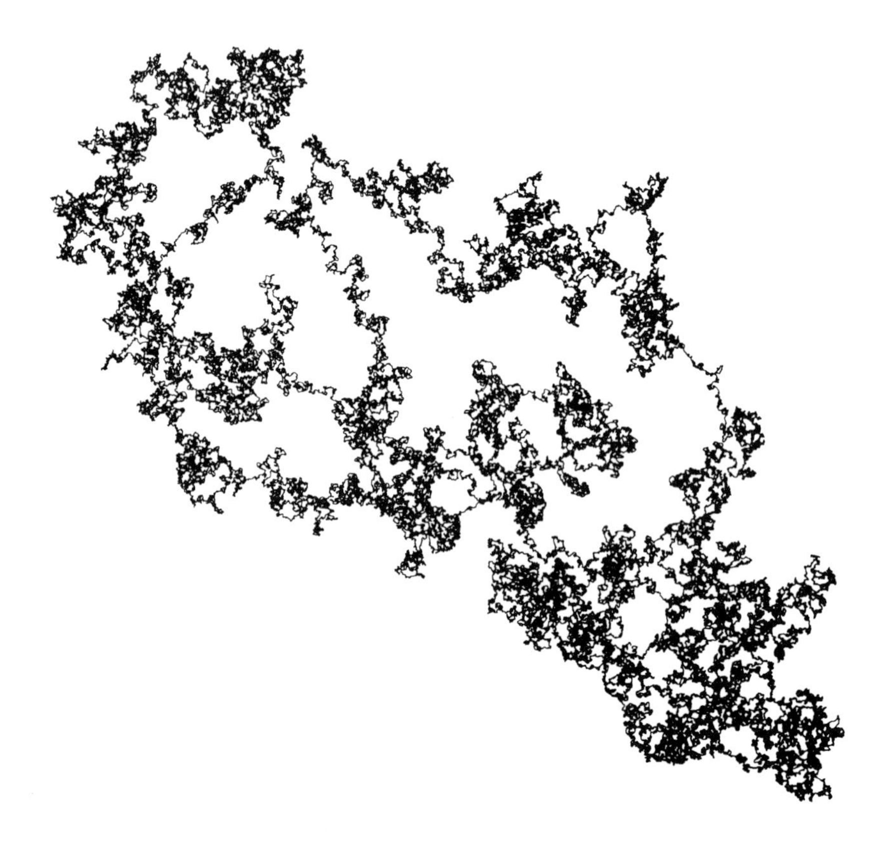

Figure 5.9. External perimeter of a percolation cluster generated by the ring-forming version of IGSAW on the square lattice. This configuration contains 194 468 sites (Ziff 1986).

The proof of (5.33) is completed by finding the fractal dimension of the hull. Assuming that the front of diffusing particles is equivalent to the perimeter of large percolation clusters, theoretical arguments (Saleur and Duplantier 1987) led to the following expression for the fractal dimension of the hull

$$D = 1 + 1/\nu_p = \frac{7}{4}, \tag{5.34}$$

where $\nu_p = \frac{4}{3}$ is the correlation length exponent for percolation known exactly. An exact derivation of (5.34) can be obtained using the formulation of percolation as a $q = 1$ Potts model, and applying a Coulomb-gas mapping technique. Furthermore, it can be shown that hulls (i.e., IGSAW) are directly related to a SAW with short range attraction, thus they are expected to be in the same universality class as chains at the θ point. IGSAW is also related to the front of diffusing particles in $d = 2$ (Sapoval *et al* 1985). We conclude the description of IGSAW by mentioning that finding an algorithm for the $d > 2$ case is far from trivial and no successful attempt has been published yet. The reason is that in $d > 2$, non-local information is required to avoid traps.

The last model discussed in this section simulates linear aggregation. It generates a one parameter family of strictly self-avoiding, indefinitely growing walks. In this model the jump probability is determined by solving the Laplace equation with appropriate boundary conditions. One of the advantages of this *Laplacian random walk* (LRW) is that it can be defined in any dimension.

LRW grows according to the following rules (Lyklema and Evertsz 1986). Imagine a configuration centred at the middle of a circle having a radius much larger than the size of the walk. For this configuration the Laplace equation $\nabla^2\phi = 0$ is solved with a boundary condition $\phi = 0$ on the walk and $\phi = 1$ on the sphere. Let us define the probability of jumping to a given neighbouring site to be proportional to the potential at this possible new position. As a result the walk is self-avoiding and never stops, since the potential is zero in the cages (because of screening) and on the walk itself.

A simple generalization can be accomplished by choosing the probability of jumping to site i equal to

$$p_i = \frac{\phi_i^\eta}{\sum_j \phi_j^\eta}, \tag{5.35}$$

where ϕ_i is the value of the field at site i, η is a parameter and j runs over the nearest neighbours. The parameter η governs the probability difference among allowed directions. For $\eta = 0$ IGSAW is recovered. In fact, this analogy provides a non-local method for growing of IGSAW-s in dimensions larger than 2. For $\eta > 0$ ($\eta < 0$) the walk is self-repelling (self-attracting), and has a smaller (larger) fractal dimension than IGSAW. The case $\eta = 1$ is analogous to a model based on linear aggregation of diffusing particles (Debierre and Turban 1986).

5.4.3. Walks on fractals

The problem of random walks on fractals has attracted great interest recently (for reviews see Aharony 1985, Havlin and Ben-Avraham 1987). Here we shall discuss only some of the specific aspects of this field which are related to fractal growth. The condition of self-avoidance will not be considered, thus we can refer to random walks on fractals as diffusion on a fractal substrate.

Consider the random motion of a particle placed onto a fractal network of dimension D made of bonds or filled sites. The walk starts at $t = 0$ and the particle jumps with equal probability to any of the nearest neighbour sites belonging to the fractal. Its mean squared distance from the origin is expected to scale with the number of steps (time) as

$$R_0^2(t) \sim t^{2/d_w} , \tag{5.36}$$

where the exponent d_w depends on the structure of the fractal in a non-trivial way. For diffusion on Euclidean lattices $D_w = 2$. But the trajectory of a diffusing particle expands slower on a fractal, and in general $d_w > 2$.

Writing (5.36) in the form $t \sim R_0^{d_w}$ suggests that d_w is a fractal dimension-like quantity. However, d_w is typically not the fractal dimension of the trail as a geometrical object, but describes the fractal scaling of mass (number of steps) within a region of radius R_0. While on Euclidean lattices the fraction of self-intersections becomes negligible in dimensions $d > 2$, in the case of fractal substrates returns to already visited sites play a relevant role in all d, except very special constructions. This is indicated by the fact that generally $d_w > D$. For example, $d_w = \ln(d+3)/\ln 2 > \ln(d+1)/\ln 2 = D$ for the d-dimensional versions of Sierpinski gaskets (Havlin and Ben-Avraham 1987), or $d_w = D + 1$ for the fractal shown in Fig. 2.1 (Guyer 1985) when defined in d dimensions. Moreover, d_w can be larger than d. This is the case, i.e., for diffusion on percolation clusters up to $d = 6$.

The anomalous diffusion law represented by (5.36) can be supported by Monte Carlo simulations, scaling arguments or solutions of the problem for simple systems. The following derivation based on scaling arguments demonstrates that (5.36) is equivalent to the assumption (O'Shaughnessy and Procaccia 1986)

$$\sigma(r) \sim \sigma_0 r^{-\theta}, \tag{5.37}$$

with $\sigma(r)$ defined through $\sigma_{tot}(r) \sim \sigma(r) r^{D-1}$ which is the total conductivity (or diffusivity) of a shell of $n(r) \sim r^{D-1}$ sites being at a distance r from the origin of walk. To see this let $p(r,t)$ denote the average probability per site at time t of finding the particle within the shell between r and $r + dr$. Then the conservation of probability requires that

$$\frac{\partial}{\partial t}[(n(r)p(r,t)] = \frac{\partial}{\partial r}\left(n(r)\sigma(r)\frac{\partial p(r,t)}{\partial r}\right). \tag{5.38}$$

Inserting (5.37) and the expression for $n(r)$ one obtains the diffusion equation in the form

$$\frac{\partial p(r,t)}{\partial t} \simeq \frac{1}{r^{D-1}}\frac{\partial}{\partial r}\left(\sigma_0 r^{D-1-\theta}\frac{\partial p(r,t)}{\partial r}\right) \tag{5.39}$$

which has the exact solution

$$p(r,t) = \text{Const}\; t^{-D/(2+\theta)} \exp\left\{-\frac{r^{2+\theta}}{\sigma_0(2+\theta)^2 t}\right\}. \tag{5.40}$$

From (5.40) one can easily obtain the main quantities of interest. First, we see that (5.40) is equivalent to (5.36) with $d_w = 2 + \theta$, since

$$R_0^2(t) = \int_0^\infty r^2 n(r) p(r,t) dr \sim t^{2/(2+\theta)} . \tag{5.41}$$

Next we observe that $N(t)$, the total number of distinct sites visited scales with time as

$$N(t) \sim t^{D/(2+\theta)} = t^{d_s/2} \tag{5.42}$$

because $p(0,t)$, the probability of returning to the origin is inversely proportional to $N(t)$. In (5.42)

$$d_s = 2D/d_w \tag{5.43}$$

denotes the so called *spectral dimension.* This quantity enters the spectral density of vibrational modes of a fractal in the form of dimension, hence its name (Alexander and Orbach 1982).

Anomalous diffusion in inhomogeneous media has been extensively studied using percolation models (Gefen *et al* 1983). It was recognized that d_s is surprisingly close to $\frac{4}{3}$ for percolation clusters in dimensions $1 < d \leq 6$. This fact led to a conjecture (Alexander and Orbach 1982) about the superuniversality of d_s, giving rise to a large number of controversial results about its validity for percolation. Scaling arguments can be used to express the spectral dimension for the infinite percolation network at the threshold through known standard exponents β, ν_p (see Section 5.1) and μ, where the conductivity scales as $\sigma \sim (p - p_c)^\mu$. The result is (Alexander and Orbach 1982)

$$d_s = 2 \frac{d\nu_p - \beta}{\mu - \beta + 2\nu_p} . \tag{5.44}$$

Finally, we discuss the fractal nature of growth sites which belong to the fractal and are nearest neighbours of the already visited sites (they are analogous to the live sites defined in Section 5.1). Let us denote the number of growth sites at time t by $G(t)$. The following relation has been proposed between $N(t)$ and $G(t)$ (Rammal and Toulouse 1983)

$$\frac{dN(t)}{dt} \sim \frac{G(t)}{N(t)} \tag{5.45}$$

expressing the assumption that the probability of access to a growth site (dN/dt) is proportional to the number of growth sites, and is inversely proportional to the number of already visited sites. Integrating (5.45) and using (5.36), (5.42) and (5.43) we have

$$G(t) \sim t^{2D/d_w - 1} \sim R_0(t)^{2D-d_w} , \tag{5.46}$$

where $2D - d_w$ is the effective fractal dimension of the set of growth sites. This result is supported by simulations of diffusion on a Sierpinski gasket-type deterministic fractal (Havlin and Ben-Avraham 1987). Plotting $\ln G(t)$ against $\ln N(t)$ the slope of the straight line fitted to the data is $x \simeq 0.53$ in good agreement with $x = 2 - 2/d_s \simeq 0.535$ obtained from (5.46), where $G(t) \sim N(t)^x$.

Chapter 6

DIFFUSION-LIMITED GROWTH

Many of the growth processes in nature are governed by the spatial distribution of a field-like quantity which is inherently *non-local*, i.e., the value of this quantity at a given point in space is influenced by distant points of the system, in addition to its immediate neighbourhood. For example, such behaviour is exhibited by the distribution of temperature during solidification, the probability of finding a diffusing particle or cluster at a given point, and electric potential around a charged conductor.

The spatial dependence of these quantities in various approximations satisfies the *Laplace equation with moving boundary conditions.* Since the concentration of diffusing particles is also described by the Laplace equation, the above mentioned class of processes is commonly called diffusion-limited growth. Diffusion-limited motion of interfaces typically leads to very complex, branching fractal objects, because of the unstable nature of growth. Thus, as a result of a self-organizing mechanism governed by the Laplace equation, structures with a rich geometry can emerge from the originally homogeneous, structureless medium. This far-from equilibrium phenomenon can be studied by approaches based on aggregation.

It is the non-local character of the probability distribution which plays an essential role in aggregation phenomena, where single particles, or clusters of particles are added to a growing aggregate. The main assumption of the related cluster models is that the particles stick together irreversibly, a condition which is satisfied in a wide variety of growth processes.

6.1. DIFFUSION-LIMITED AGGREGATION (DLA)

Consider an electrolyte containing positive metallic ions in a small concentration, and a negative electrode. Whenever a randomly diffusing ion hits the electrode or the already deposited metal on its surface, it stops moving (sticks to the surface rigidly) because of the electrostatic attraction. This experiment results in a complicated, tree-like deposit with scale-invariant structure.

The model called *diffusion-limited aggregation* (DLA) was introduced by Witten and Sander (1981) to simulate in a computer phenomena related to the above mentioned process. The rules of the model are simple: One puts a seed particle at the origin of a lattice. Another particle is launched far from the origin and is allowed to walk at random (diffuse) until it arrives at a site adjacent to the seed particle. Then it is stopped, and another particle is launched which stops when adjacent to the two occupied sites, and so forth. In this way large clusters can be generated whose structure is expected to be characteristic for objects grown under diffusion-limited conditions. Indeed, the experiments discussed in Chapter 10 support this expectation.

Figure 6.1 shows a typical DLA cluster of 3000 particles. It demonstrates that these objects i) have a randomly branching, open structure, ii) look stochastically self-similar, and iii) this special geometry is likely to be due to the effects of screening. By stochastic self-similarity here we mean the following: shrinking a large branch and omitting the finest details one obtains a structure which has the same appearance as a smaller branch. In the case of DLA screening is manifested through the fact that the tips of most advanced branches capture the incoming diffusing particles most effectively. Thus, small fluctuations are enhanced, and this instability together with the randomness inherent in the model leads to a complex behaviour (Witten and Sander 1981, 1983).

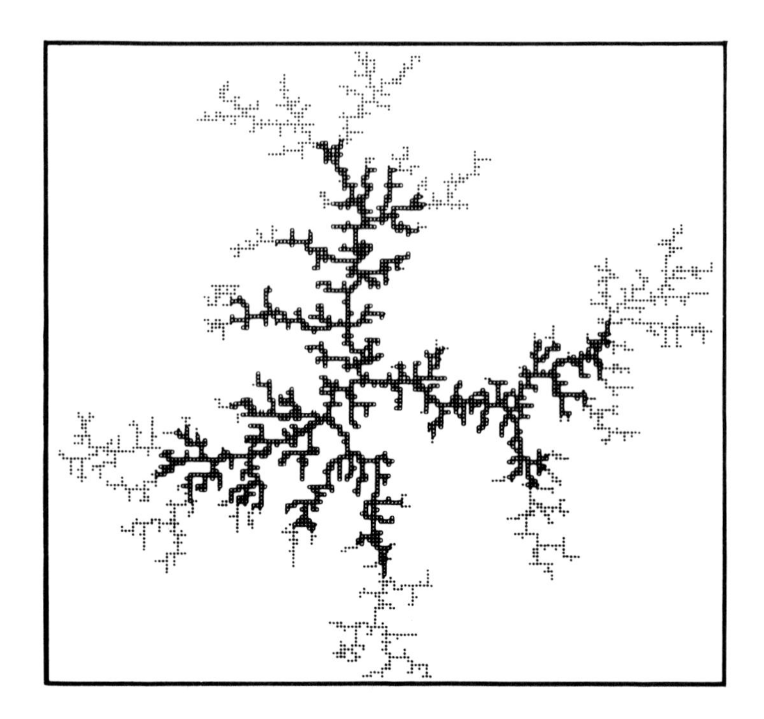

Figure 6.1. A relatively small DLA cluster consisting of 3000 particles. To demonstrate the screening effect the first 1500 particles attached to the aggregate are open circles, while the rest are dots (Witten and Sander 1983).

In the actual simulations the rules are changed in such a way that the resulting process is equivalent to the original version, but it can be realized much more efficiently in a computer. For example, the particles can be launched from a circle having a radius which is only a little larger than the distance of the furthermost particle (belonging to the cluster) from the origin. This can be done because a particle released very far from the cluster arrives at the points of a circle centred at the origin with the same probability. However, as soon as a diffusing particle enters this circle, its trajectory has to be followed until it either sticks to the cluster or diffuses far away. Only in the latter case can it be put back onto the launching circle again. There are additional relevant improvements in the algorithm (Meakin 1985) which are described in Appendix A.

6.1.1. Fractal dimension

The fractal dimension of diffusion-limited aggregates can be estimated by methods described in Section 4.2. As discussed, a possible way to determine the fractal dimension defined by the expression

$$N(R) \sim R^D \tag{6.1}$$

is to calculate the density-density correlation function $c(r)$ (2.14). Figure 6.2 shows $c(r)$ obtained for DLA clusters grown on a square lattice as a function of the distance r between the particles (Meakin 1983a). The slope of the straight line fitted to the data on this double logarithmic plot indicates that the density correlations within the clusters decay according to a power law

$$c(r) \sim r^{-\alpha} \tag{6.2}$$

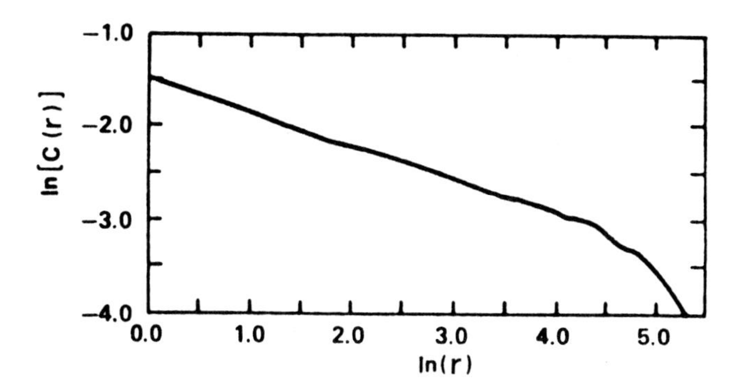

Figure 6.2. Double logarithmic plot of the density-density correlation function $c(r)$ (2.14) for a DLA cluster of 11260 particles generated on the square lattice (Meakin 1983a).

with $\alpha \simeq 0.3$ in $d = 2$. An alternative method is to determine the radius of gyration of the clusters $R_g(N)$ (4.12) as a function of the number of particles N. Plots of this kind demonstrate that (Meakin 1983a)

$$R_g(N) \sim N^{\nu}\,, \tag{6.3}$$

where $\nu = 1/D \simeq 0.585$. These results are in good agreement with the expression (2.18) (Witten and Sander 1981) for the fractal dimension $D = d - \alpha \simeq 1.7$. Therefore, the mass (number of particles) within a region of radius R of a diffusion-limited aggregate scales as $N \sim R^D$ which is equivalent to expression (2.4).

Dependence of the fractal dimension of DLA clusters on the embedding dimension and the sticking probability was extensively studied by Meakin (1983). The results for $2 \leq d \leq 6$ are summarized in Table 6.1. Several conclusions can be made from this table. First, it appears that for all d considered, the following inequalities hold

$$d - 1 < D < d\,. \tag{6.4}$$

These values are in good accord with the mean-field prediction $D = (d^2 + 1)/(d+1)$ to be discussed in Section 6.1.3.

Table 6.1. The fractal dimension (D) of DLA clusters grown on $2 \leq d \leq 6$ dimensional hypercubic lattices. The mean-field prediction $D = (d^2+1)/(d+1)$ is also shown for comparison (Meakin 1983a).

d	D	$(d^2+1)/(d+1)$
2	1.70 ± 0.06	1.667
3	2.53 ± 0.06	2.500
4	3.31 ± 0.10	3.400
5	4.20 ± 0.16	4.333
6	5.3	5.286

The numerical result (6.4) can be supported by an argument providing a *lower bound* for the fractal dimension of diffusion-limited aggregates (Ball and Witten 1984a). Consider a system of randomly diffusing particles making one step in unit time, whose number density is ρ. Let us imagine the trajectories of these particles as they would have been in the absence of the aggregate. The density of individual steps after time t is ρt, while the average number of virtual contacts with the aggregate is $N(t)\rho t$, where $N(t)$ is the actual number of particles in the DLA cluster. Next we want to estimate the average number of contacts of one trajectory with the aggregate. To obtain this quantity we recall that the trail of a randomly walking particle is a fractal of dimension 2 for $d \geq 2$ (Section 5.4.1). According to the related rule given in Section 2.3.1, the fractal dimension of the intersection of two fractals of dimensions D and 2 embedded into a d dimensional space is

$$D_\cap = D + 2 - d\,. \tag{6.5}$$

Correspondingly, the average number of contacts per trajectory goes as $R^{D_\cap}$, where R is the radius of the aggregate.

If $D_\cap > 0$, a typical trajectory intersects the aggregate many times. The number $A(t)$ of first contacts between trajectories and the aggregate is the total number of contacts divided by the number of contacts per trajectory

$$A(t) \sim \frac{N(t)\rho t}{R^{D+2-d}} \sim \rho t R^{d-2}\,. \tag{6.6}$$

The increase of $N(t)$ in unit time is the same as the total flux onto the aggregate which is equal to the time derivative of $A(t)$. Thus, $dA(t)/dt = dN(t)/dt = (dN(t)/dR)(dR(t)/dt)$. Inserting (6.6) and $N \sim R^D$ into this identity one gets

$$\rho R^{d-1-D} \sim dR(t)/dt\,. \tag{6.7}$$

Because of causality the growth speed $dR(t)/dt$ has to remain finite in the limit $R(t) \to \infty$, implying the inequality (6.4) we wanted to derive. If the assumption $D_\cap > 0$ we made earlier is violated, the aggregate is transparent to the particles, and the growth occurs nearly equally over the entire aggregate. In this way the density has to increase up to a point when $D_\cap$ becomes larger than zero and the above considerations hold.

In addition to its dependence on d, the fractal dimension may be affected by other factors. It is well known from the theory of critical phenomena that the exponents describing the singular behaviour of quantities at a second order phase transition are not changed under the influence of irrelevant parameters such as anisotropy, further neighbour interactions, type of lattice, etc. This property of the exponents, called universality, is of special importance. The *question of universality* of the fractal dimension has been addressed in the context of DLA as well, by investigating modifications of the original model.

As a first approximation to this problem the following versions of diffusion-limited aggregation were considered (Witten and Sander 1983, Meakin 1983).

i) DLA with *sticking probability* less than 1. In this variation the particles stick to the surface with a probability p_s, and continue to diffuse with a probability $1 - p_s$.

ii) DLA with *next-to-nearest neighbour* interaction. In this version the particles stop moving as soon as they arrive at a site which is next nearest neighbour to the aggregate.

iii) *Off-lattice* DLA. During the simulations of this variant the centre of a diffusing spherical particle is moved with the same probability to any point within a distance equal to the diameter of the particles. If a particle is found to overlap with another one, the particle is moved back to the position where it first touched the cluster and is incorporated into the aggregate. Figure 6.3 shows an off-lattice DLA cluster of 50,000 particles. This picture demonstrates *stochastic self-similarity* of diffusion-limited aggregates when one compares it with the much smaller aggregate in Fig. 6.1.

The results are given in Table 6.2 (Meakin 1983a). It is clear from the comparison of Tables 6.1 and 6.2 that the above mentioned modifications are irrelevant from the point of view of fractal dimension, at least for the sizes considered. However, as we shall see later, anisotropy plays a relevant role in the structure of aggregates. Among other effects, this will be manifested in the dependence of the large scale structure of aggregates on the type of lattice which is used in the simulations, making the question of universality a delicate problem.

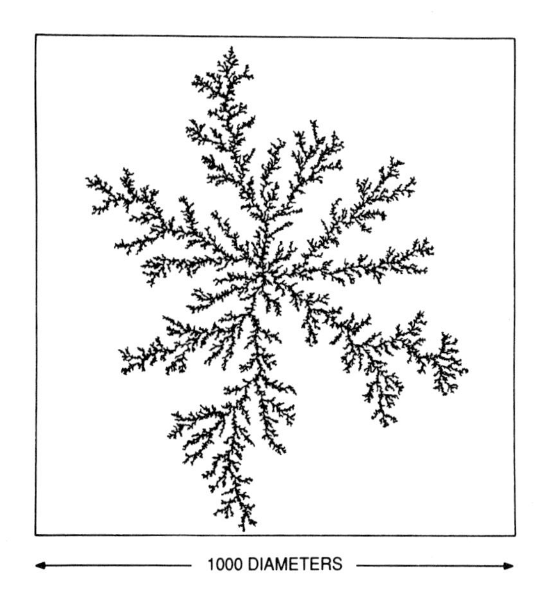

Figure 6.3. A typical off-lattice DLA cluster of 50,000 particles. A comparison with Fig. 6.1 showing a much smaller aggregate illustrates the stochastic self-similarity of diffusion-limited aggregates (Meakin 1985b).

Table 6.2. Fractal dimension (D) of DLA clusters grown using modified versions of the diffusion-limited-aggregation model. The notation is the following: p_s – sticking probability, *o-l* – off-lattice, d – embedding dimension and *nnn* – sticking at next nearest neighbours. If not indicated, $p_s = 1$ and nearest neighbour interaction is used on the square and simple cubic lattices. These results should be compared with those presented in Table 6.1 (Meakin 1983a).

Model	D
d=2, p_s=0.25	1.72 ± 0.06
d=2, o-l	1.71 ± 0.07
d=2, nnn	1.72 ± 0.05
d=3, p_s=0.25	2.49 ± 0.12
d=3, o-l	2.50 ± 0.08

The above results seem to be consistent with the picture of a perfectly self-similar DLA cluster. However, several observations (some of them discussed in the next section) indicate that the structure of diffusion-limited aggregates is more complex and the fractal dimension itself is not enough to characterize its scaling properties. The first result suggesting deviations from a standard behaviour was related to the width of the region (active zone) where the newly arriving particles are captured by the growing cluster. The *active zone* can be well described in terms of the growth probability $P(r,N)$ which is the probability of the event that the Nth particle is attached to a cluster at a distance r from the origin (Plischke and Rácz 1984).

According to the simulations, $P(r,N)$ can be well approximated by a Gaussian characterized by two parameters $\bar{r}_N$ and ξ_N, representing the average deposition radius and the width of the active zone, respectively. For self-similar growth one expects that for $P(r,N)$ a scaling form analogous to (5.12) holds, i.e., $\bar{r}_N$ and ξ_N diverge with the same exponent as $N \to \infty$. Instead, simulations of DLA up to cluster sizes $N \sim 4000$ suggest that (Plischke and Rácz 1984)

$$\bar{r}_N \sim N^{\nu} \qquad \text{and} \qquad \xi_N \sim N^{\nu'} \tag{6.8}$$

with $\nu = 1/D$ as expected, but with $\nu < \nu'$ in contradiction with the assumption of a single characteristic scaling length typical for the cutoff behaviour of a finite-size self-similar object. This conclusion may not hold in the asymptotic limit, since simulations of considerably larger off-lattice aggregates indicated that ν' tends to approach ν as $N \to \infty$ (Meakin and Sander 1985). Therefore, it appears that an extremely *slow crossover* takes place in the surface structure of DLA clusters as they grow. Similar observations have been made in the context of anisotropy of cluster shapes induced by the underlying lattice.

Finally, we briefly discuss numerical results obtained for the fractal dimension from a *Monte Carlo position-space renormalization* (PSRG) group approach (Montag *et al* 1985). This method is reviewed in Section 4.3 for growth processes in general. When applying the renormalization transformation (4.22) one has to determine the sum of probabilities $P_{N,i'}$ taken over

i' from Monte Carlo simulations, where $P_{N,i'}$ is the probability that the i'th configuration consisting of N particles has a radius of gyration R_g such that $\kappa R_g = b/2$. Here κ is an optimization parameter (Vicsek and Kertész 1981) and b is the linear size of the cell renormalized into a single site. In practice, DLA clusters are grown from the centre of a cell and one records the number of sites in the cluster when its radius of gyration becomes equal to the cell size. Then the number of such occurrences is determined as a function of the number of sites in the cluster at that time. The data are fitted to a Gaussian, and the resulting curve is integrated to obtain Monte Carlo estimates of the coefficients in the renormalization transformation. Having determined $\sum_{i'} P_{N,i'}$ the eigenvalue (λ) of the transformation (4.22) is calculated numerically, and the fractal dimension is obtained from (4.25). Figure 6.4 demonstrates that

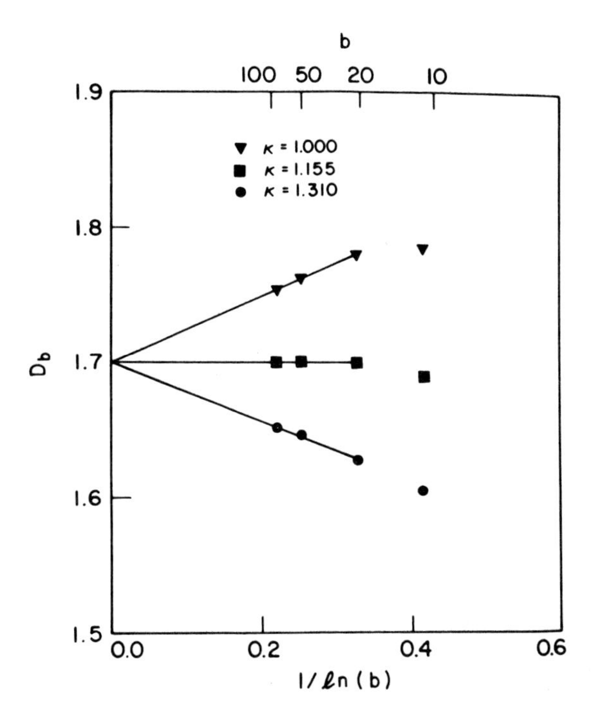

Figure 6.4. Extrapolation of the fractal dimension estimates, D_b, obtained from the phenomenological renormalization method to the large cell size ($b \gg 1$) limit. The calculations were carried out for three different values of the optimization parameter κ (Montag *et al* 1985).

an extrapolation of the fractal dimension values D_b obtained for a given cell size b to the $b \to \infty$ limit provides accurate estimates of D in spite of the relatively small cluster sizes used in the calculations.

6.1.2. Anisotropy in DLA

While an ordinary stochastic fractal has an isotropic internal structure, this is not true for diffusion-limited aggregates which are isotropic only in a crude approximation. The determination of fractal dimension alone does not allow us to get an insight into the structural details of DLA clusters, although they are of interest, since one expects that the diffusion-limited mechanism has a very specific impact on the correlations and interrelation of branches inside an aggregate.

To see this one calculates the tangential correlation function $c_R(\theta)$ defined in $d=2$ by (Meakin and Vicsek 1985, Kolb 1985)

$$c_R(\theta) = \frac{1}{N} \sum_{\theta'} \rho_R(\theta + \theta')\, \rho_R(\theta'), \tag{6.9}$$

where N is the number of particles in the aggregate, $\rho_R(\theta) = k$ if there are k particles in a box of size $R\Delta\theta\Delta R$ at the point (R, θ) and $\rho_R(\theta) = 0$ otherwise. The summation in (6.9) is taken over θ' values gradually increased by a fixed small $\Delta\theta'$ from $\theta' = 0$ to $\theta' = \pi$. According to (6.9), $c_R(\theta)$ describes the density-density correlations in a layer of width ΔR being at a distance R from the origin (Fig. 6.5) as a function of the angle θ measured from the origin of the clusters, so that θR is the distance separating two particles in the layer.

The results obtained for off-lattice aggregates are shown in Fig. 6.6. A finite size scaling analysis shows that asymptotically the tangential correlation function for $\theta \ll 1$ scales as a function of θ with an exponent $\alpha_\perp \simeq 0.41$. This exponent is definitely different from $\alpha \simeq 0.29$ which describes the algebraic decay of the ordinary radial correlation function (2.14).

It follows from $\alpha_\perp > \alpha$ that the density correlations around a particle being at point $\vec{R}$ depend on both r (denoting the distance from the given

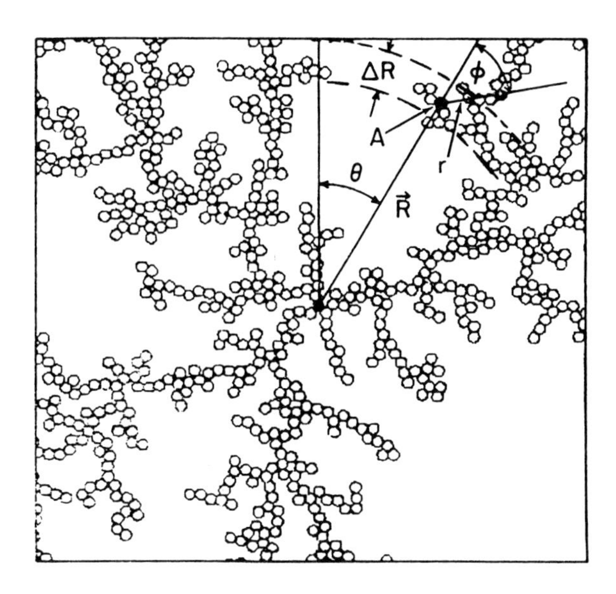

Figure 6.5. Central part of an off-lattice DLA cluster. The tangential correlations as a function of the angle θ are determined in a layer of width δR being at a distance R from the centre (Meakin and Vicsek 1985).

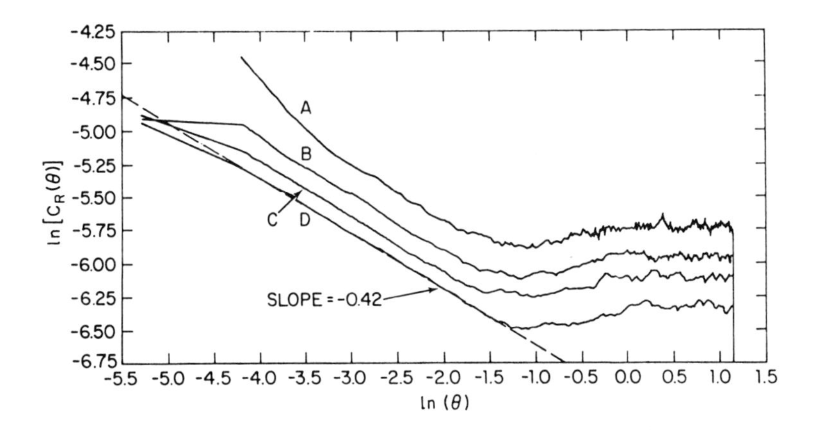

Figure 6.6. Tangential correlations in off-lattice DLA clusters of 50,000 particles. The results were obtained by averaging over the interval $\delta R = R \pm 0.05R$, where for the curves $A - D$ the radius R was respectively equal to 75, 150, 225 and 300 (Meakin and Vicsek 1985).

particle) and the local angle φ, where $\varphi = 0$ in the radial direction $\vec{R}$ as it is shown in Fig. 6.5. Let us assume that the density correlation function has the form

$$c(r, \varphi) \simeq \rho_0 r^{-\alpha_\parallel} \cos^2 \varphi + \rho_1 r^{-\alpha_\perp} \sin^2 \varphi\,, \tag{6.10}$$

where ρ_0 and ρ_1 are constants and $\alpha_\parallel$ describes the decay of $c(r, \varphi)$ in the radial direction. The relation (6.10) provides a simple example for a function with a power law decay consistent with the numerical data. Using (6.10) for the calculation of the fractal dimension in a manner analogous to (2.17) we get

$$N(a) \sim \int_0^a r dr \int_0^{2\pi} c(r, \varphi) d\varphi \sim \pi a^{2-\alpha_\parallel} (\rho_0 + \rho_1 a^{\alpha_\parallel - \alpha_\perp})\,, \tag{6.11}$$

where $N(a)$ is the number of particles within a circle of radius a. The fractal dimension is given by

$$D = \lim_{a \to \infty} \frac{\ln N(a)}{\ln a} = \lim_{a \to \infty} 2 - \alpha_\parallel + D_1(a) = 2 - \alpha_\parallel \tag{6.12}$$

with $D_1 = [\ln \pi(\rho_0 + \rho_1 a^{\alpha_\parallel - \alpha_\perp})]/\ln a$ representing a slowly decaying correction to scaling (Meakin and Vicsek 1985).

The main conclusions one can draw on the basis of the above results are the following: i) diffusion-limited aggregates are *not isotropic self-similar* fractals, instead, they possess a special kind of self-affinity with direction dependent scaling of the density correlations, ii) The fractal dimension is determined by the exponent describing the decay of radial correlations, and finally, iii) the observed slow convergence of D to its asymptotic value is due to the correction term D_1 related to $\alpha_\perp$.

The above discussed anisotropy originates in the fact that DLA clusters grow by developing branches oriented away from a fixed origin. This type of symmetry shows up in the behaviour of the *three point correlation function* as well (Halsey and Meakin 1985). The results of the related simulations indicate that while the decay of $c(r)$ is determined by the same exponent throughout the cluster the structure is not homogeneous. In particular, the amplitude of the power law decay at points close to the origin is larger than elsewhere.

Another kind of anisotropy is manifested in the studies of the *overall shape* (or envelope) of DLA clusters. Results concerning the response of diffusion-limited aggregates to anisotropy show that the anisotropy of both the underlying lattice and the sticking probability represents a relevant parameter changing the general appearance of DLA clusters drastically.

One way to impose anisotropy is to make the sticking probability direction dependent (Ball *et al* 1985). This can be achieved on a square lattice by differentiating between two possible cases: i) the particle sticks with probability 1 if the left or right nearest neighbour sites to its actual position are occupied, ii) otherwise, it only has a probability $p_s < 1$ of sticking. This modified version of DLA leads to highly elongated clusters, as it can be seen in Fig. 6.7. According to the simulations the characteristic lengths of the aggregate X and Y in the easy (x) and hard (y) direction of growth increase as $X \sim N^{\frac{2}{3}}$ and $Y \sim N^{\frac{1}{3}}$ in the limit $N \to \infty$. Since $X \times Y \sim N$, the area covered by the cluster grows linearly with N which means that for any applied uniaxially anisotropic sticking probability the cluster will eventually grow into an object homogeneous on a large scale. All of these findings are in accord with a theoretical approach discussed in the next section.

Figure 6.7. A representative DLA cluster grown using anisotropic sticking probability (Ball *et al* 1985).

Although relatively small DLA clusters grown on various lattices of the same dimension were found to have the same radius of gyration exponent ν, simulations carried out on a larger scale indicated that the symmetry of the underlying lattice may affect the asymptotic behaviour of aggregates. Growth on the square lattice was studied in more detail from this point of view, and the results confirmed the *relevant role of the lattice anisotropy* (Meakin *et al* 1987).

In order to see how the crossover to the $N \to \infty$ regime takes place particularly large clusters (consisting of 4×10^6 particles) have to be generated using the algorithm described in Appendix A. The anisotropy of the resulting cross-like structures can be demonstrated by plotting only the regions where the last 2×10^5 particles were added to the cluster. Figure 6.8 shows a typical configuration of these active places visualizing the envelope of a very large diffusion-limited aggregate. It is apparent from this figure that as DLA clusters grow larger their shape becomes more similar to that associated with conventional *dendritic growth* which is known to be governed by the anisotropy of the surface tension.

The above observation can be made quantitative by determining the characteristic sizes of the four main arms of the cross-like structure (Meakin *et al* 1987). Let us define the exponents $\nu_{\|}$ and $\nu_{\perp}$ through the scaling of the average length

$$l \sim N^{\nu_{\|}} \tag{6.13}$$

and average width

$$w \sim N^{\nu_{\perp}} \tag{6.14}$$

of the arms as a function of the number of particles in the clusters, N. The quantity l can be estimated from the maximum of the cluster radius, while w can be associated with the mean deposition distance from the nearest of the lattice axes (crossing at the origin). The crossover to the behaviour corresponding to (6.13) and (6.14) is particularly *slow*. However, approximating

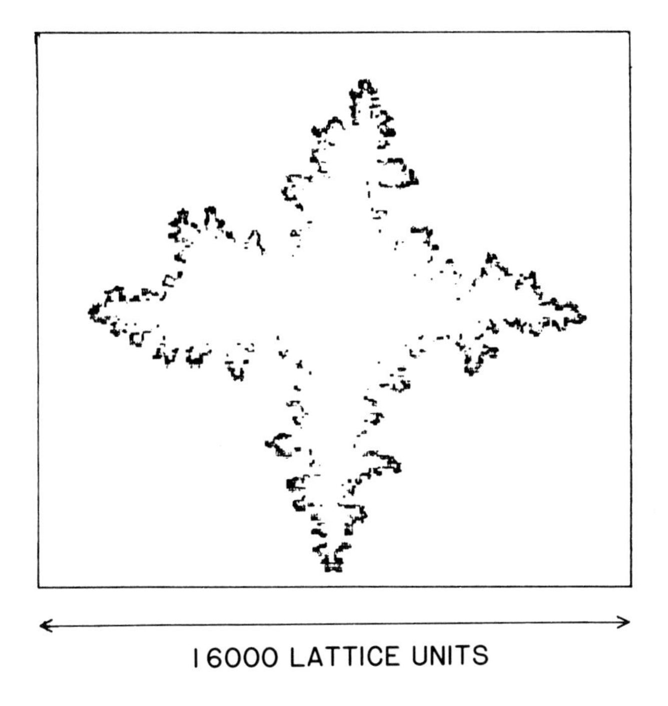

Figure 6.8. Envelope of a very large DLA cluster (consisting of 4×10^6 particles) generated on the square lattice. The effect of the lattice anisotropy is shown by plotting the last 2×10^5 particles attached to the cluster (Meakin *et al* 1987).

the behaviour of R_{max} with a curve of the form

$$R_{max} = aN^{\nu_{\parallel}}(1 + bN^{-\Delta}) \tag{6.15}$$

provides a relatively good fit to the data with $\nu_{\parallel} = \frac{2}{3}$ and with some constants a, b and Δ not relevant from the present point of view. A similar analysis for w gives $\nu_{\perp} \simeq 0.48 \simeq \frac{1}{2}$. These results correspond to at least two independent length scales describing the large scale behaviour of DLA clusters and, correspondingly, one expects that

$$w/l \to 0 \qquad \text{for} \qquad N \to \infty\,. \tag{6.16}$$

The above expression means that the angle at the tip of an arm should go to zero in the asymptotic limit. This angle plays an important role in a theoretical approach (next section) based on solving the Laplace equation for a region surrounding a tip of idealized geometry. According to this theoretical model the rate of growth at the maximum radius is determined by the angle θ through $dR_{max}/dN \sim R_{max}^{-\pi/2\varphi}$, where $\pi - \varphi$ is half of the tip angle. The dependence of φ on $\ln N$ as determined from the simulations is surprisingly linear, but it is not inconsistent with limiting values of $\leq 180°$ (thin cross) for large N and $\simeq 27°$ ($D \simeq 1.70$) for relatively small N. However, it should be pointed out that the investigation of the noise-reduced version of DLA (Section 9.2.2) indicates that (6.16) may be violated for sizes which are beyond those accessible by direct simulation.

The discussion of lattice induced anisotropy raises the question of the asymptotic shape of DLA clusters grown on other lattices. According to the simulations on the triangular and hexagonal lattices no signs of a crossover to a star-like overall form could be observed up to cluster sizes of 80,000 particles. The values obtained for the fractal dimensions indicate that diffusion-limited aggregates generated on these lattices have fractal dimensions numerically indistinguishable off-lattice DLA clusters (Meakin 1987).

These results suggest that during the process of aggregation there is a *competition* between anisotropy (provided by the lattice) and randomness (due to the stochastic motion of the particles) both having a relevant impact on the asymptotic behaviour (more about such questions can be found in Section 10.1.2). The triangular lattice is not anisotropic enough (it has too many axes) to force an arm to grow in a given direction. Consequently, there seems to exist an upper limit for the number of distinct main arms growing out from the central region of a diffusion-limited aggregate. This problem can be attacked by introducing various kinds of anisotropies and estimating the number of arms which can grow in a more or less stable fashion. The present consensus is that in the asymptotic limit diffusion-limited aggregates are not likely to have more than 5 main arms in two dimensions. In particular, off-lattice DLA clusters might acquire the shape of a 4 or 5 fold star as $N \to \infty$ (Ball 1986, Meakin and Vicsek 1987). However, up to sizes simulated so far, the overall shape of an off-lattice aggregate (like those generated on the triangular and hexagonal lattices) does not seem to deviate from a circle significantly.

6.1.3. Theoretical approaches to DLA

Diffusion-limited aggregation is a far from equilibrium phenomenon with no standard theory founded on first principles only. The approaches discussed in this section are based on various assumptions depending on the particular theoretical model considered. The most relevant difficulty is to take into account the spatial fluctuations characteristic for a DLA cluster in an appropriate way. This problem is usually treated by assuming some kind of average behaviour for such properties of a cluster as the distribution of its density, the penetration length of an incoming particle, or the aggregate's envelope.

There are two main types of *mean-field* approach to DLA. The first class is analogous to that introduced by Flory (1971) to describe the structure of linear polymers. In the Flory approximation one calculates the free energy of a cluster as a function of N and its linear size neglecting density fluctuations. Then the fractal dimension is obtained from the condition that the free energy has to be minimal. For a far from equilibrium system the construction of free energy and its minimalization do not represent well founded principles. However, following this line it is possible to obtain for the fractal dimension a simple expression (Tokuyama and Kawasaki 1984)

$$D = \frac{d^2 + 1}{d + 1} \tag{6.17}$$

which gives values surprisingly close to the simulation results.

An alternative method is to express the growth rate of a DLA cluster through the *screening or penetration length* ξ, where ξ is the average depth of the layer at the surface of a cluster which is accessible for an incoming particle (Muthukumar 1983). In fact, the result (6.17) was first obtained by this approach. As an example of such mean-field theories let us consider the following heuristic argument (Honda *et al* 1986). We assume that the particles follow a trajectory of dimension D_w (it can be a Levy walk or walk on a fractal, see previous chapter) and for convenience denote the number of particles in a layer of width dr being at a distance r from the origin by

$$\rho(r)dr = \frac{dn(r)}{dr}dr \sim r^{D-1}dr\,, \tag{6.18}$$

where $n(r)$ is the number of particles within a sphere of radius r. Suppose that $\Delta N \ll N$ new particles are added to the cluster. Then the cluster radius increases by ΔR so that

$$N + \Delta N \sim (R + \Delta R)^D\,. \tag{6.19}$$

Since $N + \Delta N \sim \int_0^{R+\Delta R}[\rho(r) + \delta\rho(r)]dr$, from (6.18) and (6.19) we get for the increment of the number of particles within a shell at a distance r from the origin

$$\delta\rho(r) \sim r^{D-1} \tag{6.20}$$

with a coefficient proportional to ΔN. Next one makes the heuristic assumption that this increase of $\rho(r)$ is proportional to the volume of the empty regions in which the particles can diffuse before deposition

$$\delta\rho(r) \sim \xi^d(r)\,. \tag{6.21}$$

To proceed we need to obtain an estimate for ξ. This can be done by taking into account that a particle makes, on average, $N_w \sim \xi^{D_w}$ steps before hitting the cluster. The number of steps on the surface of this "cloud" of steps is then proportional to $N_{w,s} \sim \xi^{D_w-1}$. On the other hand, the average density of particles belonging to the cluster grows with r as $\sigma(r) \sim r^{D-d}$. From the condition that the actual deposition takes place when $N_{w,s}\sigma(r) \simeq 1$ one gets

$$\xi(r) \sim r^{(d-D)/(D_w-1)}\,. \tag{6.22}$$

The final result is obtained by comparing (6.20), (6.21) and (6.22)

$$D = \frac{d^2 + D_w - 1}{d + D_w - 1} \tag{6.23}$$

which reduces to (6.17) for $D_w = 1$. For $d_w = 1$ (ballistic aggregation) the above expression yields $D = d$ in accord with other results. The comparison of (6.23) with estimates obtained by other approaches for various d and d_w values leads to reasonable agreement as well.

Investigation of the *growth probability scaling near the tips* of the clusters represents an important and far reaching contribution to the theoretical description of diffusion-limited aggregation (Turkevich and Scher 1985, Ball *et al* 1985). In this approach the analogy between the probability of finding a particle at a given point close to an array of traps and the distribution of electrostatic potential ϕ around a conductor (see Section 9.1) is utilized.

Consider a DLA cluster of N particles having a maximum radius R. The probability of finding a diffusing particle in a certain site $\vec{r}$ outside the cluster satisfies the Laplace equation $\nabla^2 p(\vec{r}) = 0$ with the boundary condition $p = 0$ on sites adjacent to the cluster. The flux of particles onto the cluster at a point $\vec{r}_0$ on its surface is proportional to $\nabla p(\vec{r}_0)$. The electrostatic analog of this problem is a charged conductor having the shape of a cluster. In particular, the local electric field $\vec{E} = -\nabla\phi$ (or the surface charge density) is the analog of the flux of particles $dR/dt = (dN/dt)/(dN/dR)$ onto that point, where dN/dt is the total flux of particles onto the cluster corresponding to the total charge, and $dN/dR \sim R^{D-1}$.

Clearly, if one is able to obtain an expression for dN/dR from electrostatics, comparison with the above relation for the same quantity should provide an estimate for D. It is quite natural to assume that the most advanced parts of a cluster (where the deposition of the particles takes place) can be represented by a cone of exterior half angle φ (Fig. 6.9). This is an idealization of the actual situation, however, it reflects one of the basic properties of DLA clusters: the active surface consists of advanced tips corresponding to a local singularity of both the geometry and the deposition probability.

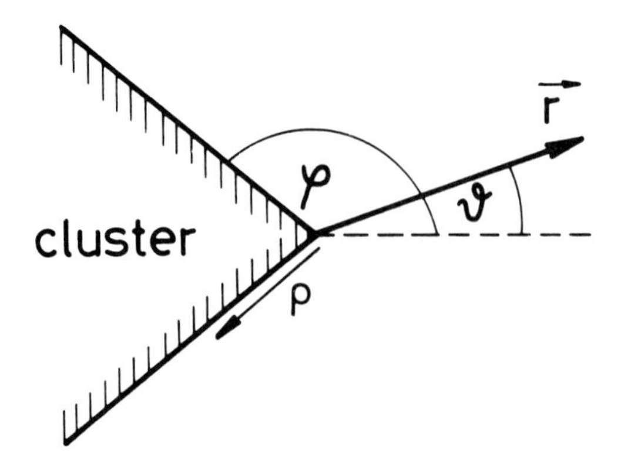

Figure 6.9. Modelling the region around the tips of DLA clusters with a cone of exterior half angle φ.

The problem of finding the solution of the Laplace equation for a conducting infinite cone can be solved exactly via conformal transformation giving

$$\phi(r, \vartheta) = C\ r^{\pi/(2\varphi)} \cos(\pi\vartheta/2\varphi)\,, \tag{6.24}$$

where C is a normalization factor. Thus the steady-state flux of diffusing particles onto the cone edge at a distance λ from its tip is

$$\nabla\phi(\lambda, \varphi) = \frac{C\ \pi}{2\varphi}\lambda^{\pi/(2\varphi)-1}\,. \tag{6.25}$$

One obtains dN/dt by integrating the above expression from $\lambda = 0$ to some large cutoff at $\lambda \sim R$ and dR/dt by integrating up to a small cutoff $\lambda \sim a$, where $a = 1$ is the size of diffusing particles or the lattice spacing. We find from (6.25) that $dN/dt \simeq C\ R^{\pi/(2\varphi)}$ and $dR/dt \simeq C$ which leads to

$$\frac{dN}{dR} = \frac{dN/dt}{dR/dt} \simeq R^{\pi/(2\varphi)}\,. \tag{6.26}$$

Integrating the above expression gives for the fractal dimension

$$D = 1 + \frac{\pi}{2\varphi} = \frac{3\pi - \theta}{2\pi - \theta} \tag{6.27}$$

which is the main result of the probability scaling approach in two dimensions (Turkevich and Scher 1985).

Let us discuss the above results considering various assumptions for θ. In the previous section DLA clusters with anisotropic sticking probability were shown to evolve into an elongated quadrangle becoming increasingly needle-like with growing N. This geometry corresponds to $\varphi_x = \pi$ and $\varphi_y = \frac{\pi}{2}$ in the limit $N \to \infty$, where φ_x and φ_y are exterior half angles of the cones oriented in the direction of easy growth (x) and hard growth (y), respectively. Then expressions analogous to (6.26) applied to the description of the tip distances in the x and y directions yield (Ball *et al* 1985)

$$X \sim N^{2/3} \qquad \text{and} \qquad Y \sim N^{1/3} \tag{6.28}$$

in complete agreement with the simulation results.

It is less clear what is the appropriate value of φ or θ in (6.24) when one is interested in D of ordinary diffusion-limited aggregates. The assumption $\theta = \frac{\pi}{2}$ corresponding to a growing square on the square lattice gives for the fractal dimension $D = \frac{5}{3}$, a good estimate coinciding with the prediction of the mean-field approach. For the unbiased off-lattice case there is no direct way to determine θ, and (6.27) can be regarded as a definition for an effective angle $\theta_{eff} \simeq 0.59\pi \simeq 106°$ through the known fractal dimension $D \simeq 1.71$.

The above method can be generalized to dimensions higher than $d = 2$ (Turkevich and Scher 1986). In this case the fractal dimension can be written in the form $D(d) = 2 + \omega$, where ω is the degree of the ordinary Legendre function, and it is determined from an implicit equation derived from the condition that the right-angled hypercone should be at zero potential. The case of d dimensional cross-shape clusters can also be described by this approach using $\theta = 0$ for the angle of the hypercone representing the tips of the needle-like arms of the aggregate. The corresponding result is

$$\omega = \begin{cases} (d-3)/2, & \text{if } 1 \leq d \leq 3, \\ d-3, & \text{if} d \geq 3\,. \end{cases} \tag{6.29}$$

Finally, it should be noted that a number of further approaches have been applied to the theoretical description of DLA. Position-space renormalization group (PSRG) (see Section 4.3) represents a standard tool for calculating non-integer exponents characterizing the singular behaviour of various quantities in systems exhibiting equilibrium phase transitions (Stanley *et al* 1982). However, its application to DLA raises a number of unresolved questions, especially, when small cells are used (Gould *et al* 1983). Small cell renormalization has also been used to calculate the multifractal spectrum of growth probabilities (Nagatani 1987a, 1987b). In fact, the actual values for D are usually not accurate, and depend too strongly on the particular (sometimes quite arbitrary) rules assumed in the course of renormalization. In a different approach first the number of main arms m in a cluster was determined from a stability analysis (Ball 1986). Then this number was used to calculate D using the probability scaling theory, assuming that the cluster has a polygonal shape with m tips. Here it is not clear whether the shape of a cluster can be simultaneously associated with both a star-like object and a convex polygon. The scaling of the length and the width of the arms with N has also been addressed using conformal transformation and scaling arguments (Szép and Lugosi 1987, Family and Hentschel 1987).

6.1.4. Multifractal scaling

According to the simulations and the theoretical arguments discussed in the previous sections the growth of a DLA cluster is governed by the distribution of the quantity $p(\vec{r}_j)$, where $p(\vec{r}_j)$ is the probability that the next growth event takes place at the site being at $\vec{r}_j$, adjacent to the cluster. This *growth-site probability distribution* (GSPD) is a very complex function changing rapidly in space due to screening (Halsey *et al* 1986, Amitrano *et al* 1986, Meakin *et al* 1986a). Let us imagine that we proceed along the surface of an aggregate and we record $p(\vec{r}_j)$ as a function of the arc length. Whenever we approach a tip in the outer region of the cluster, the growth probability associated with the actual position sharply increases since an advanced tip captures the diffusing particles with a large probability. Leaving this region one may get into a deep fjord which is almost completely screened by the surrounding branches, here $p(\vec{r}_j)$ is practically equal to zero. Getting close to another tip the growth probability becomes much larger again.

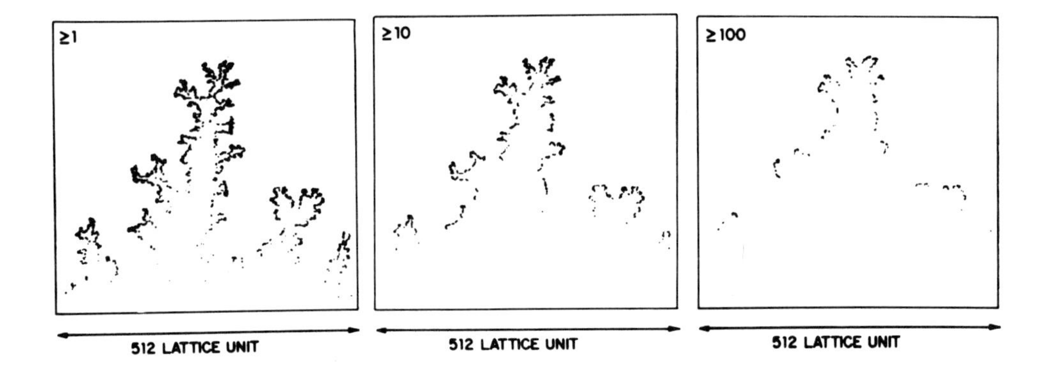

Figure 6.10. This figure shows those sites of a diffusion-limited deposit (grown on a strip of width 512 lattice units) which were contacted by the 5×10^5 random walk probes at least 1, 10 and 100 times (Meakin 1987a).

This is demonstrated in Fig. 6.10, where only those perimeter sites (sites adjacent to the aggregate) are shown which were hit by randomly walking probe particles at least 1, 10 and 100 times (Meakin 1987a). In general, the exponent describing the singular increase of $p(\vec{r}_j)$ depends on the local configuration close to a given tip. Therefore, it is quite natural to look at GSPD as a *fractal measure* (see Chapter 3) with infinitely many types of singularities. Probing the surface with many random walks is equivalent to the estimation of the corresponding solutions of Laplace's equation (Section 9.1) which are usually called harmonic functions. Consequently, the name harmonic measure is also used for GSPD.

In addition to describing the distribution of growth probabilities, the harmonic measure is relevant to the physical properties of a fractal. In the preceding section it was discussed that $p(\vec{r}_j)$ is proportional to the local charge density on a DLA, assuming that the aggregate is a charged electrical conductor. Similarly, the absorption rate of a DLA shaped catalyzer representing a sink for diffusing particles can be interpreted in terms of GSPD. Other physical processes may depend on the structure of the cluster in a different manner giving rise to fractal measures of various types associated with the aggregate. However, in most of the cases the freshly grown parts of the aggregate determine its physical properties and those are the the corresponding fractal measures which provide a detailed description of this surface region.

To calculate the harmonic measure and its characteristic properties one can follow two numerical methods. i) After having generated a DLA cluster one releases further particles (Halsey *et al* 1986, Meakin *et al* 1986a) whose diffusional motion is simulated by the same technique (Appendix A) which is used for growing the aggregate. These probe particles, however, are eliminated when they arrive at the surface, and a record is kept of how many times each of the surface sites is contacted in this way. The normalized number of contacts is then regarded as the growth probability. The main disadvantage of this method is that it can only be used to obtain information about the harmonic measure in those regions where the measure is large enough (places with $p(\vec{r}_j) \ll 1$ are not visited by a sufficient number of trajectories). Consequently, only quantities determined by the positive moments of the probability distribution can be calculated.

Therefore, more complete data can be obtained by ii) solving the Laplace equation $\nabla^2\phi = 0$ with the boundary conditions $\phi = Const$ on the cluster and $\phi = 0$ far from it. Then the growth probabilities are given by $p(\vec{r}_j) \sim |\nabla\phi(\vec{r}_j)|$ on the basis of the electrostatic analog (preceding section). For small clusters the Green's function method can be used to solve Laplace's equation (Amitrano *et al* 1986) yielding GSPD free of any effects caused by the finite distance of the boundary with $\phi = 0$ from the aggregate. For larger clusters, which are usually needed to see the true scaling behaviour, it is more practical to solve the discrete version of the Laplace equation by relaxation methods (Hayakawa *et al* 1987).

After having determined the set of $p(\vec{r}_j)$ values, the generalized dimensions D_q and the $f(\alpha)$ spectrum of fractal dimensions corresponding to the singularities of strength α can be calculated using the expressions (3.11), (3.13) and (3.16) given in Chapter 3. For this purpose one has to cover the cluster with boxes of size l and sum up the $p(\vec{r}_j)$ values within the ith box to obtain the accumulated probability p_i associated with it. Then the exponent describing the scaling of the qth moment of the harmonic measure is given by

$$D_q = \lim_{l/L \to 0} \frac{1}{q-1} \frac{\ln \sum_i [p_i(l/L)]^q}{\ln(l/L)}, \tag{6.30}$$

where L is the linear size of the aggregate. To evaluate (6.30) one can either change l for a fixed cluster, or keep $l = 1$ and consider the growth probabilities for increasing L values. Note, that in principle both of the conditions $a/l \ll 1$ and $l/L \ll 1$ should be satisfied during the calculations to produce results exactly corresponding to the multifractal spectrum as defined for finite fractals (with no lower cutoff length scale). In practical calculations these conditions can not be satisfied because of computer time and memory limitations. Due to (6.30) the log-log plots of $\sum_i p_i^q$ versus l/L have a slope $(q-1)D_q$ providing an esimate for the generalized dimensions. Figure 6.11 shows the results for two-dimensional off-lattice aggregates (Hayakawa *et al* 1987).

The obtained estimates can be examined using a few theoretical predictions. The value D_1 called information dimension has particular importance. It is equal to the fractal dimension of the set of boxes which give the dominant contribution to the first moment, i.e., to the sum of the box probabilities. According to a recent mathematical theorem, in $d = 2$ the information dimension of the harmonic measure is equal to 1 (Makarov 1985), therefore, most of the measure is concentrated on a fractal of dimension $D_1 = 1$. Furthermore, the exponent of the $q = 0$th moment has to be equal to the fractal dimension of the substrate on which the measure is defined. These predictions are consistent with the numerical data of Fig. 6.11: $D_0 \simeq 1.64$ ($D_{DLA} \simeq 1.7$) and $D_1 \simeq 1.04$. Finally, $D_3 \simeq 0.85$ is in good agreement with a recent theoretical result implying for DLA $2D_3 = D$ (Halsey 1987).

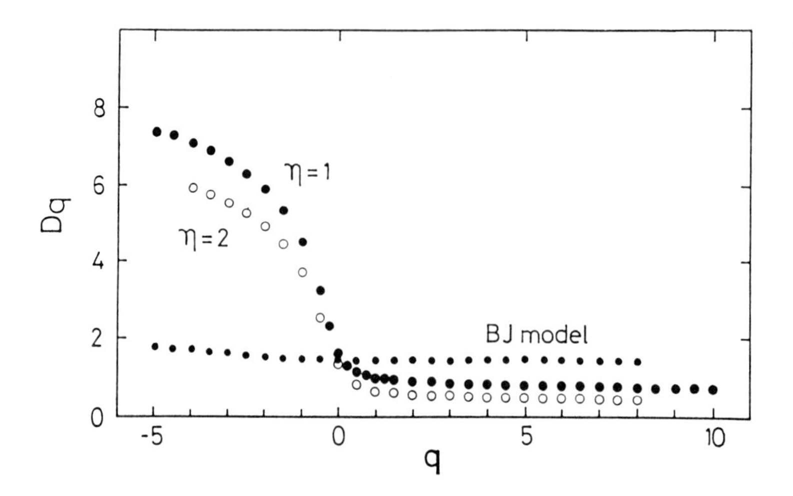

Figure 6.11. The generalized dimensions D_q calculated for off-lattice DLA clusters consisting of 3000–70000 particles ($\eta = 1$). The data obtained for the dielectric breakdown model ($\eta = 2$) and the disaggregation model (BJ) will be discussed in Sections 6.2 and 6.3 (Hayakawa *et al* 1987).

The $f(\alpha)$ spectrum can be determined through the Legendre transformation of D_q according to the expressions (3.13–3.15). The result is presented in Fig. 6.12. This continuous function demonstrates that the harmonic measure divides a DLA cluster into interwoven fractal subsets with dimensions between $0 < f(\alpha) < D_0 = D$ each characterized by the corresponding singularity of strength $\alpha_\infty < \alpha < \alpha_{-\infty}$.

The top of the curve in Fig 6.12 is at $D_0 = D \simeq 1.62$ in reasonable agreement with the simulations aimed at calculating the fractal dimension only. Another possibility for comparison with theoretical results is provided by expression (6.27) connecting the fractal dimension of the cluster with the effective angle of the cone representing the tip of an advanced branch. Writing (6.27) in the form $D = 1 + x$ and (6.25) as $\nabla\phi \sim l^{x-1}$ one can see that the integral of $\nabla\phi \sim p_i$ (which is the accumulated probability in a box of size l placed onto a tip) scales with the exponent x. Thus, assuming that the fractal dimension is determined by tips having the strongest singularity

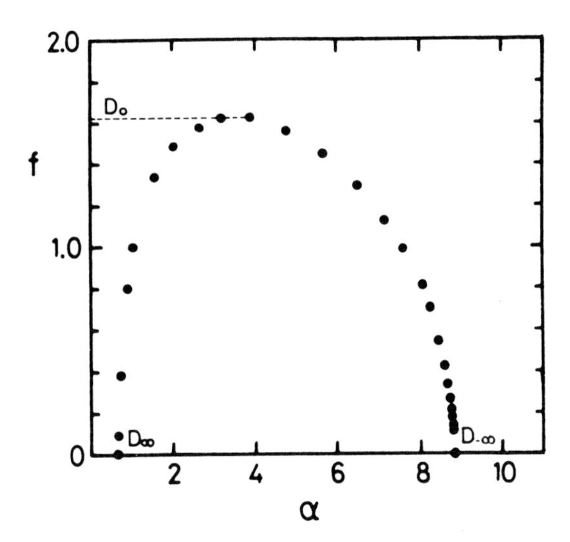

Figure 6.12. The spectrum of fractal dimensionalities, $f(\alpha)$, for the growth site probability distribution of off-lattice DLA clusters (Hayakawa *et al* 1987).

one obtains

$$D = 1 + \alpha_\infty \,. \tag{6.31}$$

The value $\alpha_\infty \simeq 0.7$ in Fig. 6.12 is in good accord with the above prediction ($D \simeq 1.7$).

The definitions (3.2) and (3.3) allows one to make an attempt to plot $f(\alpha)$ directly from the data obtained for the growth probabilities $p(\vec{r}_j)$. Assuming $l = a = 1$ and using as a measure of the linear size $M^{1/D} \sim L$ instead of L one can combine (3.2) and (3.3) to give

$$\frac{\ln[pN(p)]}{\ln M} \sim f\left(\frac{\ln p}{\ln M}\right), \tag{6.32}$$

where $pN(p)d\ln p = [dN(\ln p)/d\ln p]d\ln p$ is the number of sites with growth probabilities between $\ln p$ and $\ln p + d\ln p$ with $dp \ll 1$. If there exists a unique scaling function $f(\alpha)$, the data obtained for $\ln[pN(p)]/\ln M$ plotted against $\ln p/\ln M$ for various M should fall onto the same curve. Of course, this procedure should give the right exponents only if the approximation $l = 1$ can be justified. Very recent results indicate that the condition $l \gg a$ has to be satisfied to get reliable results, where a is the lattice constant.

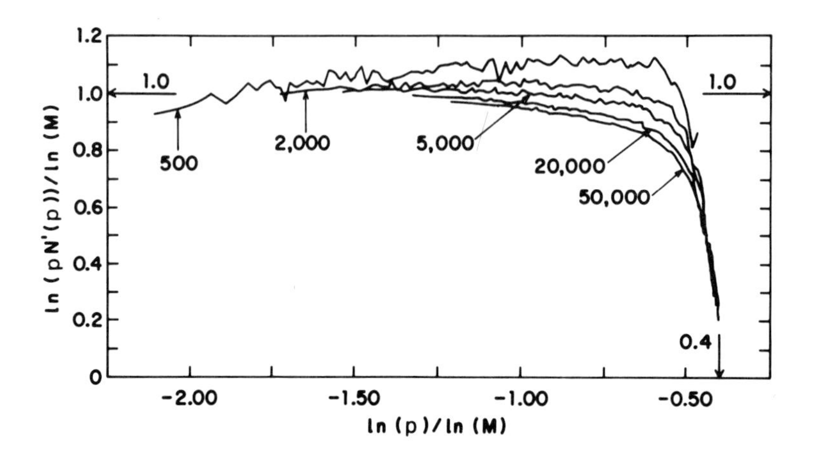

Figure 6.13. Result of an attempt to rescale the data for the growth probability distribution, $N(p)$, into a single curve corresponding to $D^{-1}f(\alpha)$ (Meakin *et al* 1986a).

In Fig. 6.13 the above scaling is tested using data for $pN(p)$ determined from probing the surface of two-dimensional DLA clusters with 10^6 diffusing particles (Meakin *et al* 1986a). The collapse of the results is not particularly good but clearly improves as the cluster mass increases. The maximum is close to 1.0 as expected, since it should be equal to $f_{max}(\alpha) = D$ when $\ln L = D^{-1} \ln M$ is used to represent the size of the system. It can also be seen in Fig. 6.13 that the maximum growth probability p_{max} scales with the size of the cluster according to

$$p_{max} \sim M^{-\delta} \tag{6.33}$$

with $\delta \simeq 0.4$ which is predicted by the probability scaling theory (preceding section) through the expression $\delta = 1 - 1/D$.

In Section 3.4 it has been shown that apart from the multifractal behaviour of a singular measure defined on a fractal, a fractal substrate itself can also exibit multifractal scaling of its mass. It is presently an open question of great interest whether randomly growing structures like DLA clusters are such geometrical multifractals. The first steps in the direction of treating this problem were concerned with the mass distribution in two-dimensional aggregates (Meakin and Havlin 1987).

To analyze the *mass distribution* one chooses a site randomly on a large off-lattice aggregate and determines $P_s(r)$ which is the probability of having s sites belonging to the cluster within a distance r of the given site. Then the fractal dimension is given by

$$\langle s \rangle = \int_0^\infty s P_s(r) ds \sim r^D . \tag{6.34}$$

Next we assume that $P_s(r)$ has the simplest scaling form yielding (6.34)

$$P_s(r) \sim \frac{1}{s} f\left(\frac{s}{r^D}\right) , \tag{6.35}$$

where the factor s^{-1} is needed to satisfy the normalization condition of $P_s(r)$

$$\int_0^\infty P_s(r) ds = 1 . \tag{6.36}$$

It is important to note that (6.35) is qualitatively different from the expression one would have for a homogeneous object, where the number of particles within a box of given size is described by the Poisson distribution. In the large r limit the Poisson distribution becomes delta function-like on a log-log plot, while (6.35) results in an invariant shape. If (6.35) is valid the moments of the distribution can be easily calculated (changing the variable of integration to $s = z/r^D$)

$$\langle s^q \rangle \sim \int_0^\infty s^{q-1} f(s/r^D) ds \sim r^{qD} \qquad \text{for } q > 0 , \tag{6.37}$$

therefore, all of the positive moments can be expressed as powers of the first one.

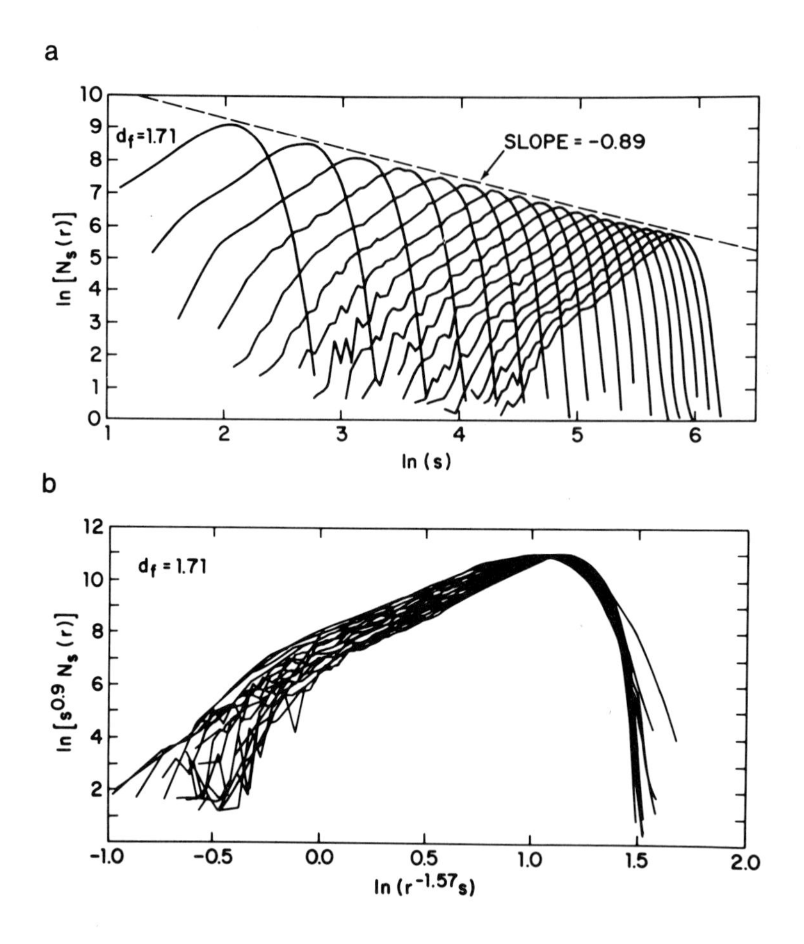

Figure 6.14. Attempt to scale the quantity $N_s(r) = N(r)P_s(r)$ determined for various r (a) into a single curve (b), where $N(r)$ is the total number of circles of radius r considered in the numerical experiment (Meakin and Havlin 1987).

However, the simulations of $2d$ off-lattice DLA ($D \simeq 1.7$) do not seem to lead to a satisfactory agreement with the above scaling picture based on the assumption of a single exponent D. Attempts to rescale $P_s(r)$ according to (6.35) onto a universal curve failed, as is shown in Fig. 6.14. The best collapse of the data was achieved using $P_s(r) \sim s^{-0.9} f(s/r^{1.57})$, which is clearly inconsistent with (6.35). In addition, the ratio $\langle s^q \rangle / \langle s \rangle^q$ was found to depend on r: another result indicating that the mass distribution within a DLA cluster can not be described by simple scaling of the type (6.35). Thus one is led to the conclusion that geometrical multifractality of diffusion-limited aggregates should be investigated further with the help of the formalism given in Section 3.4.

6.2. DIFFUSION-LIMITED DEPOSITION

The deposition of materials onto surfaces to form layers with specific properties has become an important technology with a very broad range of applications. In practice, deposition is carried out under conditions which allow various complex physical and chemical processes to occur. Diffusion-limited deposition (Meakin 1983b, Rácz and Vicsek 1983) represents a relevant limiting case, and its investigation is expected to be helpful in understanding more complicated systems used in commercial processes. The main difference between free aggregation and deposition is in the boundary condition: In the latter case a d_s dimensional surface of nucleation sites is present instead of a single seed particle. As the simulations show (Fig. 6.15) the presence of the surface and the competition of the incoming particles result in a forest of tree-like structures. For our purposes a cluster can be defined as a collection of particles connected to the same nucleation site through nearest neighbours.

Let us first summarize the results of simulations concerning the global structure of deposits grown in two dimensions along a linear substrate of length L. The distribution of particle density is very inhomogeneous in the direction perpendicular to the substrate. This can be studied by calculating the normalized number of particles at a height h, $\rho(h) = L^{-1}\sum_x \rho(h,x)$, where $\rho(h,x) = 1$ if the lattice site at (h,x) is filled and is equal to zero otherwise. The plot of $\ln \rho(h)$ versus $\log h$ suggests that for $h \ll L$ the density $\rho(h)$ behaves as (Meakin 1983b)

$$\rho(h) \sim h^{-\alpha_\parallel} \tag{6.38}$$

with $\alpha_\parallel \simeq 0.29$. Here $\parallel$ denotes the direction parallel to the direction

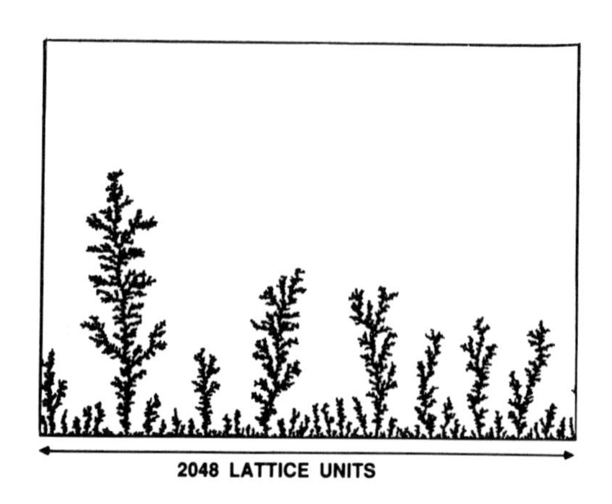

Figure 6.15. Forest of clusters grown on the square lattice along a 300 lattice unit long straight substrate. Because of screening, diffusion-limited deposition leads to a power law distribution of tree sizes (Meakin 1983b).

of growth. An effective fractal dimension can also be defined for deposits through the relation

$$N(h) \sim h^{D_s - d_s} \tag{6.39}$$

where $N(h)$ is the number of deposited particles within a distance h from the substrate. Since $N(h) \sim \int_0^h \tilde{h}^{-\alpha_\|} \tilde{h}^{d-d_s-1} d\tilde{h}$, the above effective dimension is $D_s = d - \alpha_\|$, in analogy with (2.18). Assuming that the correlations in the deposit decay in the same way as in a DLA cluster, we conclude that the effective dimension of deposites coincides with that of diffusion-limited aggregates ($D_s = D_{DLA}$).

The density correlations within a layer of a two-dimensional deposit provide relevant information about the internal structure of the deposit. To investigate the correlations along the lateral direction x (parallel to the deposition line) one can use the expression (Meakin *et al* 1988)

$$c_h(x) = \frac{1}{L}\sum_{x'} \rho(h, x + x')\rho(h, x') \,. \tag{6.40}$$

The results for the lateral correlation function are shown in Fig. 6.16a. The $c_h(x)$ curves exhibit a number of interesting features. For all h values they have a well pronounced minimum followed by a less apparent maximum. The position of the minima $x_{min}(h)$ depends on the height at which the correlation function was calculated. The corresponding log-log plot supports that the position of the minima scales with h according to an exponent $\simeq 0.8$.

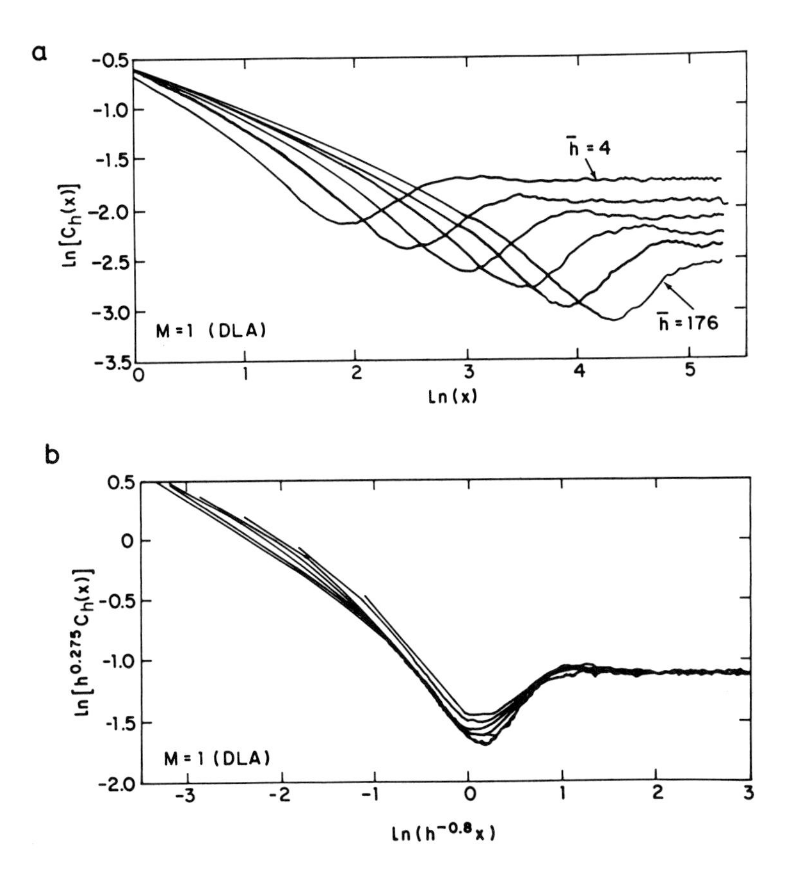

Figure 6.16. (a) Density correlations $c_h(x)$ in the lateral (x) direction within a layer of a two-dimensional deposit being at a distance h from the substrate. (b) The data for various h are shown to collapse into a single curve using the scaling form 6.41 (Meakin *et al* 1988).

One can also attempt to scale the correlation function $c_h(x)$ measured at different heights (h) onto a common curve. Figure 6.16b shows that the correlation function can be described quite well in terms of the scaling form

$$c_h(x) \sim h^{-\alpha_\parallel} f(x/h^\delta)\,, \tag{6.41}$$

where the exponents $\alpha_\parallel$ and δ have values of 0.275 and 0.8, respectively.

Finally, the behaviour of $c_h(x)$ is nontrivial for $x \ll h$. The slope of the curve seems to approach the limiting value $\alpha_\perp \simeq 0.42$ which indicates that the decay of correlations in the lateral directon is faster than in the direction parallel to the growth, and

$$c_h(x) \sim x^{-\alpha_\perp} \qquad \text{for} \qquad x \ll h\,. \tag{6.42}$$

The above results suggest that the trees are *anisotropic* and that $\alpha_\parallel$ and $\alpha_\perp$ are close to the analogous exponents determined for the radial DLA clusters. Therefore, it is of interest to calculate the dependence of mean tree height (H) and width (W) on the number of particles s in the trees. The simulations indicate that for $s \gg 1$,

$$H \sim s^{\nu_\parallel} \qquad \text{and} \qquad W \sim s^{\nu_\perp} \tag{6.43}$$

with the effective exponents $\nu_\parallel \simeq 0.65 \simeq 2/3$ and $\nu_\perp \simeq 0.56$.

An important aspect of the deposition process is that it produces a statistical *ensemble of aggregates* in a natural way. This forest of trees can be characterized by the *cluster size distribution function,* $n_s(N)$ which is the number of trees containing s particles after N particles have been deposited. Both $n_s(N)$ and N are normalized quantities and obtained from the corresponding total number of clusters and particles after a division by the area of the d_s dimensional surface onto which the deposition takes place. In large DLA clusters the density correlations decay algebraically similarly to the decay of magnetic correlations within large droplets in an equilibrium system near its critical point. Thus we expect that $n_s(N)$ exhibits a scaling behaviour analogous to that of the droplet size distribution in thermal critical phenomena (Fisher 1967) or to the cluster size distribution in percolation models (Stauffer 1985).

The related simulations (Rácz and Vicsek 1983, Meakin 1984) suggest that, indeed, the size distribution in diffusion-limited deposits can be well represented by the scaling form

$$n_s(N) \sim s^{-\tau} f\left(\frac{s^\sigma}{N}\right), \tag{6.44}$$

where $f(x)$ is a cutoff function with $f(x) \simeq 1$ for $x \ll 1$ and $f(x) \simeq 0$ for $x \gg 1$. The above scaling behaviour is assumed to be valid for large s and N values. The expression (6.44) can be used for the derivation of a scaling law between the exponents τ and σ. The total number of particles per nucleation site, N, can be calculated through $n_s(N)$ as follows

$$\begin{aligned} N = \sum_{s=1}^{\infty} s n_s(N) &\sim \int^{\infty} s^{1-\tau} f(s^\sigma/N) ds \\ &= \sigma^{-1} N^{(2-\tau)/\sigma} \int^{\infty} z^{(2-\tau-\sigma)/\sigma} f(z) dz \sim N^{(2-\tau)/\sigma} \end{aligned} \tag{6.45}$$

which leads to the scaling law (Rácz and Vicsek 1983)

$$\tau = 2 - \sigma\,. \tag{6.46}$$

In contrast to the cluster size distributions determined for homogeneous equilibrium systems (where $\tau > 2$), for diffusion-limited deposition the inequality $\tau < 2$ must hold in order to have a diverging sum in Eq. (6.45) if N goes to infinity. Because of $\tau < 2$ the main contribution to the sum comes from the $s \gg 1$ clusters and the use of the integral for its evaluation is justified.

Assuming that the mean cluster size

$$S = \frac{\sum_{s=1}^{\infty} s^2 n_s(N)}{\sum_{s=1}^{\infty} s n_s(N)} \tag{6.47}$$

scales with N as $S \sim N^{\gamma}$ the substitution of (6.44) into (6.47) results in $\gamma = 1/(2-\tau)$ which is again different from the expression $\gamma = (3-\tau)/\sigma$ valid for percolation systems.

An important step in completing the description of scaling in diffusion-limited deposition is finding relationships connecting the exponents D_s, $\alpha_{\|}$, $\alpha_{\perp}$, $\nu_{\|}$, $\nu_{\perp}$ characterizing the geometry of deposits and the exponents σ and τ describing the scaling behaviour of the cluster size distribution. The following argument based on scaling assumptions can be used to establish a relation connecting the two types of exponents. Since the number of clusters containing s particles decays as $n_s \sim s^{-\tau}$ one can write for the number of trees having a height larger than h_0

$$n_{h>h_0} \sim n_{s>s_0} \sim h_0^{(1-\tau)/\nu_{\|}}, \tag{6.48}$$

where s_0 is the number of particles in a tree of height h_0. Using (6.48) the density of the deposit at a distance h from the substrate can be expressed as

$$\rho(h_0) \sim n_{h>h_0} m(h_0), \tag{6.49}$$

where

$$m(h_0) \sim \int_0^{h_0^{\nu_{\perp}/\nu_{\|}}} c_h(x) x^{d_s-1} dx \sim h_0^{\nu_{\perp}(d_s-\alpha_{\perp})/\nu_{\|}} \tag{6.50}$$

is proportional to the number of particles in a layer of a tree at a distance h_0 from the substrate, and for the correlation function describing the density decay (6.42) was assumed. Substituting (6.38), (6.48) and (6.50) into (6.49) we get

$$\tau = 1 + \alpha_{\|}\nu_{\|} + (d_s - \alpha_{\perp})\nu_{\perp} \tag{6.51}$$

which is a scaling relation among the exponents introduced earlier. For $\alpha_{\|} = \alpha_{\perp} = \alpha = d - D_s$ and $\nu_{\|} = \nu_{\perp} = 1/D_s$ the expression (6.51) leads to (Rácz and Vicsek 1983)

$$\tau = 1 + \frac{d_s}{D_s} \tag{6.52}$$

which relates the exponent describing the power law decay of the cluster size distribution to the fractal dimension for the case of deposits with isotropic scaling.

The above theoretical predictions have been tested by various approaches. The simulations of diffusion-limited deposition led to a cluster size distribution decaying as a power law for intermediate values of s and decreasing much faster for s larger than a characteristic value depending on N. According to large scale numerical experiments (Meakin 1984) the value of τ for $d_s = 1$ and $d = 2$ is $\tau \simeq 1.55$ in good agreement with both (6.51) and (6.52), since they predict $\tau \simeq 1.5$ (with $\alpha_{\parallel} = 0.3$, $\alpha_{\perp} = 0.4$, $\nu_{\parallel} = 2/3$ and $\nu_{\perp} = 1/2$) and $\tau \simeq 1 + 1/1.7 \simeq 1.59$. For $d_s = 2$ and $d = 3$ the value $\tau \simeq 1.84$ was obtained in the simulations, while (6.52) would give (with $D_s \simeq 2.5$) $\tau \simeq 1.8$.

Another possibility for checking the validity of the scaling form (6.44) and the scaling laws (6.46) and (6.52) is to calculate the corresponding quantities using a deterministic fractal model for diffusion-limited deposition (Vicsek 1983). According to this approach the deposit is generated by a recursion whose stages are demonstrated in Fig. 6.17. The resulting structure has a geometry analogous to the deterministic fractal shown in Fig. 2.1a. In order to obtain the cluster size distribution, one should note that the largest clusters generated in the k-th stage of the deposition process contain $s_{max}(k) \sum_j^k 5^{j-1} = 5^k/4$ particles and the number of these clusters per nucleation site is 3^{-k}. This means that the cutoff function in (6.47) for this case is the step function $\theta(1 - 4s/5^k)$. Therefore, the cluster size distribution can be written in the form

$$n_s(N) \sim \frac{1}{3^l 5^l} \theta \left(1 - \frac{4s}{5^k} \right), \tag{6.53}$$

where we took into account that the normalized number of clusters of size $5^l/4$ ($l < k$) is $2/3^{l+1}$ and that these delta function-like peaks are separated from each other by a distance proportional to 5^l. Since $s_l \sim 5^l/4$ and $N =$

$(5/3)^k$ one recovers the scaling form (6.44) with $\tau = 1 + \ln 3/\ln 5$ and $\sigma = 1 - \ln 3/\ln 5$, so that the scaling law (6.46) is satisfied. Since $D = \ln 5/\ln 3$, the scaling relation (6.52) is also fulfilled in the deterministic model.

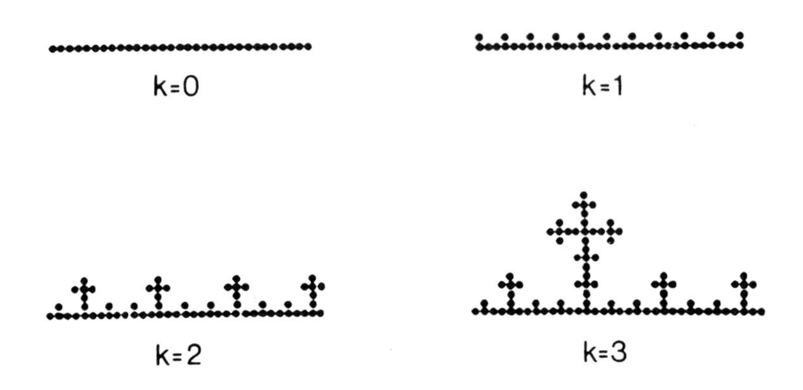

Figure 6.17. Deterministic fractal model for diffusion-limited deposition (Vicsek 1983).

6.3. DIELECTRIC BREAKDOWN MODEL

This model was introduced in order to simulate a variety of dielectric breakdown phenomena which range from atmospheric lightning to electric treeing in polymers. Although the actual physical processes can be quite different in these phenomena, the global properties of the resulting discharge patterns are very similar: they have a randomly branching, open structure resembling DLA.

Before describing the model let us outline the phenomenology of the discharge process. If an insulating material is exposed to an electric field exceeding a critical value, a conducting phase is created because a large field produces mobile charge carriers. The motion of the interface is controlled by the electric field and it is more or less stochastic in time. In the dielectric breakdown model (DBM) (Niemeyer *et al* 1984) the complicated details of the physical processes occurring at the tips of the discharge pattern are ignored, and the corresponding equations are replaced by the assumptions i) $\phi = \phi_0 = 0$ in the conducting phase, where ϕ is the electric potential satisfying Laplace's equation

$$\nabla^2 \phi = 0\,, \tag{6.54}$$

and ii) the growth velocity is stochastically proportional to some power η of the local electric field $\vec{E} = -\nabla\phi$. The apparent analogy between DLA and DBM can be understood on the level of the equations which determine the behaviour of the two models. If the exponent $\eta = 1$ the growth probability in both growth models is proportional to the local value of the gradient of a distribution satisfying (6.54), since the probability of finding a diffusing particle at a given point is also given by the Laplace equation (Witten and Sander 1981).

The actual model is formulated on a d-dimensional hypercubic lattice, and the Laplace operator, ∇^2, is replaced with its discrete version. For example, in two dimensions (6.54) takes the form

$$\phi_{i,j} = \frac{1}{4}(\phi_{i-1,j} + \phi_{i+1,j} + \phi_{i,j-1} + \phi_{i,j+1})\,, \tag{6.55}$$

where $\phi_{i,j}$ is the value of ϕ in the grid site i, j. The boundary conditions are the following

$$\phi_{i,j} = 0 \qquad \text{for sites belonging to the cluster}\,, \tag{6.56}$$

and $\phi_{i,j} = -1$ for sites on a large circle of radius r_0 centred at the origin. The boundary condition describing the motion of the interface is represented by an expression for the growth probability at the site i, j adjacent to the cluster

$$p_{i,j} = C\nabla\phi^{\eta}_{i,j} = -C\phi^{\eta}_{i,j}\,, \tag{6.57}$$

where the normalization factor is given by $1/C = \sum \phi_{i,j}^{\eta}$ with the summation running over all of the nearest neighbour sites to the cluster. It is the exponent η which is an important extra property of DBM with regard to diffusion-limited aggregation, because η turns out to be a relevant parameter from the point of view of the fractal dimension of the patterns (Niemeyer *et al* 1984). There is another slight difference in the boundary conditions. The absorbing boundary condition used in DLA corresponds to a zero potential (probability) in the sites adjacent to the aggregate (not on the cluster itself as is the case according to (6.56) for DBM).

The simulation starts with a seed particle at the origin of a lattice. The potential for each site of the lattice within a circle of radius r_0 is calculated using relaxation methods. (6.55) represents a system of linear algebraic equations (one equation per site) which can be solved by iteration. Relatively good convergency can be achieved by the Gauss-Seidel over-relaxation scheme which on a square lattice has the form

$$\phi_{i,j}^{k+1} = \phi_{i,j}^{(k)} + \omega \left[\frac{1}{4} (\phi_{i-1,j}^{(k+1)} + \phi_{i,j-1}^{(k+1)} + \phi_{i+1,j}^{(k)} + \phi_{i,j+1}^{(k)}) - \phi_{i,j}^{(k)} \right] \tag{6.58}$$

if one sweeps the sites in such a way that i and j increase as one goes to the next site. In (6.58) ω is the over-relaxation parameter. Finding an optimal value for ω by trial and error may speed up the convergence considerably. Next a perimeter site is chosen randomly, and a random number r drawn from a set of random numbers uniformly distributed between 0 and p_{max}, where p_{max} is the largest growth probability. If $r < p_{i,j}$, the perimeter site i, j is filled, and the whole procedure starts again by calculating the distribution ϕ in the presence of the new configuration.

This procedure for growing an N-site cluster requires much more computer time than generating a diffusion-limited aggregate of the same size since one has to solve the Laplace equation within a large region of radius r_0. Correspondingly, the data for the fractal dimension were obtained for clusters consisting of about 10000 particles. The simulations for $\eta = 1$ led to clusters with $D \simeq 1.70$ in good agreement with the expectation that DLA and DBM with $\eta = 1$ belong to the same universality class ($D_{DLA} \simeq 1.7$ in $2d$).

Varying η results in a non-trivial change of the fractal dimension (Wiesmann and Pietronero 1986), a property which makes the DBM model particularly interesting from a theoretical viewpoint (a direct connection between η and physical quantities has not been established). Table 6.3 shows the numerical estimates for D.

Table 6.3. Fractal dimension for several values of the probability exponent η of the dielectric breakdown model. The data were obtained for clusters consisting of N particles and generated on d-dimensional lattices (Wiesmann and Pietronero 1986).

d	η	D	N
2	0	2.00	20000
2	0.5	1.92	30000
2	1	1.70	10000
2	2	1.43	3000
3	0	3.00	20000
3	0.5	2.78	10000
3	1	2.65	4000
3	2	2.26	1500

The mean-field type argument (Section 6.1.3) leading to the expression (6.23) for the fractal dimension of DLA clusters can be generalized to take into account the effects of the growth exponent η. The result is given by (Matsushita et al 1986)

$$D(\eta) = \frac{d^2 + \eta(D_w - 1)}{d + \eta(D_w - 1)} \tag{6.59}$$

which should be compared with the results given in Table 6.3. The agreement is good, e.g., for $d = 3$ and $\eta = 2$ expression (6.59) predicts $D = 2.2$, while the numerical value is $D \simeq 2.26$.

Finally, the distribution of the growth probabilities $p_{i,j}$ represents a fractal measure, just like in the case of DLA. The corresponding hierarchy of exponents can be determined numerically (see Fig. 6.11), leading to results being in reasonable agreement with the expectations.

6.4. OTHER NON-LOCAL PARTICLE-CLUSTER GROWTH MODELS

DLA has attracted great interest because of its obvious relevance to a large class of important physical processes. The success of diffusion-limited aggregation models has prompted a rapid increase in the number of various non-local cluster growth models leading to fractal structures. Many of these new constructions have been shown to exhibit remarkable scaling behaviour. However, none of them has such a direct relation to any significant physical phenomenon as DLA has to diffusion-limited growth. Therefore, the investigation of these models is important from a didactical point of view rather than for describing real processes.

The *screened growth* model (Rikvold 1982, Meakin *et al* 1985a) is truly non-local, since the probability of adding a particle to the cluster at site i is determined by the position of all the other particles in the aggregate. This probability for the ith perimeter site of an N particle cluster is given by

$$p_i = \frac{\prod\limits_{j=1}^{N} e^{-Ar_{ij}^{-\lambda}}}{\sum\limits_{k=1}^{N_s} \prod_{j=1}^{N} e^{-Ar_{kj}^{-\lambda}}} = \frac{P_i}{\sum\limits_{k=1}^{N_s} P_k}, \qquad (6.60)$$

where N_s is the total number of perimeter sites, r_{ij} is the distance of the jth particle to the ith perimeter site and A is a constant which does not affect the scaling behaviour. Because of the long-range, independent multiplicative nature of the contributions coming from the particles already belonging to the aggregate, a cluster grown according to (6.60) is a fractal. Computer simulations and theoretical considerations suggest that for this model

$$D = \lambda\,. \qquad (6.61)$$

The following heuristic argument (Meakin *et al* 1985) supports the above relationship. Let us imagine that the cluster grows by adding sites at a rate P_i given in (6.60), so that $\sum_k P_k$ is the average number of sites created in unit time. It is possible to estimate the D-dependence of the rate P_i by changing the summation to integration when calculating the exponent describing the behaviour of P_i as a function of the distances r_{ij}

$$\sum_{j=1}^{N} r_{ij}^{-\lambda} \sim \int_a^R r^{D-1-\lambda} dr \sim BR^{D-\lambda}\,, \tag{6.62}$$

where a is a short distance cutoff length of the order of the lattice unit, B is a constant and $R \gg 1$ is the cluster radius. Supposing that $D > \lambda$ we have $P_i \sim e^{-BR^{D-\lambda}}$ and

$$\sum_{k=1}^{N_s} P_k \leq R^D e^{-BR^{D-\lambda}} \ll 1\,. \tag{6.63}$$

The above expression implies that the growth slows down dramatically everywhere with R becoming large. This is not expected to occur, since the formation of any branches protruding from the more compact structure can grow much faster than the other regions. On the other hand, such unstable perturbations decrease the fractal dimension D. Accordingly, a state with $D < \lambda$ has to be reached. However, it is the largest $D < \lambda$ which will be most favourable on combinatorial grounds: the number of different possible ways to obtain a given cluster shape by successive addition of sites increases with the fractal dimension of the cluster.

Studying the properties of the screened growth model requires large amounts of computer time because the growth probabilities p_i have to be updated for all of the perimeter sites after each growth event. The reward for the extra cost is a knowledge of p_i with a high accuracy, making it possible to carry out a detailed scaling analysis of the growth probability measure.

As was discussed in Section 6.1.4, if the distribution of growth probabilities can be described in terms of a fractal measure characterized by a spectrum of singularities of strength

$$\alpha = -D \ln p / \ln N\,, \tag{6.64}$$

then the quantity

$$f(\alpha) = D \ln[pN(p) \ln N] / \ln N \tag{6.65}$$

corresponds to the fractal dimension of the set of singular parts with strength α. Here $pN(p)d\ln p$ is the number of sites with probabilities in the range $\ln p$ to $\ln p + \ln dp$ expressed through $N(p)$ which is the number of sites with p between p and $p + dp$ divided by dp. Figure 6.18 shows the results for $f(\alpha)$ determined for $\lambda = 1.5$ (Meakin 1987). The fact that the plots obtained for various cluster sizes N collapse onto a single curve demonstrates the existence of a unique $f(\alpha)$ spectrum.

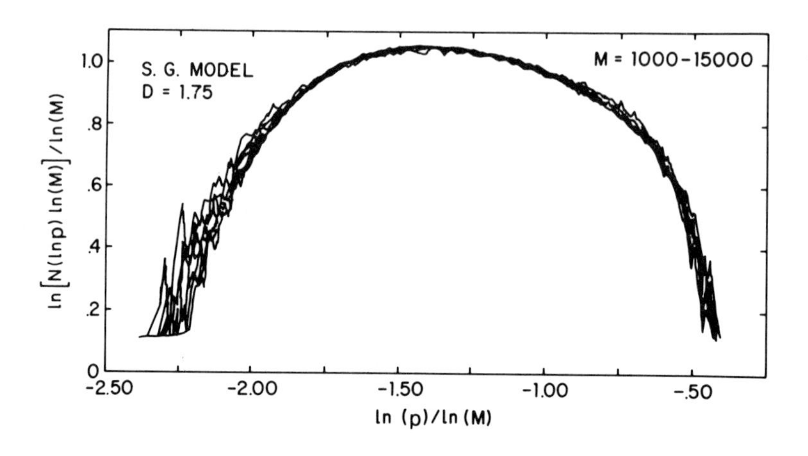

Figure 6.18. Scaling of the growth probability distribution for screened growth clusters with a fractal dimensionality of 1.75 (Meakin 1987a).

Diffusion-limited aggregation with *disaggregation* (Botet and Jullien 1985) is an interesting modification of the original DLA model, because during its simulation an equilibrium type regime is attained. In this lattice model a cluster is defined as a set of particles connected by filled bonds (not all of the bonds of the underlying lattice are filled). Given a cluster of N particles and the corresponding bond configuration at time $t_0 = 0$, one chooses with the same probability one of the particles which is linked to the rest of the cluster by only one bond. This particle is then allowed to diffuse away, and as in DLA it sticks to the cluster again forming a single new bond when it arrives at an adjacent site to the aggregate. At this point ($t = 1$) a new single connected particle is chosen randomly, and so on. In case the particle gets too far from the aggregate, it is eliminated and a new one is simultaneously released from a randomly selected point on a sphere centred at the origin of the cluster and having a radius just exceeding that of the aggregate.

This model does not seem to represent any realistic physical process because the strength of a bond depends in a rather unusual way on the actual configuration. Nevertheless, it is of interest to compare the resulting structures with those generated by other aggregation or equilibrium models. Figure 6.19 shows the process of approaching the steady-state configuration from two qualitatively different initial cluster shapes. The fractal dimension determined from the simulation of diffusion-limited aggregation with disaggregation on a square lattice is $D \simeq 1.54$, definitely smaller than the fractal dimension of DLA clusters. It is, however, very close to the fractal dimension $D \simeq 1.56$ of lattice animals representing a classical example of clusters existing in an equilibrium model (lattice animals are all the possible connected configurations of N particles each considered with the same weight). Obviously, in the present model two different configurations may have different weights. This was seen in the related simulations.

The growth probability distribution for diffusion-limited aggregation with disaggregation is expected to be uniform because of the equilibrium nature of the model. This can be checked by leaving a given configuration unchanged while releasing particles from the single connected places. Counting the number of trajectories terminating at the perimeter sites the retrapping probability distribution and the corresponding D_q spectrum can be determined from (6.30). According to the simulations (Hayakawa *et al* 1987) D_q is independent of q, as expected (see Fig. 6.11).

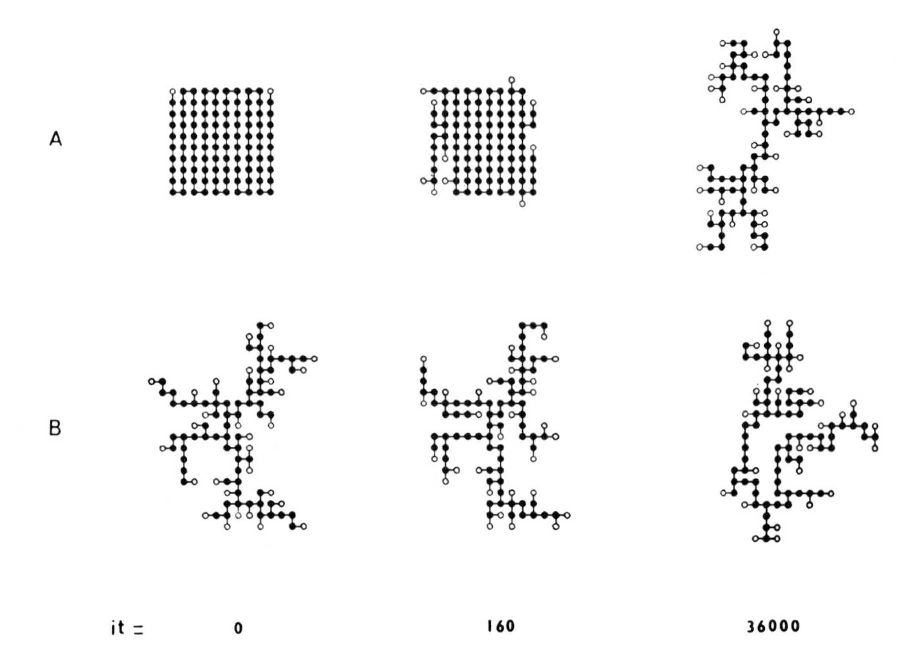

Figure 6.19. The process of aggregation with disaggregation results in qualitatively equivalent clusters even if it starts from very different initial configurations (Botet and Jullien 1985).

All models discussed in Part II so far have been stochastic concerning both the generated structures and the rules used to grow an object. This is quite understandable: the overwhelming majority of the structures existing in nature have a random geometry (it is man who prefers to produce regular shapes). Non-local growth enhances the fluctuations which leads to an increase of the already present randomness (see Chapter 9). However, the degree of randomness can be rather different depending on the particular process considered. There are growing objects which have a simple overall shape (but still a complex surface, next chapter), like a water droplet in the supersaturated vapor of the atmosphere. Under somewhat different circumstances, instead of a droplet a snow crystal starts growing and the resulting structure usually has a complicated regular structure.

Deterministic aggregation models lead to clusters with perfect symmetry corresponding to the lattice on which the aggregation takes place. To obtain a noiseless structure which is grown in the presence of a diffusion field one can consider the following deterministic variation of the dielectric breakdown model. The process starts with a seed particle. The distribution of the electric potential satisfying (6.54) is calculated with the boundary condition $\phi = 0$ on the cluster and $\phi = -1$ on a circle (in $d = 2$) far from it.

From this point the two algorithms are different. In the deterministic model (Garik 1985, Family *et al* 1987) each surface site is considered for occupation simultaneously. At a given time step, all perimeter sites for which $\nabla\phi_i = \phi_i > p$ are filled, where ϕ_i denotes the potential at the ith perimeter site and $0 < p < 1$ is a fixed parameter. Next the potential distribution is calculated for the new configuration, more particles are added, and so

Figure 6.20. A Laplacian carpet generated on the square lattice using a deterministic version of the dielectric breakdown model. This configuration was obtained for $p = 0.45$.

on. Although simple, this method produces regular *Laplacian carpets* with a fractal dimension which can be tuned by changing p. Figure 6.20 shows a typical example. Note, that while in the case of dielectric breakdown model a branch advances with a probability linearly proportional to the local gradient, in the present model this dependence is a deterministic step function.

Chapter 7

GROWING SELF-AFFINE SURFACES

Many growth processes lead to space filling objects with a trivial dimension coinciding with the dimension of the space d in which the growth takes place. However, the surface of these objects may exhibit special scaling behaviour. The three basic possibilities are the following: the surface may be i) smooth, having a trivial dimension $d_s = d - 1$, ii) fractal with $D < d$ and iii) self-affine, characterized by an anisotropic scaling of the typical sizes. In this chapter cluster growth models producing the third kind of interfaces will be discussed.

Irreversible growth phenomena rarely result in smooth surfaces. Determining the fractal dimension of a self-similar interface is an important step in characterizing its properties. There are a number of known examples for such surfaces, including shore lines and the surface of silica colloid particles or materials used for catalysis. *Self-similarity*, however, implies the presence of "*overhangs*" in the surface structure: to satisfy isotropic scale invariance all possible directions should be represented equally.

During the growth of compact (non-fractal) objects the motion of the interface is directed outward, and this orientation plays a special role.

Typically, the interface can be well approximated by a *single valued* function of $d-1$ variables, e.g., one can describe the properties of the surface by examining only those points of the object which are the farthest from the centre of the structure in a given direction. The scaling properties of such surfaces (with irrelevant overhangs) are *direction dependent*: parts of various sizes can be rescaled into an object with the same overall behaviour using a rescaling factor in the direction parallel to the growth which is different from that needed to rescale the perpendicular lengths.

Mathematical examples for self-affine surfaces invariant under anisotropic rescaling of distances were discussed in Section 2.3.2. It was shown that for these objects there exists a crossover scale x_c separating two regimes. For example, the local fractal dimension of a self-affine fractal embedded in two dimensions can be observed only for length scales $x \ll x_c$, while for sizes $x \gg x_c$ the object has a global fractal dimension $d = 1$. It is important to realize that for cluster growth models $x_c = a$, where a is the lattice constant. Since a cluster of particles of size a does not have any detail on a length scale smaller than a, we conclude that *no local fractal dimension* can be associated with growing self-affine surfaces generated on a lattice.

7.1. EDEN MODEL

Perhaps the simplest cluster growth model was introduced by Eden in 1961 to simulate the growth of tumors. In addition to its biological applications, this model has relevance to many other types of stochastic growth phenomena with stable or marginally stable interfaces. When growing an Eden cluster one of the empty sites next to the aggregate (perimeter sites) is chosen randomly, and it is added to the cluster. A large cluster is obtained after having repeated this procedure many times. The particular method by which a site is selected for occupation is slightly different in the three basic variants A, B and C of the Eden model.

In the simulations of the most common version A, a single perimeter site is filled with probability $1/N_p$, where N_p is the total number of perimeter sites. Therefore, each nearest neighbour site to the cluster has the same probability to be occupied at the given time step. In version B one of the free bonds is occupied by a particle with a probability $1/N_b$, where N_b is the number of bonds on the lattice connecting a filled and an empty site. In this way a perimeter site connected to the cluster through more than 1 bond has more chance to become occupied than in version A. Finally, it is possible to define a method in which all occupied surface sites of the cluster (sites with empty nearest neighbours) have the same probability to have a new neighbour in the next step. In this version i) a surface site is chosen with a probability $1/N_s$, where N_s is the number of surface sites. Then, ii) the new particle is added to one of the adjacent empty sites picked randomly. All three versions are expected to have the same scaling properties, but the rate of approaching the asymptotic behaviour can strongly depend on the particular variation used. In general, version C exhibits *faster convergence* than A and B (Jullien and Botet 1985).

Eden growth from a single seed leads to compact, d-dimensional objects which are nearly spherical and have a non-trivial surface (Fig. 7.1).

Figure 7.1. This Eden cluster consisting of 5000 particles was grown from a single seed by occupying randomly selected perimeter sites (version A).

The properties of the surface of a large Eden cluster can be investigated by determining $N_s(r, N)$, the number of perimeter or surface sites in a layer of width dr at a distance r from the origin. For a self-similar distribution of surface sites with a fractal dimension D one expects an expression of the form (5.11) $N_s(r, N) \sim N^{\delta - 1/D} f(r/N^{1/D})$, where δ is the exponent describing the increase of the total number of surface sites $N_S \sim N^\delta$ with N. According to the simulation and theoretical results (see e.g., Leyvraz 1985), for the Eden growth the exponent δ has a trivial value

$$\delta = (d-1)/d\,. \tag{7.1}$$

In two dimensions $\delta = 1$, in complete agreement with our earlier statement about the global dimension of a self-affine surface.

Because of the self-affine nature of the surface, the above quoted expression (5.11) *does not hold* for the growth site distribution in Eden clusters. Numerical evidence shows that there is more than one relevant length scale determining the behaviour of $N_s(r, N)$ and, correspondingly, one has to use a more complex form for the scaling function $f(x)$ in (5.11) than for the self-similar case. In particular, the width $\sigma(N)$ of the distribution $N_s(r, N)$ was found to scale differently from its average radius $R(N)$ (Plischke and Rácz 1984).

There are two factors determining the asymptotic behaviour of the surface site distribution of Eden clusters grown on a lattice with a single seed particle. In addition to the self-affine geometry of the surface zone, very large clusters become distorted because of the anisotropy of the underlying lattice. In order to see the weak *anisotropy* of the asymptotic overall *shape* of Eden clusters grown on a square lattice, aggregates containing 5×10^7 particles have to be generated. For such sizes the deviation from a circular shape is about 2% (Freche *et al* 1985, Zabolitzky and Stauffer 1986).

There is a simple reason for the observed asphericity of single-seed Eden clusters. Imagine that one grows a cluster on a square lattice filling all perimeter sites simultaneously at each time step. It is easy to see that this procedure leads to a perfect diamond shape, since the distances of the layers perpendicular to the axes are larger by a factor $\sqrt{2}$ than the distances of layers perpendicular to the diagonals. Correspondingly, the "velocity" of the interface depends on the local orientation of the surface. During random Eden growth this trivial anisotropy is not manifested in a direct way because the local direction of the surface changes randomly. However, the interface always has parts with an average direction which results in differences in the growth velocity (Wolf 1987). In fact, it is possible to show analytically that Eden clusters grown on a hypercubic lattice in $d > 54$ dimensions must be anisotropic (Dhar 1985).

Instead of investigating the complex scaling of $N_s(r, N)$ it is more effective to concentrate on the properties of the *width of the surface region* $\sigma(N)$. Furthermore, as an alternative to the geometry corresponding to a single seed particle, Eden growth in a *strip geometry* can be studied to provide less biased data (Jullien and Botet 1985). This means that, for example, in two dimensions the seed is a line of L occupied sites and the growth is confined within a strip of width L using periodic boundary conditions. Simulation of the process in a $d-1$-dimensional "strip" has the advantage that the two parameters L and N controlling σ can be well separated.

Let us characterize the average height of an Eden deposit by

$$h \sim \bar{h} = \frac{1}{N_S} \sum_{i=1}^{N_S} h_i \,, \tag{7.2}$$

where $h = N/L$ is the average number of particles per column, N_S is the total number of surface sites and h_i is the distance of the ith surface site from the substrate. (7.2) expresses the fact that the vertical size grows linearly with the number of time steps (number of deposited particles), because of the compactness of the structure. In the single-seed case the number of particles N controls both the height, i.e., the mean radius $R \sim N^{1/d}$, and the "strip width" which corresponds to the circumference at the mean radius, $2\pi R \sim N^{1/d}$. Consequently, one expects that the scaling properties of the single-seed Eden clusters can be identified with those in strip geometry provided $h \sim \bar{h} \sim L \sim N^{1/d}$.

Let us define the surface width as

$$\sigma(L,h) = \left[\frac{1}{N_S}\sum_{i=1}^{N_S}(h_i - \bar{h})^2\right]^{1/2}, \tag{7.3}$$

where $\bar{h}$ is given by (7.2). One expects that in the strip geometry there are two separate scaling regimes: i) for $h \ll L$ the fluctuations in the shape of the surface grow as some power of h, while ii) for $h \gg L$ the surface becomes stationary and its width depends on L algebraically. This behaviour can be described by the following scaling form (Vicsek and Family 1985, Jullien and Botet 1985)

$$\sigma(L,h) \sim L^{\alpha} f\left(\frac{h}{L^z}\right), \tag{7.4}$$

where the exponents α and z correspond to the stationary and "dynamic" scaling of the interface width, respectively. The scaling function $f(x)$ is such that

$$f(x) \sim \begin{cases} \sim x^{\beta} & \text{for } x \ll 1 \\ \simeq \text{constant} & \text{for } x \gg 1 . \end{cases} \tag{7.5}$$

This is equivalent to

$$\sigma(L,h) \sim h^{\beta} \qquad \text{for} \qquad h/L^z \ll 1 \tag{7.6}$$

and

$$\sigma(L,h) \sim L^{\alpha} \qquad \text{for} \qquad h/L^z \gg 1. \tag{7.7}$$

Comparing (7.5) and (7.6) we find that

$$z = \alpha/\beta . \tag{7.8}$$

The scaling assumption (7.4) forms the basis of the investigations of Eden surfaces. Together with (7.7) it expresses the self-affine nature of the surface of Eden clusters, since in the $h/L^z \to \infty$ limit the size of the surface in the direction perpendicular to the substrate diverges as L^α, i.e., with a different exponent than its size L along the substrate. This is a situation analogous to the graph of the one-dimensional Brownian motion, where the distance of the particle at time t is plotted along the vertical axis versus the number of steps t (see Section 2.3.2). For such a plot the average width of the curve (which is the square root of the mean squared displacement) scales as the square root of t representing the horizontal size of the sample.

Numerical investigation of the scaling properties of $\sigma(L, h)$ is surprisingly difficult. In order to see the precise asymptotic behaviour, an extremely large number of particles is needed because in the intermediate region the dependence of the effective exponent β is unusually complex as a function of the deposition height. The main reason for the slow convergence has its origin in the so called *intrinsic width* σ_i which is independent of L. According to this picture the surface width contains two additive terms: i) width coming from the relevant, long wavelength fluctuations, and ii) a size independent width σ_i which corresponds to the local stochastic configuration of the particles in a narrow zone at the surface. This means that for $h/L^z \to \infty$ one has $\sigma_\infty^2 = L^{2\alpha} f_\infty^2 + \sigma_i^2$ (Kertész and Wolf 1988), and the second, correction to scaling term of the right-hand side strongly affects the numerical results, especially in dimensions higher than 2. Here the quadratic summation arises naturally if σ_∞ is regarded as the width of the convolution of two Gaussian distributions, one describing the long wavelength fluctuations and the other the intrinsic width.

The actual value of α and z has been investigated by a number of simulations. The situation for $d = 2$ seems to be settled: the numerical estimates are close to $\alpha = \frac{1}{2}$ and $z = \frac{3}{2}$. To observe scaling for $d = 3$ and $d = 4$ the noise-reduction method was applied which reduces the intrinsic width (Wolf and Kertész 1987a). During the application of this method to Eden growth, counters are placed on the perimeter sites and initially are set to have a value 0. Each time a perimeter site is selected for occupation, the counter is incremented by one, and the site is actually occupied only if the counter reaches a prescribed value m called noise-reduction parameter. As a result, "old" perimeter sites are occupied preferentially so that the holes, high steps and overhangs which are mainly responsible for the intrinsic width are suppressed (Fig. 7.2).

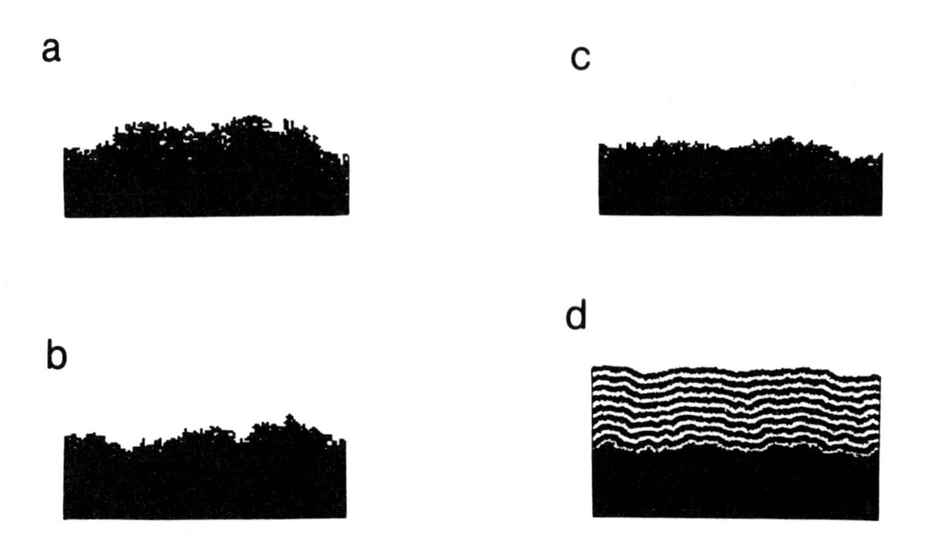

Figure 7.2. Eden deposits of 25000 particles generated on a substrate of width $L = 160$ using the noise reduction algorithm with $m = 1, 2, 4$ (a–c). Figure 7.2d shows the time evolution of a cluster for which at $N = 10000$ noise reduction with $m = 16$ was switched on (following a growth with $m = 1$) (Kertész and Wolf 1988).

For a given noise-reduction parameter the surface width is expected to scale as (Kertész and Wolf 1988)

$$\sigma^2(L, h, m) = [a(m)L^\alpha f(h/L^z m^y)]^2 + \sigma_i^2(m)\,, \tag{7.9}$$

where $a(m)$ is an m dependent constant and y is some exponent. It is clear from the above relation that the exponents α and z are not affected by m and simulations with $m > 1$ can be used to estimate their values. The results are summarized below (Wolf and Kertész 1987b)

$$\begin{array}{lll} d = 2 & \beta = 0.33 \pm 0.015 & \alpha = 0.51 \pm 0.025 \\ d = 3 & \beta = 0.22 \pm 0.02 & \alpha = 0.33 \pm 0.01 \\ d = 4 & \beta = 0.15 \pm 0.015 & \alpha = 0.24 \pm 0.02\,. \end{array} \tag{7.10}$$

The above values are in good agreement with the scaling relationship $\alpha+z = \alpha+\alpha/\beta = 2$ following from theoretical arguments (Section 7.4). Furthermore, they suggest that $\alpha = 1/d$, in contradiction with theoretical results predicting either superuniversality ($\alpha = 1/2$ for all d) or $\alpha = 0$ for $d > 2$ (see Section 7.4). The best way to visualize the results of computer simulations is to plot the scaling function $f(x)$ using $\alpha = 1/d$ and $z = 2 - \alpha$. Figure 7.3 shows a reasonable collapse of data for various values of L.

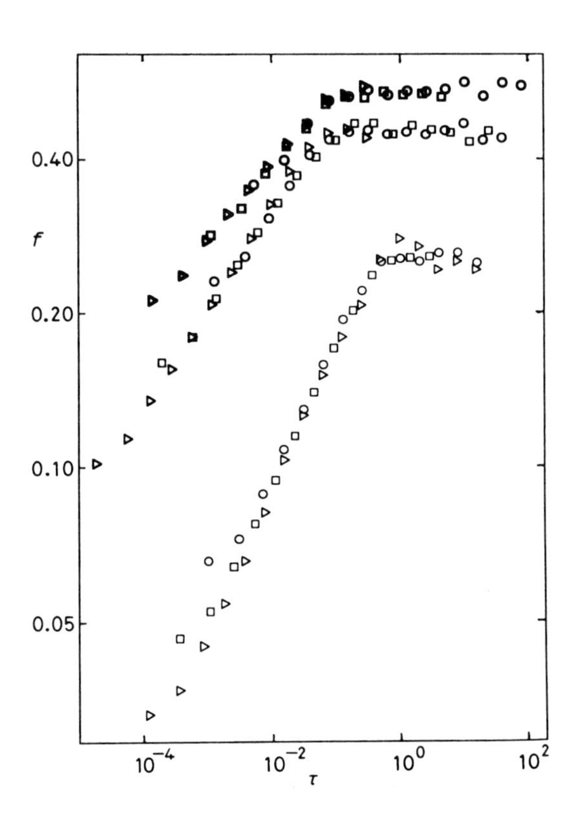

Figure 7.3. Scaling plot for $d = 2$ (lower curve, $L = 60$ (○), 120 (⊓), and 240 (▷)), for $d = 3$ (middle curve, $L_1 \times L_2 = 10 \times 10$ (○), 30×32 (⊓) and 120×128 (▷)), and for $d = 4$ (upper curve, $L_1 \times L_2 \times L_3 = 4 \times 4 \times 4$ (○), $9 \times 10 \times 10$ (⊓), and $30 \times 32 \times 32$ (▷)). The noise-reduction parameter is $m = 8$ for all data (Wolf and Kertész 1987b).

7.2. BALLISTIC AGGREGATION

In ballistic aggregation models the particles move along straight trajectories until they encounter the growing aggregate and stick to its surface irreversibly. This kind of kinetics is typical for experimental situations when molecules move in a low density vapour. Therefore, ballistic aggregation can be useful for the interpretation of technologically important processes, such as vapour deposition on a cold substrate (Vold 1963, Sutherland 1966, Leamy *et al* 1980).

Two basic versions of the model have been considered recently. In the first case the particles move along randomly oriented straight lines, while trajectories are assumed to be parallel in the second type of models. In addition, the geometry of the substrate can also affect the results and, accordingly, ballistic aggregation on both a single seed (this section) and surfaces (next section) has been investigated.

Figure 7.4a shows a ballistic aggregate grown using a two-dimensional off-lattice model with a single seed particle and randomly oriented trajectories. In spite of the simplicity of the rules the structure of large ballistic aggregates is far from trivial. There are large, elongated holes and open streaks of various sizes in the cluster, however, the structure does not appear to be self-similar. Thus ballistic aggregates are not fractals. This statement can be made more quantitative by carrying out simulations (Meakin 1985, Family and Vicsek 1985, Liang and Kadanoff 1985) and also follows from a simple theoretical argument (Ball and Witten 1984) given below.

For ballistic aggregates the considerations resulting in a causality bound for the fractal dimension of DLA clusters (see Section 6.1.1) take a considerably simpler form. In the case of straight trajectories and a fixed small density of the moving particles, the number of particles in the aggregate, $N(t)$, grows in time linearly with the cross section of the cluster

$$\frac{dN(t)}{dt} \sim R(t)^{d-1}, \tag{7.11}$$

where $R(t)$ is the radius of the aggregate at time t. On the other hand,

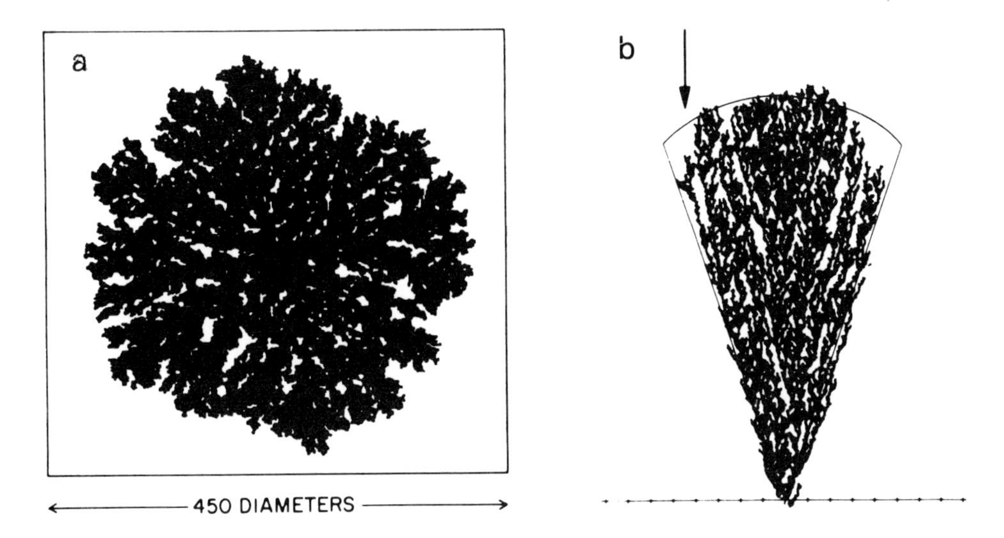

Figure 7.4. Off-lattice ballistic aggregates. (a) This cluster consisting of 180,000 particles was generated by simulating randomly oriented trajectories (Meakin 1985b). (b) Randomly positioned vertical trajectories lead to a fan-like structure when a single seed particle is used (Ramanlal and Sander 1985).

$$\frac{dN(t)}{dt} = \frac{dN(t)}{dR(t)}\frac{dR(t)}{dt} \sim R^{D-1}v\,. \tag{7.12}$$

Since v, the growth rate of the radius $R(t)$, can not increase indefinitely (in other words it must remain smaller than some v_{max}) we have

$$R^{d-D} \le v_{max}\,, \tag{7.13}$$

therefore, $D = d$ has to be satisfied to guarantee $v_{max} < \infty$ when $R(t) \to \infty$ ($D > d$ is excluded for obvious reasons).

Although the dimension of ballistic aggregates turns out to be the same as the Euclidean dimension of the space they are grown in, they exhibit non-trivial scaling behaviour. Consider the following version of the model. A seed particle is put into the origin of a square lattice and the rest of the particles move parallel to the y axis with random x coordinates till they either disappear or stick to the growing cluster (the latter takes place if a particle arrives at a site adjacent to the aggregate). The resulting cluster looks like a *fan* with higher density in the central region and long holes near its edge. This process can be formulated easily for the off-lattice case as well, leading to a similar structure (Fig. 7.4b).

The scaling behaviour of the *density distribution* $\rho(r, \theta)$ within the fan can be described as a function of the distance from the seed, r, and the angle θ measured from the y direction. Here the density is defined as the number of particles within the area $r\Delta r\Delta\theta$. According to the simulations $\rho(r, \theta)$ can be well represented by the scaling form (Liang and Kadanoff 1985)

$$\rho(r, \theta) \sim r^{-\mu} f\left(r^{\vartheta}(\theta - \theta_c)\right), \tag{7.14}$$

where the scaling function $f(x) \simeq$ constant for $x \ll 1$ and $f(x) \ll 1$ for $x \gg 1$, θ_c is a critical angle corresponding to the region close to the edges of the fan, $\theta > \theta_c$, and μ and ϑ are scaling indices. This assumption is supported by the fact that the data points for ρr^{μ} as a function of the renormalized angle $r^{\vartheta}(\theta - \theta_c)$ fall onto the same universal curve for various r and θ. In the case of off-lattice aggregates the best collapse is obtained using $\theta_c = (15.5 \pm 0.7)^{\circ}$, $\mu = 0.13 \pm 0.05$ and $\vartheta = 0.39 \pm 0.05$ (Joag *et al* 1987). These values for the exponents agree within the error limits with those obtained for the on-lattice case. The critical angle, however, is considerably less than $\theta_c \simeq 32^{\circ}$ found in the simulations carried out on the square lattice.

Let us now investigate the average density in the region $|\theta| < \theta_c$ as a function of r. For such a fixed angle the density within a large aggregate approaches a constant as $r \to \infty$. This means that the fan has a trivial dimension $D = d$ as was mentioned before. On the other hand, the approach to the constant density is very slow, and follows a power law of the form (Liang and Kadanoff 1985)

$$\rho(r) = \rho_{\infty} + A r^{-\beta}, \tag{7.15}$$

where ρ_∞ and A are constants and β is a non-trivial correction to the dimensionality. The above relation is analogous to the expression (2.37), assuming that $l \sim 1/r$. Therefore, (7.15) can be regarded as an indication of the *fat fractal* character of ballistic aggregates. The exponent β was determined from the slopes of the plots $\ln(\rho - \rho_\infty)$ versus $\ln r$ for both the off-lattice and the square lattice cases giving respectively $\beta \simeq 0.55$ and $\beta \simeq 0.66$. Thus the metadimension β appears to be non-universal.

7.3. BALLISTIC DEPOSITION

In this section, ballistic aggregation of particles onto surfaces will be discussed. We assume that the particles are released from randomly chosen launching points and move along parallel straight lines till they are attached to the aggregate. In general, the *angle of incidence* ϑ (the angle between the trajectories and the normal vector to the surface) can be varied from 0 (vertical trajectories) to $\pi/2$. This situation is common in a number of technologies used to control electrical, optical and other physical properties of surfaces by means of vapour deposition.

In an actual simulation of ballistic deposition the *strip geometry* with periodic boundary conditions is used. The linear size of the substrate is denoted by L. The particles are launched at randomly selected positions at a height of $h_{max} + 1$ which is the maximum height of any particle in the deposit. Then the particles follow a straight trajectory with a prescribed angle of incidence until they contact either a particle in the deposit or reach the original surface. At the point of the first contact they are stopped and become part of the growing deposit. In the lattice version of the model with $\vartheta = 0$ it is easy to see that there is only one active perimeter site (a site which has the possibility to be filled in the next step) in each column. Since one needs to record only the height of these L sites such an algorithm is fast and does not require much computer time. The off-lattice case with $\vartheta \neq 0$ can be simulated effectively by choosing a suitable underlying lattice to find particles near the trajectory of a moving particle.

Both the experiments and computer simulations showed that in the case of non-zero incidence angle, *columnar structures* grow on the surface. This columnar morphology is most distinctive for large ϑ, and the angle ζ between the growth direction of the columns and the normal to the surface is less than the angle of incidence. Investigations of vapour and sputter-deposited aluminium and rare-earth-metal thin films suggested the *empirical relationship* (Nieuwenheuzen and Haanstra 1966, Leamy *et al* 1980)

$$\tan \zeta \simeq \frac{1}{2} \tan \vartheta \tag{7.16}$$

known as the "tangent rule". It is easy to see why ζ should be less than ϑ: Particles passing the "high" side of an existing column can be caught and cause the column to tilt towards the normal. However, (7.16) has not been shown to hold by any rigorous theoretical argument.

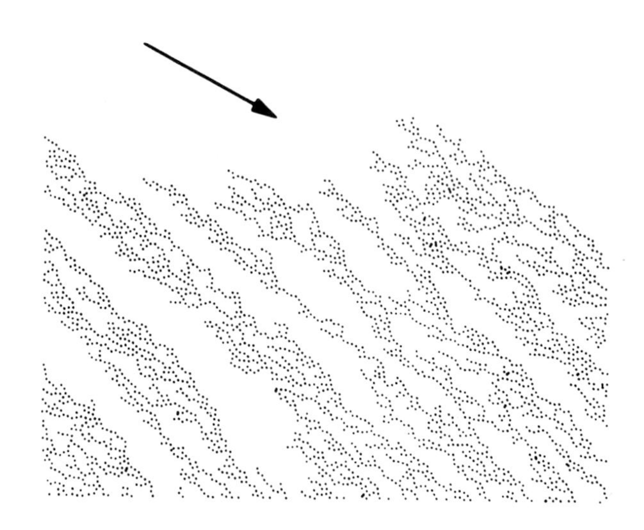

Figure 7.5. Off-lattice ballistic deposit obtained for a fixed angle of incidence $\vartheta = 60°$ (Ramanlal and Sander 1985).

Figure 7.5 shows a typical configuration obtained in a relatively small scale simulation of two-dimensional off-lattice ballistic deposition with a fixed angle of incidence $\vartheta = 60°$. The deposit has a columnar structure and the deviation of the direction of growth from the direction of the incident beam is qualitatively consistent with the tangent rule. To check whether the tangent rule is qualitatively valid for ballistic deposition one needs a quantitative determination of the growth direction ζ. This can be achieved by noting that for off-lattice aggregates the incoming particles make contact with only a single particle belonging to the deposit. Thus the deposit can be considered as a set of trees which consist of particles connected to the same root particle at the surface. For large ϑ these trees can be identified with the columns. The growth direction of a tree is given by the expression (Meakin et al 1986b)

$$\tan\zeta(j) \simeq \frac{\sum_{i=n_j}^{n_j+\Delta n_j} x_i - x_{j0}}{\sum_{i=n_j}^{n_j+\Delta n_j} y_j}, \tag{7.17}$$

where x_i and y_i are respectively the horizontal and vertical coordinates of the ith particle in the jth tree, x_{j0} and $y_{j0} = 0$ are the positions of the root particles and the summation is taken over Δn_j particles added to the tree already consisting of $n_j \gg 1$ particles. The average growth direction can be obtained by averaging over the angles corresponding to the individual trees.

The results of simulations for the ϑ dependence of ζ, and $\zeta' = \tan^{-1}(\frac{1}{2}\tan\vartheta)$ (which is the prediction of the tangent rule) are displayed in Table 7.1. This table demonstrates that the deviations from the tangent rule are quite strong. It is evident from this table that the simplest two-dimensional ballistic aggregation models can not be used to explain the tangent rule on a quantitative basis.

Let us now consider the structure of the surface of deposits generated on a square lattice with $\vartheta = 0$. Figure 7.6 shows part of the surface of a large deposit. The apparent similarity of this plot to Fig. 2.13 exhibiting the graph of the one-dimensional Brownian motion indicates the self-affine nature of the surface. This analogy can be made more quantitative by studying the variance of the heights of the active perimeter sites. According to the simulations to be discussed later, in analogy with the Eden model in a strip

Table 7.1. Dependence of some characteristic quantities on the angle of incidence ϑ obtained from simulations of two-dimensional off-lattice deposition. ζ – mean angle of tree growth, ζ' – prediction of the tangent rule, and $\vartheta - \zeta$ – a quantity which appears to saturate close to 16° for large angles of incidence. All angles are shown in degrees (Meakin *et al* 1986b).

ϑ	ζ	ζ'	$\vartheta - \zeta$
10	11.55	5.04	-1.55
20	16.17	10.31	3.83
30	23.94	16.10	6.06
40	31.02	22.76	8.98
50	39.46	30.79	10.54
60	47.13	40.89	12.87
70	55.46	53.95	14.54
80	63.93	70.57	16.07

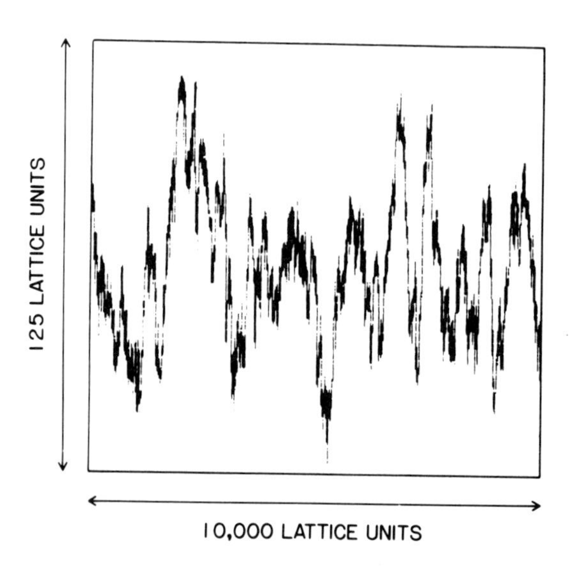

Figure 7.6. Section from the surface of a deposit grown on the square lattice to an average height of 5000 lattice units along a base line of length 2^{18} units (Meakin *et al* 1986b).

geometry, the width of the surface defined by (7.3) scales as $\sigma(L,h) \sim h^\beta$ for $h \ll L$ (7.6) and $\sigma(L,h) \sim L^\alpha$ if $h \gg L$ (7.7) (Family and Vicsek 1985).

The scaling of σ as a function of L for $h \gg L$ indicates the *self-affine* nature of the surface of ballistic deposits. In Section 2.3.2 it was discussed that if the root mean square distance of a one-dimensional random motion scales with the number of steps t as $\langle X^2(t)\rangle^{1/2} \sim t^H$, then the graph of the actual distance of the walker from the origin as a function of t is a self-affine function with a local fractal dimension $D = 2 - H$. Furthermore, $X(t)$ has the self-affine property $X(t) \sim b^{-H}X(bt)$, where b is a rescaling factor and H is a characteristic exponent. In the present case, for $d = 2$ the role of t is played by the actual distance x along the substrate, and $X(t)$ corresponds to $\tilde{h}(x) = h(x) - \bar{h}$, where $h(x)$ is the height of the active perimeter site in column x (there is only one such site in each column). Therefore, in the $h \to \infty$ limit the surface of the deposits satisfies

$$\tilde{h}(x, \bar{h} \to \infty) \sim L^{-\alpha}\tilde{h}(Lx, \bar{h} \to \infty)\,. \tag{7.18}$$

To incorporate the $\sigma \sim h^\beta$ behaviour for $h \ll L$ one assumes that (Meakin *et al* 1986b)

$$\tilde{h}(x, \bar{h}) \sim \bar{h}^\beta f(x/\bar{h}^{1/z})\,, \tag{7.19}$$

where the scaling function $f(y)$ is assumed to be a randomly changing function with $|f(y)| < Const$ for $y \gg 1$, and

$$f(y) \sim L^{-\alpha}f(Ly) \qquad \text{for} \qquad y \ll 1\,. \tag{7.20}$$

The latter condition is needed to satisfy (7.18). Since the width of a self-affine function defined by (7.19) scales as y^α, we see that $\sigma \sim L^\alpha$ for $h \ll L$ is also satisfied if $\alpha/z = \beta$. Expressions analogous to (7.18) and (7.19) are expected to be valid in dimensions higher than 2 as well both for ballistic deposits and Eden clusters generated in the strip geometry.

One consequence of (7.18) is that the intersection of the surface and a plane placed at a distance $\bar{h}$ from the substrate has the fractal dimension

$$D = d - 1 - \alpha \, . \tag{7.21}$$

To see this we recall rule f) of Section 2.2 relating the fractal dimension of the intersection of two fractals of dimensions D_A and D_B through the expression $D_\cap = D_A + D_B - d$. For a self-affine deposit the local fractal dimension is $D_A = d - \alpha$ (see Section 2.3.2, where D_{local} was shown to be equal to $2 - \alpha$ for $d = 2$) and the dimension of the plane parallel to the substrate is $D_B = d - 1$. Although the deposits do not have a well defined local fractal dimension because of the coincidence of the crossover scale and the lattice spacing (this question was discussed at the beginning of this chapter), the global behaviour of the cross section is not affected by the crossover scale and (7.21) is expected to hold.

The *scaling of the surface width* has been extensively studied for a number of deposition models. In the case of the square lattice large scale simulations (Meakin *et al* 1986b) led to numerical values for the exponents α, β and z close to the theoretical prediction (Kardar *et al* 1986)

$$\alpha = 1/2, \qquad \beta = 1/3 \quad \text{and} \quad z = 3/2 \, . \tag{7.22}$$

The surface width can also be determined for off-lattice ballistic deposits as a function of the angle of inclination ϑ. For off-lattice aggregates the surface sites are not well defined and the surface thickness should be calculated from a modified expression (Meakin and Jullien 1987)

$$\sigma(L, h) = \frac{1}{N} \sum_{i=M}^{M+N} |h_i - h_{i+1}| \, , \tag{7.23}$$

where h_i is the minimum distance from the substrate of the point at which the ith particle is deposited and $N - M$ is the increment in the deposit mass over which the characteristic quantities are determined. According to the $2d$ simulations the L dependence of $\sigma(L, h)$ is described by the exponent

$\alpha = 1/2$, for all ϑ. The exponent β, however, was found to depend on the angle of inclination. This is demonstrated by the following few selected values: $\beta \simeq 0.343$ ($\vartheta = 0$), $\beta \simeq 0.281$ ($\vartheta = 45°$) and $\beta \simeq 0.402$ ($\vartheta = 80°$).

In addition to the ϑ dependence of β, its value is changed when the deposition process is modified to take into account *surface restructuring.* Consider a lattice model with angle of inclination equal to zero, in which a particle after having contacted the deposit is moved to a new surface site with the smallest height within a given region surrounding the point of first contact. Such and analogous relaxation rules tend to smooth out irregularities and result in estimates of β close to 1/4 (Family 1986, Meakin and Jullien 1987).

Large scale simulations of ballistic deposition in three dimensions seem to support the scaling assumption (7.4) for the width of the surface. The numerical results $\alpha \simeq 0.33$ and $\beta \simeq 0.24$ obtained for the corresponding exponents (Meakin *et al* 1986b) suggest that the exact value of α in $d = 3$ might be equal to $1/3 = 1/d$. In addition, $\beta = 1/5$ would be consistent with the scaling relation $\alpha + \alpha/\beta = 2$. The situation seems to be similar to that of Eden growth: there is only one independent exponent describing the scaling of the surface width and it is found to be close to $\alpha = 1/d$ (Wolf and Kertész 1988). While the situation is settled for the two-dimensional case, the theoretical predictions for higher dimensions are not consistent with the above numerical results. It is not clear whether there exists an upper critical dimension for the problem of growing self-affine surfaces above which $\alpha = 0$.

7.4. THEORETICAL RESULTS

The theoretical treatment of growing self-affine interfaces is based on constructing a continuum differential equation for describing the motion of the interface. We shall consider first the general case of a random interface evolving in the strip geometry with a $d-1$ dimensional substrate of linear size L. Since the exponents α and β characterizing the width of the interface σ were found to be less than unity, both σ/L and σ/h go to zero as $\sigma \to \infty$. It is convenient to ignore overhangs so that h can be considered as a single valued function of $\vec{x}$. Therefore, one can assume that local coarse scale derivatives dh/dx exist. Let us now express the velocity of the interface $h(\vec{x},t)$ as a function of its local gradient. To take into account the stochastic nature of the growth one writes down the simplest non-linear Langevin equation for $\tilde{h} = h - vt$ (v is the velocity normal to the surface) in the form (Kardar *et al* 1986)

$$\partial\tilde{h}(\vec{x},t)/\partial t = \gamma\nabla^2\tilde{h}(\vec{x},t) + \lambda[\nabla\tilde{h}(\vec{x},t)]^2 + \eta(\vec{x},t)\,, \tag{7.24}$$

where the time variable t is associated with the average deposition height $\bar{h}$. In the above equation the first term describes the relaxation of the interface due to the surface tension γ. Its meaning is quite obvious; protrusions (places with local curvature $\nabla^2\tilde{h}$ less than zero) tend to disappear under the influence of the smoothing effect of surface tension. Such effects are expected to be particularly important if *surface restructuring* is allowed.

The second term in (7.24) is the lowest-order non-vanishing term in a gradient expansion. Its inclusion can be justified by the following argument. Consider, for example, the growth of an Eden cluster in two dimensions. In general, the growth takes place in a direction locally normal to the interface. When a particle is added, the increment projected onto the h axis is $\delta h = [(v\delta t)^2 + (v\delta t\nabla h)^2]^{1/2}$ (Fig. 7.7) which leads in the weak gradient limit to

$$\partial h/\partial t = v[1 + (\nabla h)^2]^{1/2} \simeq v + (v/2)(\nabla h)^2 + \ldots\,, \tag{7.25}$$

where v is the velocity normal to the interface. The above expression reduces to the second term in (7.24) after transformation to the comoving frame and is supposed to play a relevant role in situations where *lateral growth* is allowed.

The third term in expression (7.24) is included to take into account the fluctuations. The noise denoted by $\eta(\vec{x},t)$ is assumed to have a Gaussian distribution, so that $\langle\eta(\vec{x},t)\rangle = 0$, and

$$\langle\eta(\vec{x},t)\eta(\vec{x}',t')\rangle = 2C\delta^d(\vec{x}-\vec{x}')\delta(t-t')\,. \tag{7.26}$$

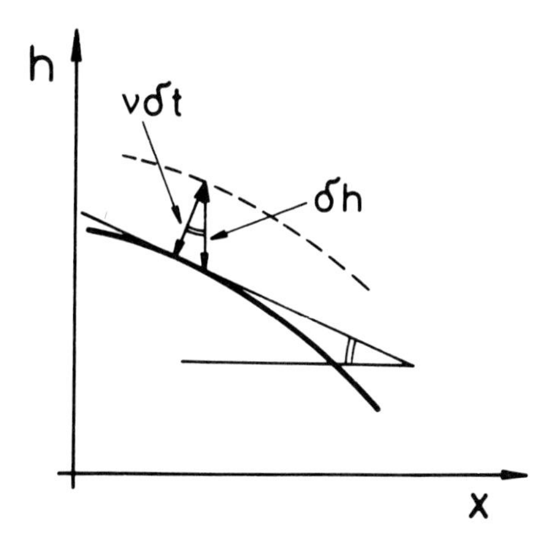

Figure 7.7. Schematic picture showing the increment of h as the growth locally occurs along the normal to the interface.

Let us use Eq. (7.24) to obtain a scaling relation among the exponents α, β and z in addition to (7.8). For this purpose we assume that for large scale solutions one can neglect the noise in (7.24). Furthermore, if the surface tension effects are small, we can omit the first term in (7.24) as well. Then we have the following simplified equation

$$\partial \tilde{h}/\partial t \simeq (\nabla \tilde{h})^2 \tag{7.27}$$

which has a solution of the form (7.19) with $\bar{h} \sim t$. This can be checked by inserting (7.19) into (7.27) (Meakin *et al* 1986b). When carrying out this substitution it is easy to see by comparing the leading terms in t that the two sides of Eq. (7.27) are equal if

$$z + \beta z = z + \alpha = 2 \tag{7.28}$$

(where (7.8) $z = \alpha/\beta$ was used), and $f(\vec{y})$ satisfies

$$\beta f(\vec{y}) - \frac{1}{z}\vec{y}\frac{\partial}{\partial y} f(\vec{y}) = \left[\frac{\partial}{\partial y} f(\vec{y})\right]^2 . \tag{7.29}$$

It can be shown by similar arguments that the scaling relation (7.28) is not affected if the surface tension term is also included into the calculations. In addition to the above heuristic arguments, more complete proofs of (7.28) can be given by application of mode coupling techniques to the full Eq. (7.24) (Krug 1987) or by mapping it to the directed polymer problem in a d-dimensional space (Kardar and Zhang 1987).

In order to determine the actual values of the exponents in (7.28) the formalism of the dynamic renormalization group can be applied to the full stochastic problem defined by Eq. (7.24) (Kardar *et al* 1986). The corresponding method was elaborated for the Burgers's equation which can be obtained from (7.24) using the transformation $\vec{v} = -\nabla\tilde{h}$. Instead of reproducing the calculations, here we discuss the results which can be summarized as follows. Equation (7.24) embodies three different universality classes, depending on the values of the parameters γ and λ.
i) The $\gamma = \lambda = 0$ case corresponds to random deposition of particles with no surface restructuring or sticking of the particles to each other. Then for $L \gg 1$ the columns grow according to the Poisson statistics describing the probability that the number of particles in a given column is equal to h if $\bar{h}$ particles per column have been deposited. Thus $\sigma \sim \bar{h}^{1/2}$ is given by the central limit theorem, and

$$\beta = 1/2\,. \tag{7.30}$$

Since the size of the substrate does not have an effect on σ, the other two exponents can be regarded to be equal to zero.
ii) If $\lambda = 0$, the evolution of the interface is dominated by surface restructuring. For this case

$$\alpha = (3-d)/2, \qquad \beta = (3-d)/4 \qquad \text{and} \qquad z = 2 \tag{7.31}$$

can be obtained by Fourier transforming (7.24) (Edwards and Wilkinson 1982).
iii) The third universality class corresponds to the general case when neither γ nor λ is equal to zero. From the dynamical renormalization group approach

the following exponents were found

$$\alpha = (3-d)/2, \qquad \beta = (3-d)/3 \qquad \text{and} \quad z = 3/2\,. \tag{7.32}$$

It is clear from (7.31) and (7.32) that the critical dimension appearing in these theories is $d = 3$, where logarithmic corrections are expected to complicate the situation.

A comparison of the above results with those obtained in the simulations leads to satisfactory agreement between the theory and numerical experiments in two dimensions. Indeed, for $d = 2$ both the Eden growth and ballistic deposition results suggest $\alpha = 1/2$ and $\beta = 1/3$ in accord with (7.32). Simulations of ballistic aggregation with surface restructuring resulted in estimates for β close to $1/4$ which is predicted by (7.31).

The situation is more controversial in higher dimensions. The simulations with no restructuring provide numerical evidence for the conjecture that $\alpha = 1/d$ and the validity of the scaling relation $\alpha + z = 2$ which allows one to obtain the rest of the exponents once α is known. These results are in disagreement with the theoretical predictions. It is apparent from (7.32) that the perturbative dynamic renormalization method suggests an upper critical dimension equal to $d = 3$. According to very recent theoretical arguments $d_c = 3$ corresponds to the weak coupling regime (i.e., to small noise), and for large $\langle \eta(\vec{x}, t)^2 \rangle$ another regime exists with a dimension independent (or superuniversal) exponent $z = 1.5$ (Kardar and Zhang 1987, McKane and Moore 1988).

As concerning ballistic deposition with surface restructuring three-dimensional simulations of the off-lattice process (Jullien and Meakin 1987) indicate that the surface width diverges logarithmically with $\bar{h}$ and L in a better agreement with the theoretical result $d_c = 3$ (7.31) obtained for this case.

Let us now concentrate on the development of a *mean-field approach* to the description of the *columnar geometry* of ballistic aggregates (Ramanlal and Sander 1985). This continuum theory which is to treat aggregation with uniaxial trajectories is based on the tangent rule (7.16) relating the angle of incidence ϑ to the angle of holes or columns ζ, where both angles are measured with respect to the normal to the envelope of the surface. It is easy to understand why $\vartheta > \zeta$: the particles arrive at the surface of the columns close to their most advanced parts and this results in a growth closer to the normal. Although (7.16) does not hold precisely, this fact does not change the general features of the theory.

The tangent rule can be used as a local prescription for the motion of the interface. According to this assumption each element of the surface moves in the direction which is determined by (7.16). For the envelope of the fan structure of Fig. 7.4b the interface in the long time limit satisfies (Ramanlal and Sander 1985)

$$\begin{aligned} \frac{\partial y}{\partial x} &= -\tan\vartheta, \\ \frac{y}{x} &= \tan(\theta)\,, \end{aligned} \tag{7.33}$$

where θ is measured from the y axis (Fig. 7.8). On the other hand, we must have $\theta = \vartheta - \zeta$, since the surface moves along a straight line originating at the seed (because of the tangent rule). The solution of the set of equations (7.16) and (7.33) can be obtained by converting to polar coordinates. Then $\partial r/\partial\theta = r\tan\zeta$ with the polar angle

$$\theta = \tan^{-1}[\tan\vartheta/(2+\tan^2\vartheta)] \tag{7.34}$$

follows from the tangent rule. The solution is

$$\begin{aligned} r &= r_0 f(\theta), \\ f(\theta) &= \frac{1}{\sqrt{2}|\sin\theta|}\,\frac{(n+1)^{1/4}(n-1)^{1/2}}{(3n-1)^{3/4}}, \\ n &= (1-8\tan^2\theta)^{-1/2}\,, \end{aligned} \tag{7.35}$$

where r_0 is a time dependent constant. $f(\theta)$ can also be written in a more compact form

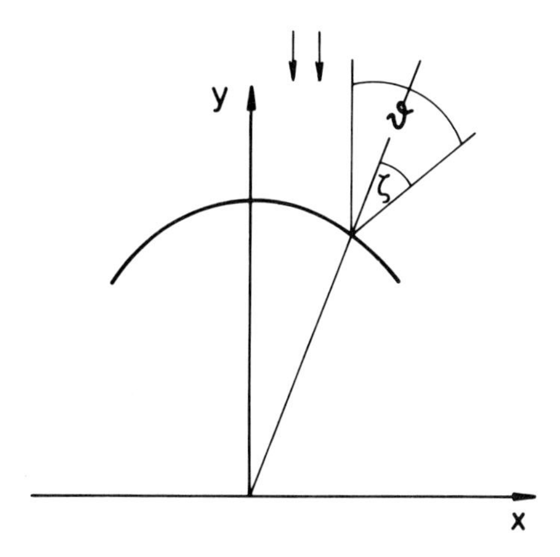

Figure 7.8. Schematic picture of the envelope of the interface (used to construct Eqs. 7.33).

$$f(\theta) = (\cos\vartheta)^{1/2}/\cos\zeta\,. \tag{7.36}$$

The most important feature of the solution appears quite naturally in (7.35). When $\tan\theta = 1/\sqrt{8}$, n diverges. This condition corresponds to an angle $\theta \simeq 19.5°$ which is in good agreement with the simulation results. One way to understand this result is to note that (7.34) has a maximum at 19.5°, thus one never expects to see a fan with an opening angle away from the incident direction larger than this value. However, the above angle is not universal since in the case of the square lattice, (discussed in Section 7.2), the opening angle was found to be about 32°.

Equation (7.36) describes the static behaviour of the surface. In order to determine the actual motion of the interface one writes for the velocity of a point of the interface whose normal is $\hat{\mathbf{n}}$

$$\vec{v} = v\hat{\mathbf{m}} = -(\hat{\mathbf{n}}\cdot\vec{J}/\rho)\hat{\mathbf{m}} = v_0 f\,, \tag{7.37}$$

where $\hat{\mathbf{m}}$ is a unit vector in a direction between the incident beam and the normal, as given by the tangent rule, $\vec{J}$ is the constant flux of incoming particles per unit area per unit time, and ρ is the local density. In the above expression f is given by (7.36) in order to be consistent with the solution obtained before for a fixed r_0. Equation (7.37) has to be completed by the condition that there is no growth if $\cos\vartheta < 0$ (the surface is in a geometrical shadow). Finally, the velocity of the interface in a given direction $\hat{\mathbf{k}}$ can be expressed as

$$\frac{v_k = \hat{\mathbf{n}} \cdot \vec{v}}{\hat{\mathbf{n}} \cdot \hat{\mathbf{k}}\,.} \tag{7.38}$$

Returning to polar coordinates with $\hat{\mathbf{k}} = \hat{\mathbf{r}}$, Eqs. (7.37) and (7.38) can be written in the form

$$\begin{aligned} \partial r(\theta, t)/\partial t &= v_0[(1 + Q^2)\cos\vartheta]^{1/2}, \\ Q &= \partial r/\partial\phi, \\ \cos\vartheta &= (\cos\theta - Q\sin\theta)/(1 + Q^2)^{1/2}\,. \end{aligned} \tag{7.39}$$

It can be checked by direct insertion of (7.35) into (7.39) that the former is an exact solution to the partial differential equation (7.39), with the initial condition of growth from a point. For other initial conditions (7.39) can be treated by either stability analysis or numerical integration.

The actual solution depends on the roughness of the initial conditions. It can be shown that the asymptotic solution (7.39) is *marginally stable*, therefore, a smooth enough initial interface does not lead to instabilities. Let us assume that initially the surface looks like (Ramanlal and Sander 1985)

$$\delta(\theta, 0) = \delta_0 \cos(m\theta) \tag{7.40}$$

representing the fluctuations in the profile due to the random distribution of incoming particles. If $\delta_0 \ll 1$, the surface remains smooth because of the marginal stability of the solutions of (7.39). However, if (for example) $\delta_0/\bar{r}(0) = 0.05$ and m=44, the long empty strips characteristic for the fan

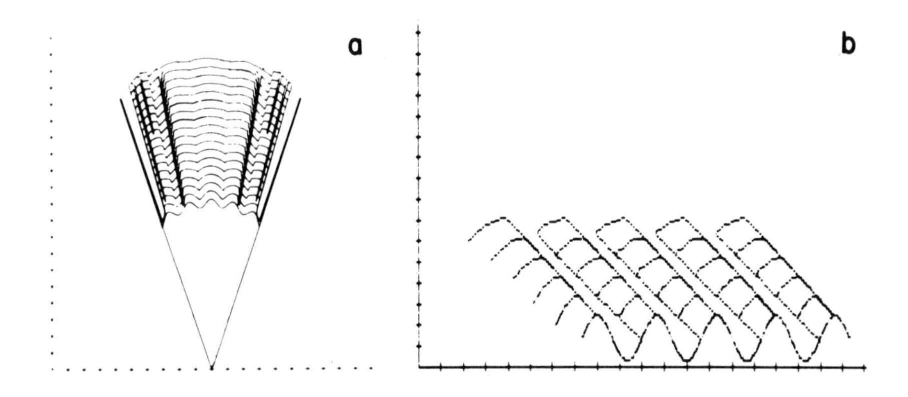

Figure 7.9. Numerical solutions of Eq. (7.39) for perturbed initial conditions. (a) Growth on a seed particle, and (b) on a line (Ramanlal and Sander 1985).

structure appear, as is demonstrated in Fig. 7.9a. The streaks and ragged edges of Fig. 7.4b probably arise in this way. This suggests that the specific shape of ballistic aggregates emerges from the interplay of fluctuations (rough initial condition) and the consequent geometrical shadowing. The same line of reasoning and a periodic initial condition give rise to the growth of columns shown in Fig. 7.9b.

The main conclusion of the above mean-field continuum theory for ballistic deposition is the following. The columns are nucleated due to the random nature of the deposition. Whenever the height per width of a cluster is large enough to produce geometrical shadow, this non-linear effect gives rise to the growth of a distinct column growing out from the given seed cluster.

Chapter 8

CLUSTER-CLUSTER AGGREGATION

Aggregation of microscopic particles diffusing in a fluid medium represents a common process leading to fractal structures. If the density of the initially randomly distributed particles is larger than zero, the probability for two "sticky" particles to collide and stick together is finite. It is typical for such systems that the resulting two-particle aggregate can diffuse further and may form larger fractal clusters by joining other aggregates (Friedlander 1977). As a result the mean cluster size increases in time and, in principle, after a sufficiently long period all of the particles in the finite system become part of a single cluster. In many cases the force between two particles is of short range and it is strong enough to bind the particles irreversibly when they contact each other. For example, such behaviour can be observed for iron smoke aggregates formed in air (Forrest and Witten 1979) or in aqueous gold colloids (Weitz and Olivera 1984).

In the above process each cluster is equivalent with regard to the conditions for their motion, i.e., there is no seed particle as in the case of DLA. Consequently, this process is called cluster-cluster aggregation (CCA) to distinguish it from particle-cluster aggregation phenomena discussed in the previous sections of Part II. CCA directly corresponds to the physical situation taking place in a system of aggregating particles, in contrast to DLA which in general should be regarded as a computer model for phenomena not necessarily involving attachment of particles.

The possibility of simulating colloidal aggregation in a computer has been recognized for a few decades (Sutherland 1967). However, large scale numerical investigation of cluster-cluster aggregation has become feasible only in recent years. Simple *computer models* for CCA (e.g., Meakin 1983c, Kolb *et al* 1983) can be successfully used to study the structure of aggregates and the dynamics of their formation. A typical two-dimensional simulation is started by randomly occupying a small fraction of the sites on a square lattice to represent particles. At each time step a particle or a cluster is selected randomly and is moved by one lattice unit in a randomly chosen direction. Two clusters stick when they touch each other. Figure 8.1 shows four stages of such a process. This figure demonstrates the most important properties of cluster-cluster aggregation. With increasing time the number of clusters decreases, and large, randomly branching aggregates appear in the system. The computer generated clusters and the real aggregates observed in many recent experiments were found to have very similar fractal scaling.

Because of the simultaneous diffusional motion of aggregates, the time is a well defined quantity in CCA (including simulations). Accordingly, the related numerical and experimental investigations have concentrated on both the *geometrical and dynamical* aspects of the aggregation process. The results suggest that in analogy with equilibrium phase transitions, non-trivial scaling can be found in both approaches. Therefore, in addition to the fractal structure of aggregates, in this Chapter we shall discuss the dynamic scaling for the cluster size distribution as well (Vicsek and Family 1984, Kolb 1984).

Most of the real cluster-cluster aggregation processes are more complex than the simple simulation described above. It is mainly the shape of the short-range interaction potential between two particles which determines the nature of the statics and dynamics of CCA. A deep minimum in the potential and a negligible repulsion part results in the so-called i) *diffusion-limited* regime, when two clusters stick rigidly together as soon as they contact. The relevant time scale in this process is the typical time needed for two diffusing clusters to approach each other. During ii) *reaction-limited* (or chemically-

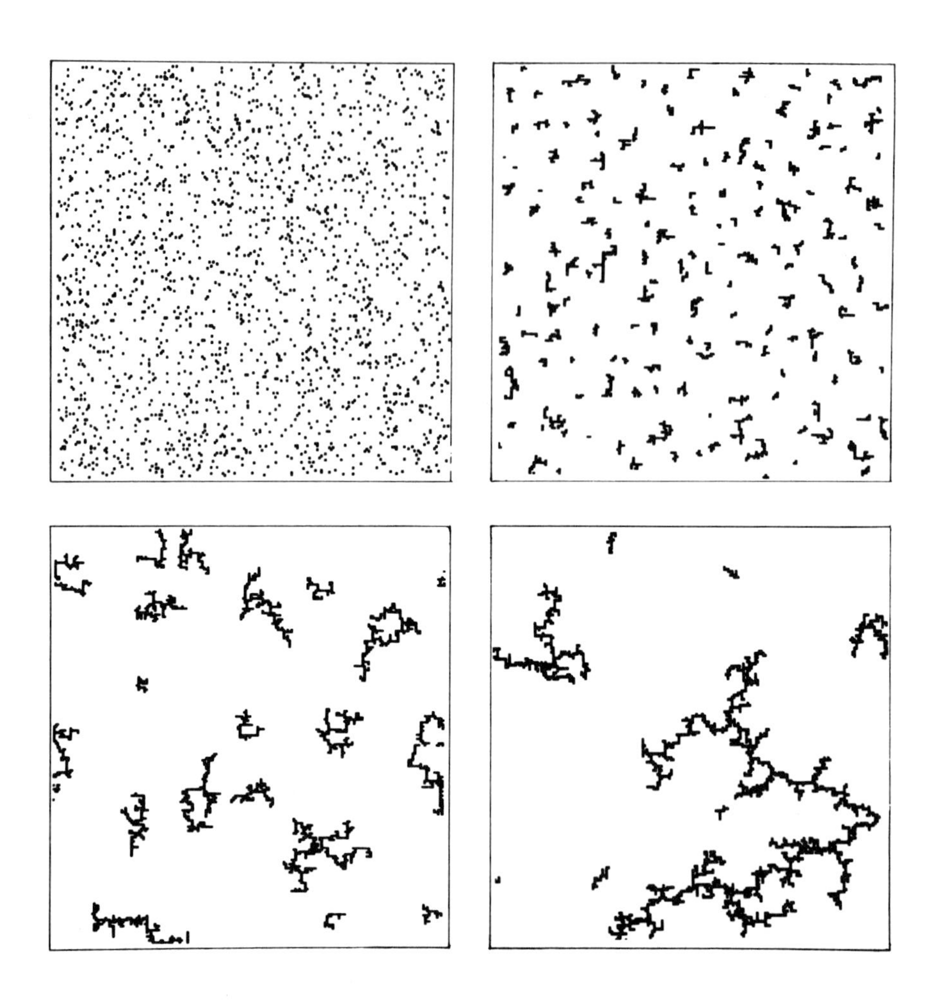

Figure 8.1. Snapshots of configurations taken at various "times" during the computer simulation of diffusion-limited cluster-cluster aggregation in two dimensions.

limited) CCA a small, but relevant repulsive potential barrier can prevent the clusters from joining each other even if they are close. However, after a number of contacts they may become joined irreversibly. In this case it is the time needed for the formation of a bond between adjacent clusters which determines the characteristic time.

If the attractive part is not deep enough, one expects that the event of aggregation of two clusters can be followed by reorganization (restructuring) or dissociation of the aggregates. In the latter case the irreversible character of the process is lost and one is led to deal with iii) *reversible* CCA. The properties of cluster-cluster aggregates are also affected by the kind of motion they undergo. The trajectory of a cluster can be Brownian or ballistic. In addition, the clusters may rotate. Many of these processes have been studied by the three main approaches (simulations, theory and experiments) to be discussed in this section.

8.1. STRUCTURE OF CLUSTER-CLUSTER AGGREGATES

Both the related experiments and simulations indicate that cluster-cluster aggregates are typically highly ramified, almost loopless structures exhibiting fractal properties. In contrast to off-lattice DLA clusters, the overall shape of cluster-cluster aggregates is not spherical. Instead, these aggregates can be characterized by a well defined *asphericity* which becomes more pronounced as the clusters become larger (Medalia 1967). This fact is a trivial consequence of the growth mechanism: the overall shape can not be spherical since joining two spherical clusters would immediately destroy the symmetry. In contract to their overall shape, however, the density correlations within cluster-cluster aggregates are isotropic (Kolb 1985).

The fractal dimension of CCA clusters can be determined by the application of methods discussed in Chapter 4. For aggregates generated by Monte Carlo simulations, the power law decay of the density-density correlation function (Eq. (2.14)), the dependence of the radius of gyration on the mass of the aggregates (4.12) or the scaling assumption (4.14) can be used to evaluate D.

8.1.1. Fractal dimension from simulations

The actual realization of a cluster-cluster aggregation model in the computer depends on the particular process to be simulated. However, the most widely used simulations are based on the following assumptions. The particles are represented by occupied sites of a d-dimensional hypercubic cell of linear size L. To make the finite-size effects smaller, periodic boundary conditions are used. Initially, $N_0 = \rho L^d$ sites are randomly filled, where $\rho \ll 1$ is the density of the particles in the system. Then the clusters are allowed to move following Brownian or ballistic trajectories. If during their motion two or more particles belonging to different clusters accidentally occupy adjacent (nearest neighbour) sites, the clusters combine to form a single new aggregate with a probability $0 < p_s \leq 1$.

In *diffusion-limited cluster-cluster aggregation* (Meakin 1983c, Kolb *et al* 1983) the clusters are assumed to undergo random walks on the lattice, and $p_s = 1$. The mobility of the clusters is presumed to depend on the number of particles s they are made of. In particular, it is assumed that the diffusion coefficient D_s of a cluster of size s is given by

$$D_s = Cs^\gamma , \tag{8.1}$$

where C is a constant and γ can be used to take into account the effects of cluster geometry. For example, in a typical physical system one expects that $\gamma \simeq -1/D$, because the mobility of a cluster in a fluid is inversely proportional to its hydrodynamic radius which for an aggregate of fractal dimension D is close to to its linear extension (e.g., de Gennes 1979). For the case $\gamma = 0$, corresponding to a mass-independent diffusion coefficient, clusters are selected randomly and moved by one lattice unit in a direction chosen randomly from the $2d$ possible directions. If $\gamma \neq 0$ the following procedure is used to decide which of the clusters should be moved next. A random number r uniformly distributed in the range $0 \leq r \leq 1$ is selected and the cluster is moved only if $r < D_s/D_{max}$, where D_s is the diffusion coefficient of the given cluster and D_{max} is the largest diffusion coefficient for any cluster in the system.

The above lattice model can be generalized to the off-lattice case in a manner analogous to that used to simulate off-lattice DLA (Section 6.1.1). Similarly to diffusion-limited aggregation, this modification is not expected to change the fractal dimension of the clusters (Meakin 1987b). Figure 8.2 shows the projection of an off-lattice diffusion-limited cluster-cluster aggregate embedded into three dimensions. Since this projection is not space filling, the fractal dimension of the aggregate itself should be less than 2 (see rule a) of Section 2.2).

Figure 8.2. Projection of an off-lattice diffusion-limited CCA cluster grown in a three-dimensional simulation (Meakin 1987b). Comparison of this figure with Fig. 8.14 shows the relevance of computer simulation results to the experiments.

One way to investigate the structure of aggregates is to calculate the density-density correlation function $c(r)$ given by (2.14) (Meakin 1987b). In addition to the determination of the fractal dimension, this method can also be used to study the effects of finite constant density of particles ρ. Results for $c(r)$ obtained from three-dimensional simulations of CCA are displayed in Fig. 8.3. For $r < r_c$ the plot of $\ln c(r)$ versus $\ln r$ is approximately a straight line with a slope $-\alpha = D - d$, where r_c is a ρ-dependent crossover scale. The power law decay of $c(r)$ indicates that the density distribution within the aggregates has a fractal scaling up to r_c. In the vicinity of the crossover scale this behaviour is changed, and for $r > r_c$ the correlation function becomes approximately constant. Such a crossover corresponds to a structure which is homogeneous on length scales larger than r_c. To estimate r_c one assumes that the large network spanning the whole cell at the final stage of the simulation is made of fractal subunits of dimension D. The number of subunits is proportional to N_0/r_c^D. Since the network fills the cell of volume L^d more or less homogeneously, and the effective volume occupied by a subunit is r_c^d, we can write

$$L^d \sim \frac{N_0}{r_c^D} r_c^d . \tag{8.2}$$

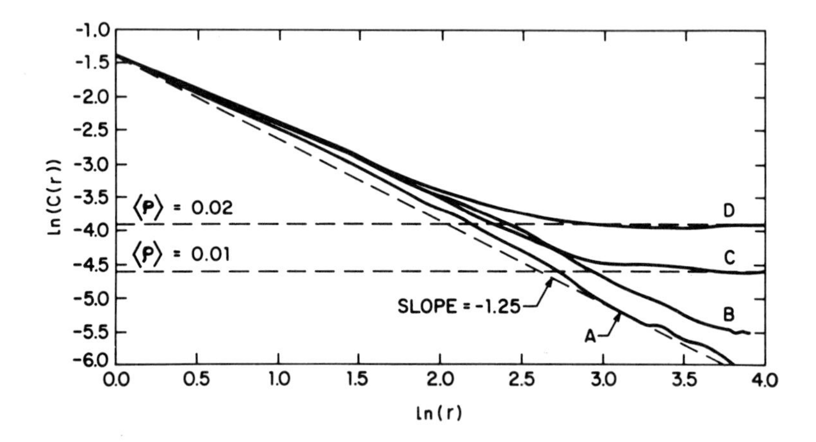

Figure 8.3. Dependence of the density correlations within diffusion-limited CCA clusters on the interparticle distance r. The crossover to the non-fractal behaviour is indicated by the horizontal part of the curves (Meakin 1987b).

From the above expression one obtains

$$r_c \sim \rho^{D-d} \sim \rho^{-\alpha} , \tag{8.3}$$

where $\alpha = d - D$ is the codimension (2.18) and $\rho = N_0/L^d$.

The d-dependence of the fractal dimension of aggregates grown by diffusion-limited CCA is presented in Table 8.1. Clearly, these aggregates have a considerably smaller D than DLA clusters generated on lattices of the same dimension. This result is quite plausible; individual particles can penetrate a DLA cluster easily enough to increase its dimension to at least $d - 1$ (6.4). Cluster-cluster aggregates do not tend to fill holes within each other, being fractal structures themselves. The values obtained for the $d = 2$ and $d = 3$ cases are in good accord with the experimental results discussed in Section 8.3.1.

Table 8.1. Dependence of the fractal dimension of diffusion-limited cluster-cluster aggregates on d as determined from the correlation function (D_α) and from the radius of gyration (D_ν).

d	D_α	D_ν	N_{max}
2	1.44±0.03	1.43±0.02	$\sim 10^4$
3	1.78±0.06	1.75±0.01	$\sim 10^4$
4	2.12±0.10	2.03±0.04	$\sim 10^4$
5	–	2.21±0.02	$\sim 10^3$
6	–	2.38±0.02	$\sim 10^3$

The question of the dependence of D on the diffusivity exponent γ arises naturally. The simulation carried out in two dimensions suggests that for $\gamma < 1$ the aggregates have approximately the same fractal dimension $D \simeq 1.45$. However, it is clear that in the limit of $\gamma \gg 1$ and low concentration CCA should become equivalent to DLA, as in this case practically only a single cluster (the largest one) is moving and it collects the rest of the individual particles during its diffusional motion. The behaviour of the density-density correlations in clusters obtained at the end of the simulations for various γ indicates a continuous change in the fractal dimension from a value of $D \simeq 1.45$ for $\gamma = 0$ to $D \simeq 1.7$ for $\gamma = 2$ (Meakin 1985c). Therefore, if $\gamma > 2$ the dimension of cluster-cluster aggregates is the same as that of DLA clusters. The nature of the *crossover* from CCA to DLA is not completely understood. The above mentioned simulations suggest a continuous change between the two regimes, while theoretical considerations imply a sudden jump from one type of behaviour into the other one as a function of γ.

Models for *reaction-limited cluster-cluster aggregation* (Jullien and Kolb 1984, Brown and Ball 1985) are constructed to represent the zero sticking probability limit of CCA. If $p_s \simeq 0$, each of the possible contact configurations of two clusters has the same probability to occur. This model is a cluster-cluster analogue of the particle-cluster model of Eden (1961), where each surface site is filled with equal probability. While for the Eden model $D = d$, in the case of reaction-limited CCA because of obvious steric constraints, the clusters can not become compact.

A possible realization of reaction-limited CCA on cubic lattices is based on placing two clusters at random positions in a large cell. The resulting configuration is accepted as a new cluster only if the two clusters are adjacent and do not overlap. There are two main possibilities for chosing the above two clusters. i) To produce a *monodisperse* size distribution (Jullien and Kolb 1984) one starts with 2^n monomers. At each iteration a dimerization is made until there are no monomers left. Next the dimers are joined to form clusters of 4 particles and so on. ii) In the *polydisperse* case (Brown and Ball 1985) the clusters are always randomly selected from the ones available at the given stage.

Based on the results for the fractal dimension, the monodisperse and the polydisperse cases are principally different. The following values were obtained

$$D = 1.53 \pm 0.01 \quad (d = 2) \quad \text{and} \quad D = 1.94 \pm 0.02 \quad (d = 3) \tag{8.4}$$

for the monodisperse distribution and

$$D = 1.59 \pm 0.01 \quad (d = 2) \quad \text{and} \quad D = 2.11 \pm 0.03 \quad (d = 3) \tag{8.5}$$

for the polydisperse system. It is clear from the above expressions that D is larger for these models than it is for diffusion-limited CCA. This is well illustrated by the most important three-dimensional case for which $D \simeq 1.75$ in the diffusion-limited and $D \simeq 2.11$ in the reaction-limited version of cluster-cluster aggregation, in good agreement with the experimental results.

A relevant quantity associated with reaction-limited CCA is the number of ways C_{s_1,s_2} in which cluster 1 and cluster 2 can be positioned adjacent to each other. The square root of this quantity for $s_1 = s_2 = s$ gives an estimate of the number of active sites (sites at which a contact can be made) of an s particle cluster. The simulations indicate that the average value of the active sites scales as (Jullien and Kolb 1984)

$$\langle C_s^{1/2} \rangle \sim s^{\delta} \,. \tag{8.6}$$

For the monodisperse case $\delta \simeq 0.37$ ($d = 2$), $\delta \simeq 0.58$ ($d = 3$) and $\delta \simeq 0.72$ ($d = 4$) were found to give the best fit to the data. It is obvious that δ must be smaller than 1 since only a fraction of the particles of a cluster is accessible when two aggregates are probed for contact. However, at the upper critical dimension d_c (if it exists) we expect $\delta = 1$, because for $d \geq d_c$ the aggregates are transparent to each other (see next Section).

In the off-lattice *ballistic cluster-cluster aggregation* model (Sutherland 1967) clusters are combined in pairs without the presence of other clusters (i.e., in the zero concentration limit). The process is started with a list of hyperspherical particles. During the simulation, particles or clusters are picked from the list in a stochastic manner, rotated to a random orientation, and are combined via randomly selected straight trajectories to form a larger cluster. The two clusters are joined at the point of first contact. After the new cluster has been formed, it is returned to the list, while the two precursors are erased. This process is continued until a single large cluster is obtained.

The clusters can be selected from the list according to a probability depending on some characteristics of the aggregates. As was discussed for reaction-limited CCA, one possibility is i) to pick a cluster completely *randomly* (polydisperse case). In the ii) *hierarchical models* (e.g., Sutherland 1967, Jullien and Kolb 1984) only clusters of the same size are joined. In this case the simulation starts with 2^k particles which are combined to form 2^{k-1} binary clusters (monodisperse size distribution). iii) An additional possibility is to choose a cluster with a *probability* depending on its *size*, e.g., in the form (8.1). In this model the probability of choosing two clusters of masses s_1 and s_2 simultaneously is proportional to $(s_1 s_2)^{\gamma}$.

Simulations of off-lattice ballistic aggregation corresponding to cases i) and ii) resulted in practically unchanged fractal dimensions for a given d (Meakin 1987b). For $d = 2$ and $d = 3$ the values $D \simeq 1.55$ and $D \simeq 1.91$ were obtained, respectively. These numbers are closer to the fractal dimension of reaction-limited CCA clusters than to that determined for diffusion-limited cluster-cluster aggregates. Model iii) provides a possibility to demonstrate the effects of cluster mobility. In Fig. 8.4 (upper row) the crossover from CCA to particle-cluster ballistic aggregation-type structures is illustrated. In the diffusion-limited case a crossover from CCA to DLA can be seen (lower row) occurring in this approach at $\gamma = 1$.

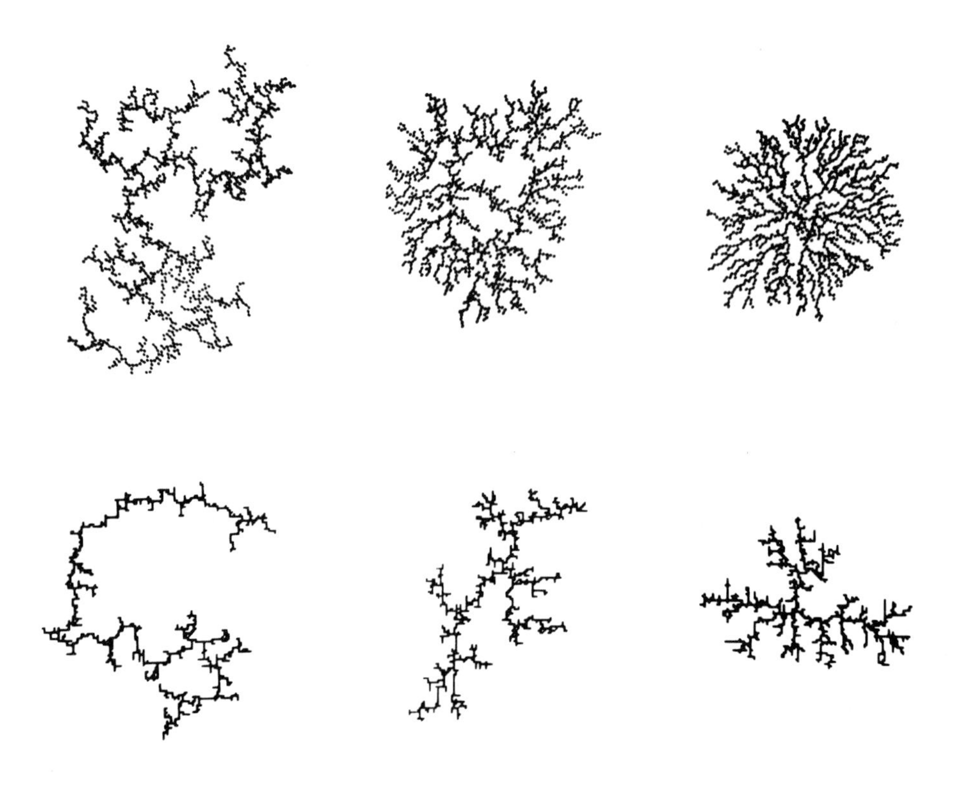

Figure 8.4. Crossover from CCA to ballistic particle-cluster aggregation as a function of γ when the objects move along straight line trajectories (upper row). In the diffusion-limited case the process crosses over to DLA for large γ (Jullien *et al* 1984).

In all of the CCA models discussed above it was assumed that the density of particles in the system is much less than unity. However, it is of interest to consider what happens for ρ close to 1. The $\rho = 1$ limit can be investigated using the following procedure (Kolb and Herrmann 1987). Initially, one particle is placed on each site of the square lattice. The particles try to move in a randomly selected direction, and form a cluster (establishing a bond) with the particle they would collide with (clusters also try to move and are combined with those they would hit). Therefore, the clusters are defined as sets of particles connected by permanent bonds formed during their attempts to move. A cluster is selected for a trial jump with a frequency given by (8.1). The results are demonstrated in Fig. 8.5. Depending on γ the following structures can be observed just before the linear size of the largest cluster becomes comparable to that of the cell: a) Compact clusters with fractal surface ($\gamma = -2$), b) fractal aggregates ($\gamma = 1$), and c) non-fractal objects ($\gamma = 2$).

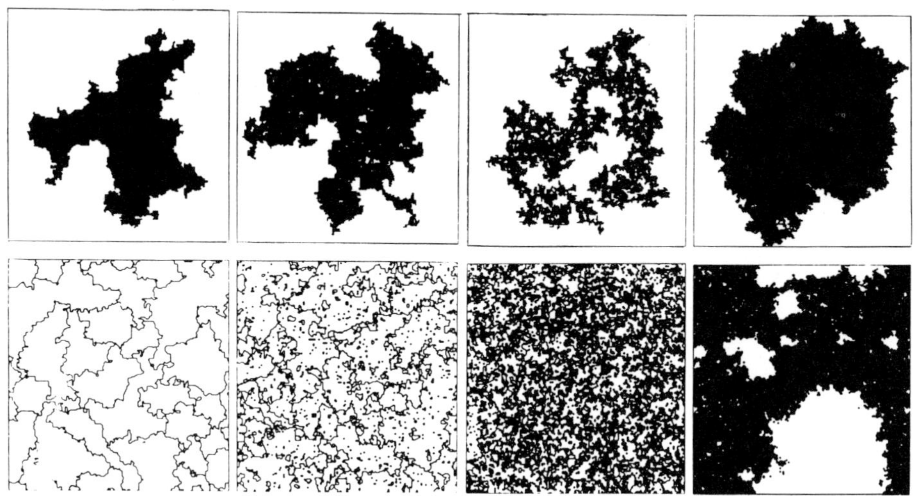

Figure 8.5. Clusters generated in the simulations of high density CCA. In these numerical experiments both surface and volume fractals could be observed. The first picture (from left) shows a non-fractal cluster having a fractal surface ($\gamma = -2$), while the third configuration represents a fractal cluster ($\gamma = 1$).

There are many more variants of cluster-cluster aggregation. A number of them lead to fractal dimensions different from the ones given above for the basic models. Non-universality was observed, for example, in ballistic CCA with a zero impact parameter (Jullien 1984) and in chain-chain diffusion-limited aggregation (Debierre and Turban 1987), where branching is not allowed.

8.1.2. Theoretical approaches

Before presenting a few theoretical results, it has to be pointed out that there is no standard theory for the fractal dimension of cluster-cluster aggregates. The theoretical methods well founded to determine the scaling behaviour in equilibrium phenomena are not applicable to diffusion-limited CCA because of its far from equilibrium nature. The situation is similar to the case of other growth processes such as DLA. However, unlike diffusion-limited aggregation, CCA can be shown to have an *upper critical dimension.* This is perhaps the most important result obtained for cluster-cluster aggregates by theoretical arguments.

To treat CCA theoretically we consider the following "Sutherland's Ghost" model (Ball and Witten 1984b). Let us imagine that the clusters are constructed according to an algorithm similar to the hierarchical version of the ballistic aggregation model (previous section). However, instead of combining the clusters using ballistic trajectories, one selects a monomer belonging to each of the two clusters, and joins the clusters by positioning these monomers on adjacent sites of the lattice (this is done in the spirit of reaction-limited CCA). The specific feature of this model is that the particles are *allowed to overlap.*

The fractal dimension of such clusters can be calculated by determining the average number of bonds $b(2s)$ (chemical distance) separating one particle from another on an aggregate consisting of $2s$ particles. If two particles are chosen at random on a $2s$-site cluster, the probability that they both belong to the A cluster or both belong to the B cluster is equal to $1/4$, where A and B denote the two s-site constituent clusters. In this case the average number of bonds is $b(s)$. With a probability $1/2$ one of the two chosen particles belongs to A, the other one to B. Then the path on A to the contact point, and on B from this point to the other chosen particle add up on average to give $2b(s)$. The three cases together give $b(2s) = 3/2b(s)$

which means that $b(s) \sim s^{\ln(3/2)/\ln 2}$. It follows from the construction (clusters of independent orientations are linked, and no overlaps occur) that the shortest path connecting two particles on the cluster behaves as a random walk (Section 5.3.1). Thus one has for the mean square distance between the particles $R^2 \sim b(s)$. From here it follows (Ball and Witten 1984b) that $s \sim R^D$ with

$$D = D_c = 2\frac{\ln 2}{\ln 3/2} \simeq 3.4 \,. \tag{8.7}$$

The above fractal dimension (which is independent of d) was obtained by letting the clusters interpenetrate freely. This crude approximation can be improved by modifying the model to take into account self-avoidance or excluded volume effects. This can be done in a computer by discarding overlapping configurations. The obtained hierarchical model is identical to reaction-limited CCA discussed above. However, if

$$2D - d < 0 \,, \tag{8.8}$$

only a small fraction of the configurations have to be removed and the two models become equivalent. Condition (8.8) is related to rule f) discussed in Section 2.2 giving the fractal dimension of the intersection of two fractals. If (8.8) is satisfied, the intersection has a fractal dimension equal to zero, consequently, overlaps occur with a probability less than 1. Therefore, excluded volume effects are negligible in embedding dimensions above $d_c = 2D_c \simeq 6.8$, which can be regarded as the upper critical dimension of reaction-limited cluster-cluster aggregation in analogy with d_c for ordinary equilibrium systems.

In fact, *reaction-limited* CCA can be considered as an *equilibrium* model since the various clusters appear with the same probability (Witten 1985). One expects that this model is closely related to lattice animals. Lattice animals are the collection of all possible connected configurations on a lattice consisting of a given number of sites taken with equal weight. The main difference between the two ensembles is that the former one is a subset of lattice animals corresponding to binary decomposable configurations. For equilibrium systems the heuristic Flory theory gives up to $d = d_c$

$$D = \frac{d+2}{2(1+1/D_c)}, \tag{8.9}$$

where D_c is the fractal dimension of clusters in d_c-dimensional space. If $d \geq d_c$, $D = D_c$. For lattice animals $D_c = 4$, and the critical dimension is $d_c = 8$. Substituting $D_c = 3.4$ into the above expression one gets the estimates for the fractal dimension of monodisperse reaction-limited aggregates $D \simeq 1.55$ ($d = 2$) and $D \simeq 1.93$ ($d = 3$) in surprisingly good agreement with the simulation results (8.4).

For *diffusion-limited and ballistic* CCA similar arguments can be used to obtain the upper critical dimension. Again, two particles can be used to link together two aggregates, but the obtained configuration has to be discarded if letting one of the clusters undergo a random walk results in an overlap during its diffusional motion (for the ballistic case straight line trajectories have to be considered). The effective dimension of the object consisting of the sites visited by the diffusing aggregate of dimension D is $D + 2$, since this object can be obtained by replacing each particle by a random walk of dimension 2. Consequently, for diffusion-limited CCA $d_c = 2D_c + 2 \simeq 8.8$. For the ballistic case we have $d_c = 2D_c + 1 \simeq 7.8$ (Witten 1985). In general, one expects that a higher upper critical dimension results in a lower fractal dimension for a given $d < d_c$. The numerical results are in good agreement with the above critical dimensions, because for a fixed d they satisfy $D_d(\text{diff. lim}) < D_d(\text{ballistic}) < D_d(\text{react. lim.})$.

8.2. DYNAMIC SCALING FOR THE CLUSTER SIZE DISTRIBUTION

The fractal dimension conveys information about the static or geometrical properties of a single aggregate. However, in a typical cluster-cluster aggregation process there are many clusters simultaneously present in the system, and the evolution of this ensemble of aggregates is of interest as well. This time dependence can be investigated by determining the *dynamic cluster-size distribution function* $n_s(t)$, which is the number of clusters in a unit volume consisting of s particles at time t.

The study of the statistics of clusters is a common approach to the description of ensembles of clusters. In many equilibrium systems n_s is known to decay as a power law at the critical point. Analogously, $n_s(t)$ has been shown to exhibit static (as a function of s) and dynamic (as a function of t) scaling in a number of close-to or far-from equilibrium systems (Binder 1976). In this Section we first treat computer simulations of the diffusion-limited and related cluster-cluster aggregation models together with the dynamic scaling picture emerging from these numerical investigations. This will be followed by a discussion of the mean-field Smoluchowski (1917) equation.

8.2.1. Diffusion-limited CCA

In the diffusion-limited cluster-cluster aggregation model (Section 8.1.1) the clusters move along Brownian trajectories and stick together on contact. Initially there are $N_0 = \rho L^d$ monomers in a cell of linear size L. To make the cluster size distribution function independent of the cell size we use the definition $n_s(t) = N_s(t)/L^d$, where $N_s(t)$ is the number of s-clusters at time t in the cell. The elapsed time is measured by increasing t by an amount Δt each time a cluster is selected to move.

The dynamic scaling of $n_s(t)$ can be well demonstrated simulating the simple version of CCA on a square lattice with mass independent cluster mobility corresponding to $\gamma = 0$ in (8.1). In this case $\Delta t = 1/n(t)$ has to be used to provide a physical time, where $n(t)$ is the normalized total number of clusters in the system at time t. The choice $\Delta t = 1$ would result in an unphysical acceleration with growing t (since $n(t) \to 0$ for $t \to \infty$, and this would lead to an increase of the number of steps per cluster per unit time). The expression $\Delta t = 1/n(t)$ yields the same diffusion coefficient for each cluster, and it corresponds to the Monte-Carlo step per spin type time definitions common in equilibrium simulations.

The results of simulations are illustrated in Figs. 8.6. Figure 8.6a shows the dynamic cluster-size distribution as a function of s for fixed times,

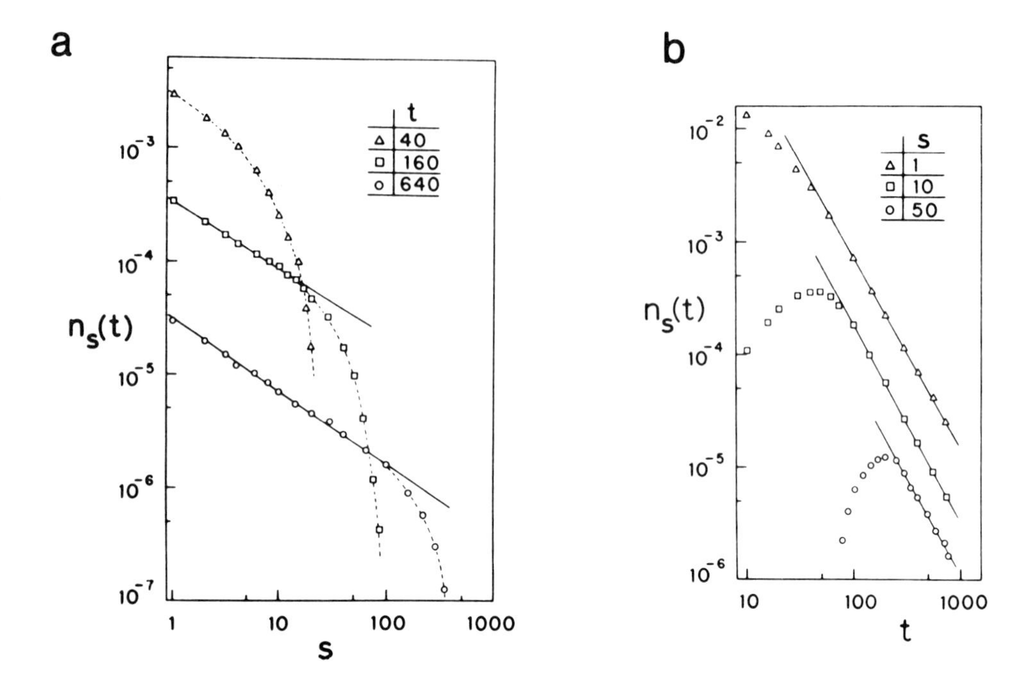

Figure 8.6. (a) Dependence of the dynamic cluster-size distribution function ($n_s(t)$) on the number of particles in the clusters (s)for fixed times (t). (b) $n_s(t)$ as a funcion of time for three selected values of s (Vicsek and Family 1984).

while $n_s(t)$ as a function of t is plotted in Fig. 8.6b for three selected values of the cluster size s. These simulations were carried out with $L = 400$ and initial particle density $\rho = 0.05$. There are a few important conclusions which can be drawn from these figures. i) The straight lines in Fig. 8.6a correspond to a power law decay of the cluster-size distribution as a function of s up to a cutoff value. This behaviour is analogous to that observed for equilibrium systems close to the critical point. However, for diffusion-limited CCA the exponent τ describing the decay of $n_s(t)$ with s, is smaller than 2, in contrast to equilibrium systems, where $\tau > 2$. ii) The position of the cutoff diverges with t. iii) According to Fig. 8.6b, $n_s(t)$ scales as a function of t (for fixed s) as well.

These observations can be well represented by the *scaling assumption* (Vicsek and Family 1984a, Kolb 1984)

$$n_s(t) \sim s^{-\theta} f(s/t^z), \tag{8.10}$$

where θ and z are constants analogous to critical exponents, and $f(x)$ is a scaling function. $f(x)$ is such that $f(x) \ll 1$ (it is exponentially small), if $x \gg 1$ and $f(x) \sim x^\delta$, if $x \ll 1$ with δ usually called the crossover exponent. The above expression can be written in an alternative form which contains the scaling of $n_s(t)$ as a function of s and t *explicitly* for small s/t^z

$$n_s(t) \sim t^{-w} s^{-\tau} \tilde{f}(s/t^z), \tag{8.11}$$

where the cutoff function is approximately a constant for $x \ll 1$ and decays faster than any power law as $x \to \infty$. The term t^{-w} corresponds to a process typical for cluster-cluster aggregation; the clusters which are much smaller than t^z gradually die out forming larger aggregates. The characteristic cluster size is determined by the denominator t^z.

The scaling assumption (8.11) and the normalization condition

$$\rho = \sum_{s=1}^{\infty} n_s(t) s \sim \int_1^\infty n_s(t) s ds \tag{8.12}$$

can be used to obtain a scaling relation among the exponents w, z and τ. Inserting (8.11) into (8.12) we have

$$\rho \sim t^{-w} \int_1^\infty s^{1-\tau} f(s/t^z) ds \sim t^{-w+(2-\tau)z} \int_{t^z}^\infty x^{1-\tau} f(x) dx\,. \tag{8.13}$$

From (8.13) it follows that (Vicsek and Family 1984a)

$$w = (2-\tau)z\,, \tag{8.14}$$

since $0 < \rho < 1$ when $t \to \infty$ and the last integral is a constant in the same limit. Obviously, $\tau < 2$ has to be satisfied, because in a physical system $w, z > 0$.

The *mean cluster size* $S(t)$ diverges for $t \to \infty$. Expressing $S(t)$ through $n_s(t)$, and using (8.11) and (8.14) we get

$$S(t) = \frac{\sum_s n_s(t)s^2}{\sum_s n_s(t)s} \sim t^z \,. \tag{8.15}$$

Similarly, for the *total number of clusters* in the system, $n(t) = \sum_s n_s(t)$ one has (Vicsek and Family 1984b)

$$n(t) \sim \begin{cases} t^{-z}, & \text{if } \tau < 1 \\ t^{-w}, & \text{if } \tau > 1 \,. \end{cases} \tag{8.16}$$

Thus the scaling of the total number of particles in time is determined by the value of the exponent τ. The simulations in $d = 2$ resulted in the estimates $w \simeq 1.7$, $z \simeq 1.4$ and $\tau \simeq 0.7$.

For $x \ll 1$ (8.10) can be written in the form $n(t) \sim t^{-z\delta} s^{-\theta+\delta}$. Comparing this with (8.11) one obtains $w = z\delta$ and $\theta - \delta = \tau$. Because of the scaling relation (8.14), from here it follows that in (8.10)

$$\theta = 2 \,. \tag{8.17}$$

The mass dependent cluster mobilities of the form (8.1) strongly influence the dynamics as well. These effects were investigated by simulating diffusion-limited CCA in two and three dimensions (Meakin *et al* 1985). To provide cluster mobilities proportional to s^γ one can use the following procedure. i) First a cluster is selected randomly. ii) Then a random number p distributed uniformly on the unit interval is generated, and the given cluster is moved in a random direction by one lattice unit only if $p < D_s/D_{max}$, where D_s is the diffusion coefficient of the selected cluster and D_{max} is the largest diffusion coefficient for any cluster in the system. iii) Finally, on each occasion when a cluster is chosen the time is incremented by $\Delta t = 1/[n(t)D_{max}]$ even if the cluster is not moved. In other words the time is incremented by Δt for each attempted move.

The dependence of the exponents w, z and τ on γ can be determined from log-log plots of the quantities $S(t)$, $N(t) = n(t)L^d$ and $N_s(t)$ for various γ. The results obtained from three-dimensional simulations are summarized in Table 8.2. This table shows that for γ smaller than some critical γ_c the shape of the cluster-size distribution function changes qualitatively. If $\gamma < \gamma_c$, $n_s(t)$ does not decay as a power law for small s, and in the scaling law (8.11) $w = 2z$.

Table 8.2. Exponents τ, w and z obtained from three-dimensional simulations of diffusion-limited CCA for various values of the diffusivity exponent γ. Note that the scaling theory predicts $w = (2 - \tau)z$ for $\gamma > \gamma_c$ and $w = 2z$ for $\gamma < \gamma_c$.

γ	z	w	τ
-3	0.33	0.64	
-2	0.45	0.90	
-1	0.85	1.60	
-1/D	1	2.2	
-1/2	1.3	2.6	$\simeq 0$
0	3	2.2	1.3
1/2	~ 100	12	1.87

Thus the behaviour of $n_s(t)$ is *non-universal* as a function of γ. For $\gamma > 0$ the large clusters move faster, and relatively many small ones do not take part in the aggregation process. In this case $n_s(t)$ is characterized by many small clusters and a gradually decreasing number of clusters for growing sizes. If $\gamma < 0$, the small clusters have a higher velocity, so they die out (aggregate) quickly, forming larger clusters. Thus there will be only a few small and very large clusters in the system, resulting in a non-monotonic, bell-shaped distribution.

The *crossover* between the monotonic and the bell-like behaviour occurs at some γ_c depending on d (Meakin *et al* 1985). For example, in $d = 2$

this qualitative change in $n_s(t)$ occurs near $\gamma_c \simeq -0.27$. This is demonstrated in Fig. 8.7. According to Table 8.2, in three dimensions $\gamma_c \simeq -0.5$. On the other hand, one expects that the mobility of clusters in a fluid is inversely proportional to their hydrodynamic radius. The latter scales the same way as the radius of gyration of the clusters, thus $D_s \sim 1/R_s \sim s^{-1/D}$, where $1/D \simeq 0.57$. Consequently, in $d = 3$ the critical mobility exponent γ_c is just in the range where γ is likely to be in a real system in which the diffusion process is controlled by shear viscosity.

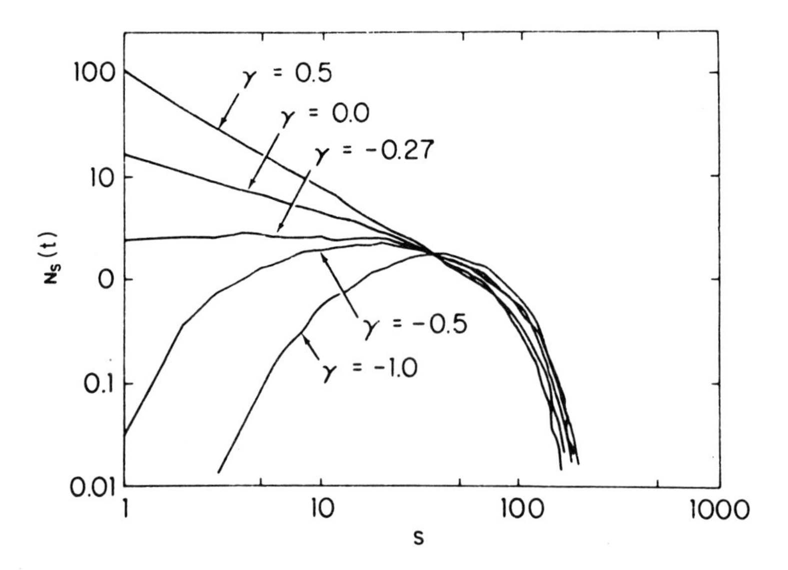

Figure 8.7. Cluster-size distribution functions obtained from simulations of diffusion-limited CCA on the square lattice. As γ is decreasing, at a critical value of the diffusivity exponent $\gamma_c \simeq -0.27$, the monotonic decay of the distribution crosses over into a different, bell-shaped behaviour (Meakin *et al* 1985).

The above discussed results for the dynamics of diffussion-limited cluster-cluster aggregation can be summarized as follows. The cluster-size distribution is described by dynamic scaling of the form

$$n_s(t) \sim s^{-2} f(s/t^z)\,, \tag{8.18}$$

where $f(x)$ depends on the mobility exponent γ. In particular,

$$f(x) \sim x^2 g(x) \qquad \text{for} \quad \gamma < \gamma_c \tag{8.19}$$

with $g(x)$ exponentially small for both $x \ll 1$ and $x \gg 1$, and

$$f(x) \sim \begin{cases} x^\delta, & \text{if } x \ll 1 \\ \ll 1, & \text{if } x \gg 1 \end{cases} \qquad \text{for} \quad \gamma > \gamma_c \,. \tag{8.20}$$

The scaling (8.18) can be checked by plotting the quantity $s^2 n_s(t)$ as a function of s/t^z. If (8.18) is valid, then the results obtained for a given γ should fall onto a single curve corresponding to the scaling function $f(x)$. Figure 8.8 demonstrates that for the simulations carried out in $d = 3$ with $\gamma = -2$ this is indeed the case.

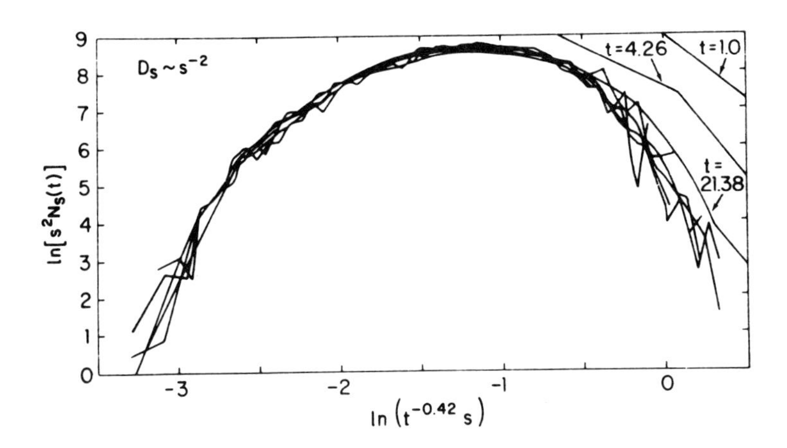

Figure 8.8. Scaling of the dynamic cluster-size distribution function. (8.18) is supported by the fact that after a relatively short time the data determined for various times fall onto a single curve. In these three-dimensional simulations $\gamma = -2$ was used (Meakin *et al* 1985).

The *region of validity* of the dynamic scaling (8.18) depends on the parameters of the problem. Equation (8.18) is expected to describe the evolution of the process in the limit when the particle density is small ($\rho \ll 1$). In addition, $s \gg 1$ and $t \gg 1$ are required in order to be in the scaling region.

For the time variable, however, there is an *upper bound* as well, for the following reasons. First, in a finite system the total number of clusters $N(t)$ after some time becomes so low that a statistical interpretation of the cluster size distribution loses its meaning. Second, a non-trivial crossover is expected to occur at a time t_g depending on the particle density ρ. As the aggregation process goes on, the number of clusters decreases and, correspondingly, the average number of particles in the clusters, $\bar{s}(t)$, increases. Assuming that the clusters are approximately of the same size (this is so for $\gamma < \gamma_c$) the average cluster radius grows in time as $\bar{R}(t) \sim [\bar{s}(t)]^{1/D} \sim [\rho/n(t)]^{1/D} \sim t^{z/D}$. In contrast, the average distance between the centres of two clusters, $\bar{r}(t)$, grows as $\bar{r}(t) \sim [n(t)]^{-1/d} \sim t^{z/d}$, i.e., slower, because $d > D$. This means that approaching the time t_g at which $\bar{R}(t)$ becomes as large as $\bar{r}(t)$, the aggregation process crosses over into *gelation* and an infinite network (gel) appears. Naturally, in this limit the dynamic scaling (8.18) breaks down. Thus, diffusion-limited CCA is a suitable model to simulate gelation in colloidal systems.

The above qualitative picture can be made more quantitative (Kerstein and Bug 1984) by using (8.18) for the determination of the effective volume, $V(t)$, occupied by the clusters. This is done by calculating

$$V(t) = \sum_s V_s n_s(t) \sim \rho \int_1^\infty s^{d/D-2} f(s/t^z) \sim \rho t^{(d/D-1)z} , \qquad (8.21)$$

where $V_s = s^{d/D}$ is the effective volume occupied by an s-site cluster and the prefactor is chosen so that $V(1) = \rho$ (at the first time step V is equal to the volume fraction of the diffusing particles). Setting $V(t_g) = 1$, (8.21) gives the estimate

$$t_g \simeq \rho^{1/[z(1-d/D)]} \qquad (8.22)$$

depending on ρ, D and z (where z itself depends on γ but it is always larger than zero). Accordingly, gelation occurs at finite time for every γ, although $t_g \to \infty$ as $\rho \to 0$. The gelation time becomes very large also for $\gamma \ll 0$ since in this case $z \ll 1$.

Dynamic scaling of the form (8.18-8.20) is a general property of many cluster-cluster aggregation processes in which the aggregates move along Brownian trajectories (Section 8.3.2). The cluster-size distribution in a number of related computer models was found to follow (8.18), including chain-chain aggregation (Debierre and Turban 1987), the particle coalescence model and aggregation of anisotropic particles (Miyazima *et al* 1987).

8.2.2. Reaction-limited CCA

Reaction-limited cluster-cluster aggregation in the form introduced in Section 8.1.1 does not allow the investigation of dynamic properties because there is no physical time scale defined in the model. One way to study the dynamics of aggregation in realistic reaction-limited processes is to use the diffusion-limited CCA model with very small sticking probabilities (Family *et al* 1985b). The formation of a permanent bond between two clusters may depend on a number of parameters, including the mass of the two colliding clusters. To be specific, in the following we shall describe the results of simulations with a sticking probability of the form

$$p_{s,s'} = p_0 s^\sigma s'^\sigma , \tag{8.23}$$

where s and s' denote the number of particles in the two clusters. In (8.23) $p_0 < 1$ and σ are constants and, naturally, if $p_0 s^\sigma s'^\sigma > 1$ one uses $p_{s,s'} = 1$. As before, the mobility of clusters depends on their size according to (8.1).

Let us first consider the $\sigma = \gamma = 0$ case to study the effects of a small, mass-independent sticking probability p_0. The behaviour of the total number of clusters in the two-dimensional simulations is shown in Fig. 8.9. According to this figure, if $p_0 \ll 1$, at the beginning $N(t)$ decreases very slowly, but after a sufficiently long time it tends to zero as t^{-z}. The value of the exponent z is independent of p_0, and is about 1.5. This fact can be interpreted as the sign of a *crossover from reaction-limited into diffusion-limited* CCA. For short times there are only small clusters in the system. If they do not stick when colliding, there is a good chance that they diffuse away without coming into contact again. The number of clusters decreases slowly, the aggregation is reaction-limited. As the time increases, larger clusters are formed which can be linked in many ways and touch each other many times while they are in the vicinity of each other. Thus if two large clusters get close, the probability that they stick becomes nearly 1, and the process crosses over into diffusion-limited aggregation.

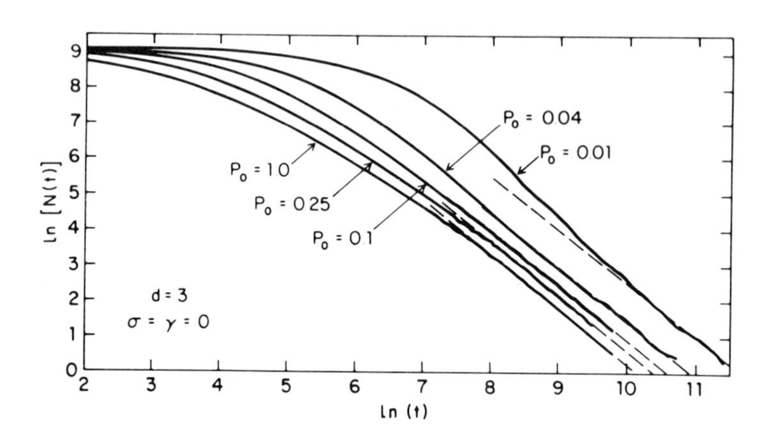

Figure 8.9. Time dependence of the total number of clusters in the three-dimensional cell for a fixed mass-independent sticking probability p_0 (Family *et al* 1985b).

Simulations with various values of the sticking probability exponent σ (Family *et al* 1985b) indicate that in analogy with the role of γ, changing σ may result in a qualitative change in the behaviour of the cluster-size distribution function $n_s(t)$. In particular, decreasing σ there exists a critical σ_c at which $n_s(t)$ crosses over from a monotonically decreasing into a bell-shaped distribution. This is demonstrated by Fig. 8.10. According to the simulations $\sigma_c \simeq -0.6$ in $d = 2$ and $\sigma_c \simeq -0.8$ in $d = 3$. Figure 8.10 allows one to draw an additional conclusion. The envelope of the cluster-size distribution function for various times is a straight line (the common tangent) having a slope approximately equal to -2. It can easily be shown that this has to be so if the time evolution of $n_s(t)$ is described by dynamic scaling of the form (8.18).

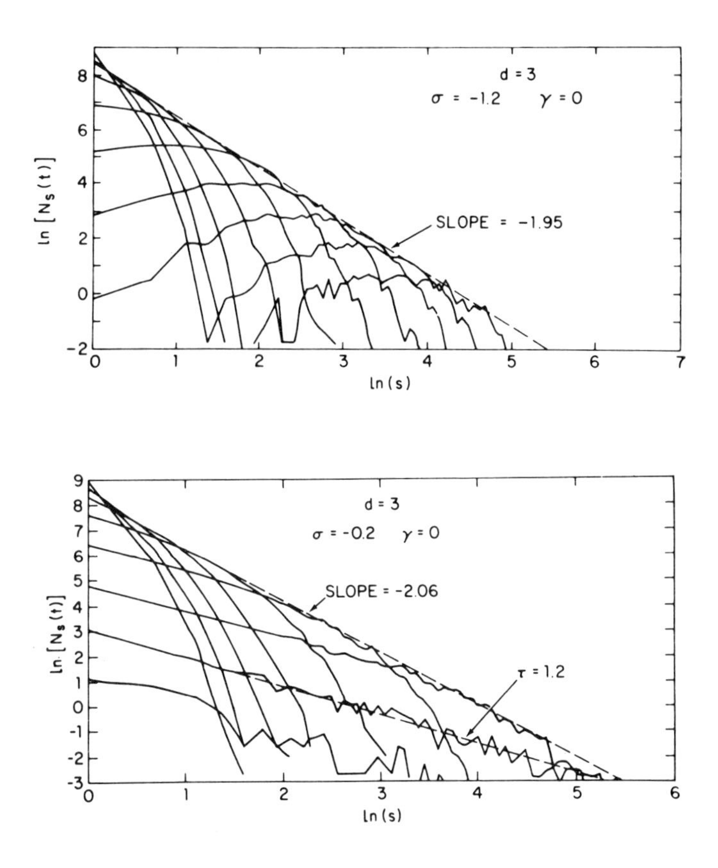

Figure 8.10. The cluster-size distribution function versus s for various times and for two selected values of the sticking probability exponent σ (Family *et al* 1985b).

8.2.3. Steady-state and reversible CCA

The above discussed cluster-cluster aggregation models describe processes with a permanent evolution in time since the number of clusters in the corresponding systems is gradually decreasing. CCA models can be easily modified in order to simulate an important process of both practical and theoretical interest, in which a far-from-equilibrium steady-state distribution of clusters develops in the system. This can be accomplished by adding the following rules to those defining CCA: i) single *particles are fed* into the system and ii) simultaneously, clusters *are removed* according to some rules. In another approach the clusters are allowed to *break up*, and this process naturally leads to a time-independent, equilibrium cluster-size distribution.

Steady-state conditions are typical in many applied fields. For example, small smoke particles fed into the atmosphere form larger aggregates by coagulation. These aggregates disappear from the air by sedimentation due to the gravitational force. In the stirred tank reactors, often used for modelling chemical reactors in industry, an analogous process takes place but the particles are removed by letting them flow out from the chamber. In addition to the possible applications to engineering, steady-state coagulation is interesting from the theoretical viewpoint because of its analogy with dynamical critical phenomena in near equilibrium systems. Steady-state coagulation has been investigated by various approaches including experiments (Madelaine *et al* 1979), numerical methods (McMurry 1980), simulations (Vicsek *et al* 1985), and studies of the Smoluchowski equation approach (White 1982, Rácz 1985a).

One of the simplest models in which the scaling behaviour of steady-state aggregation can be investigated is the following. The process is the same as the original diffusion-limited CCA model, except that at every unit time kL^d particles are added to the system at different sites selected randomly, where L is the linear size of the system and k is a small parameter. In addition, a cluster is discarded as soon as it becomes larger than a previously fixed number s_r. This latter rule is an extreme version of the situation in which larger clusters leave the system with a higher probability. In this way both the total number of clusters per unit volume, $n(t)$, and the number of particles in a unit volume, $m(t)$, in the system go to a k-dependent constant value (n_∞ and m_∞) for long times. The *relaxation time*, $t_r(k)$, corresponding to the characteristic time scale of the relaxation towards the steady-state is also expected to depend on the feed rate k.

The results of the related simulations indicate that both the total number of clusters in the steady-state, n_∞, and the relaxation time, t_r, scale as a function of k. Correspondingly, for the case when $k \ll 1$ and the initial number of particles is very small the data for $n(t)$ can be well represented by the following *scaling form* for the number of clusters in the system at time t (Vicsek *et al* 1985)

$$n(t) \sim k^{\delta} f(k^{\Delta} t), \tag{8.24}$$

where $f(x)$ is a scaling function with $f(x) \sim x$ for $x \ll 1$ and $f(x) = 1$ for $x \gg 1$. The actual shape of $f(x)$ may depend on the parameters γ or s_r but for a fixed set of these numbers $n(t)$ can be expressed through the scaling form (8.24). The scaling behaviour represented by (8.24) is demonstrated in Fig. 8.11, where the $N(t)$ curves obtained in the three dimensional simulations for various feed rates are scaled into one universal function ($N(t)$ is the total number of particles in the L^d cell).

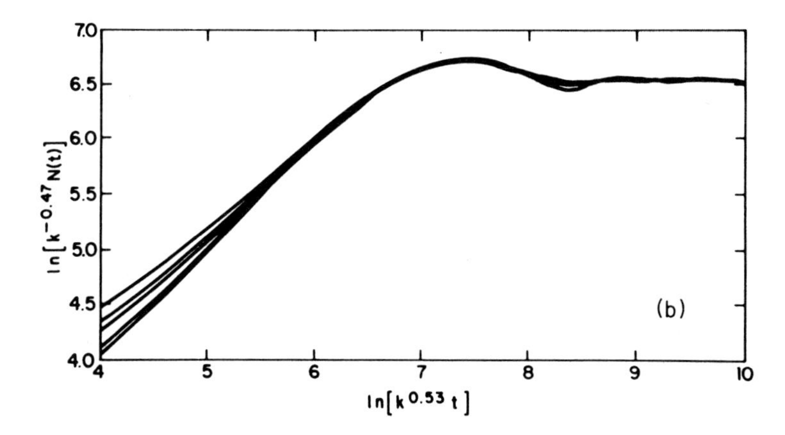

Figure 8.11. This figure shows how the scaling form (8.24) for the total number of clusters in the three-dimensional model of CCA with sources and sinks can be used to scale the data into a single function. The deviations from scaling apparent on the left-hand side of the plot is caused by the non-zero initial particle concentration used in the simulations (Vicsek *et al* 1985).

The values of the exponents δ and Δ can be determined from the slopes of the straight lines drawn through the data on the log-log plots of N_∞ and t_r versus k. It was found that in one dimension $\delta = 0.33 \pm 0.02$ and $\Delta = 0.65 \pm 0.03$. The two-dimensional simulations gave $\delta = 0.40 \pm 0.04$ and $\Delta = 0.58 \pm 0.05$ without, and $\delta = 0.52 \pm 0.04$ and $\Delta = 0.46 \pm 0.05$ with the logarithmic corrections taken into account, while in three dimensions $\delta = 0.47 \pm 0.05$ and $\Delta = 0.54 \pm 0.05$ were obtained.

The one-dimensional case can be treated both exactly and using an approximate rate equation for the number of clusters. Here we briefly discuss the latter approach because, in spite of its simplicity, the equation

$$\frac{dn(t)}{dt} = k - bn^3(t) - F(k, s_r, t) \tag{8.25}$$

reproduces the exact values for the exponents δ and Δ. In (8.25) b is a constant and the term $-F(k, s_r, t)$ describes the removal of clusters. The term $n^3(t)$ is included because of the following consideration. The rate of change of $n(t)$ due to coagulation is proportional to the number of clusters itself and to the average collision frequency of the clusters, ν. In one dimension, because of the diffusional motion, this frequency is inversely proportional to the square of the average distance X between the clusters. On the other hand, $X = 1/n(t)$, therefore, $\nu \sim n^2(t)$. Although the above arguments are not rigorous, they are quite plausible, and therefore (8.25) is expected to provide the right asymptotic behaviour of $n(t)$.

From (8.25) it can be seen that in the steady-state for $s \to \infty$ the number of clusters is equal to $(k/b)^{1/3}$, thus $\delta = 1/3$. In order to get an expression for the relaxation time one integrates (8.25) for $s_r \gg 1$. Keeping only the term which becomes singular if $n(t) \to (k/b)^{1/3}$ we get from (8.25)

$$n(t) \simeq (k/b)^{\delta} - \lambda e^{-t/t_r}\,, \tag{8.26}$$

where the relaxation time is $t_r = k^{-\Delta}/3b^{1/3}$ with $\Delta = 2/3$ and $\delta = 1/3$. In (8.26), λ is a constant depending on the initial conditions. These values for δ and Δ satisfy the scaling law $\delta + \Delta = 1$ following from a scaling analysis of the Smoluchowski equation (next section) and are in good agreement with the simulation results. A rigorous derivation (Rácz 1985b) using an analogy with the domain wall dynamics in the kinetic Ising model gives the same values for the exponents, but a different value for λ.

Having determined δ and Δ in $d = 1, 2$ and 3 we have the necessary data to discuss the question of the upper critical dimension for the dynamics of cluster-cluster aggregation under steady-state conditions. The simulations indicate scaling as a function of the feed rate with exponents $\delta = \Delta = 1/2$ for $d \geq 2$, if we assume the existence of logarithmic corrections in $d = 2$. Below two dimensions different values have been found. These results are consistent with an *upper critical dimension* $d_c = 2$ for the *kinetics* of steady-state CCA. In addition, the theoretical prediction $\delta + \Delta = 1$ is fulfilled for all cases considered.

Reversible cluster-cluster aggregation takes place if the bonds connecting the particles within a cluster can break. This is clearly a relevant process in a number of situations (e.g., Barrow 1981, Ziff and McGrady 1985, Ernst and van Dongen 1987); for example, as the clusters grow in size, the possibility of breakup increases. In reversible CCA coagulation decreases the number of clusters, whereas breakups increase $n(t)$. There exists a relaxation time t_r such that after a sufficiently long time, $t \gg t_r$, a balance is established between the two processes leading to an *equilibrium state* in which $n_s(t)$ is independent of time.

Let us consider the scaling behaviour of diffusion-limited CCA models in which the breakup probability for a particular bond that breaks the cluster of size $s = i + j$ into two clusters of masses i and j is

$$F_{ij} = h\Phi_{ij} \, , \tag{8.27}$$

where h is the breakup constant, Φ_{ij} is a function determining the dependence of the fragmentation rate on the cluster sizes, and $\Phi(1,1) = 1$. To describe the behaviour of the steady-state cluster-size distribution function $n_s(h, \infty)$ we generalize the scaling assumption (8.24) to reversible aggregation. In (8.24) the argument of the scaling function was equal to the ratio of the cluster size s to the mean cluster size $S(t)$ (8.15). We expect that the value of the mean cluster size in reversible CCA for $t \to \infty$ scales with the fragmentation rate as (Family *et al* 1986)

$$S(h, \infty) \sim h^{-y} \, . \tag{8.28}$$

The above expression combined with (8.18) and (8.24) provides the following scaling assumption

$$n_s(h,\infty) \sim s^{-2} f(sh^y) \tag{8.29}$$

for the number of clusters of size s in a unit volume in the long time limit. The scaling function $f(x)$ is expected to behave as $f(x) \sim x^{2-\tau}e^{-cx}$, where the exponent τ and the constant c depend on the actual choice for F_{ij}. The scaling assumption (8.29) implies that $n(h,\infty) \sim k^y$ for the total (normalized) number of clusters.

The scaling form (8.29) can be checked by simulations. In order to avoid the complexities originating in the geometrical properties of the clusters it is practical to use the particle coalescence model in which clusters are represented by single sites (Kang and Redner 1984). However, when two clusters of masses i and j meet they coalesce into a heavier single-site cluster of mass $i+j$. The breakup probability (8.27) is chosen to be $F_{ij} = h(i+j)^\alpha$.

According to the *simulations* of the above model (Family *et al* 1986) the exponent y determined for various α is independent of the dimension of the lattice on which the aggregation process takes place for $1 \le d \le 3$. In addition, the scaling law $y = 1/(\alpha + 2)$ resulting from the mean-field Smoluchowski equation approach (next section) is satisfied in all cases as well. Thus we conclude that in contrast to irreversible aggregation, the spatial fluctuations in the density of the particles are compensated by the fragmentation effect down to at least one dimension, i.e., *the upper critical dimension* is $d_c \le 1$. The agreement with the scaling assumption (8.29) for the cluster-size distribution is demonstrated in Fig. 8.12, where the one-dimensional results for the quantity $s^2 N_s(h,\infty)$ obtained for $\alpha = 1$ and various values of h are scaled into a universal function when plotted against sk^y. Similar agreements exist for $d = 2$ and $d = 3$ and for other values of α.

8.2.4. Mean-field theories

The classical understanding of aggregation kinetics is given by the *rate equation approach* proposed by Smoluchowski (1917). The basic assumptions of

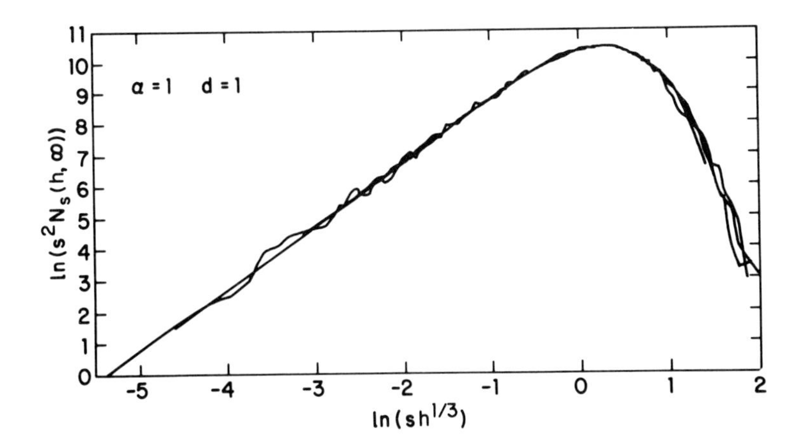

Figure 8.12. Scaling of the steady-state cluster-size distribution function $N_s(h, \infty)$ obtained from one-dimensional simulations carried out with breakup constants $k = 10^{-4}$, 10^{-5}, 10^{-6} and 10^{-7} (Family *et al* 1986).

his theory are the following. i) The reaction rate K_{ij} for two clusters of masses i and j is about the same for any pair of clusters having masses i and j, respectively. ii) The concentration of clusters with a given mass can be represented by its spatial average. Thus the space dependence of all quantities is neglected, consequently, this approach is an intrinsically mean-field theory. iii) Finally, it is assumed that the system is sufficiently diluted so that the reaction rate between two types of clusters is not influenced by the presence of other clusters.

With these assumptions the rate of aggregation of i-clusters with j-clusters to form a cluster of mass $i + j$ is proportional to the concentration of reactants $Rate \sim K_{ij}n_i(t)n_j(t)$, where K_{ij} is called the collision matrix or kernel. Smoluchowski's equation is then found by writing the population balance equation, taking into account both the gains and losses due to collisions

$$dn_s(t)/dt = \frac{1}{2} \sum_{i+j=s} K_{ij}n_i(t)n_j(t) - n_s(t) \sum_{j=1}^{\infty} K_{sj}n_j(t) . \tag{8.30}$$

Thus the above coagulation equation constitutes an infinite set of coupled non-linear rate equations which have to be solved for a given initial distribution $n_s(t=0)$. Eq. (8.30) has been studied extensively for decades, and there exists a vast number of papers in the literature devoted to the description of its properties. Here only those aspects of the Smoluchowski approach will be discussed which are closely related to fractal aggregation and its simulation.

The question of validity of (8.30) for *diffusion-limited CCA* arises naturally. It can be investigated by checking the collision rates and the evolution of the cluster-size distribution during the simulation of CCA and relating the results to those obtained from the Smoluchowski theory. From the point of view of Eq. (8.30) the question translates into finding the form of the collision matrix when the clusters are fractal. Smoluchowski showed that for diffusing, simple spherical clusters in $d = 3$ the coagulation kernel is proportional to $K_{i,j} \sim \left(i^{1/3} + j^{1/3}\right)\left(i^{-1/3} + j^{-1/3}\right)$, where the first term of the right-hand side is related to the effective cross-section proportional to the sum of the cluster radii $R_i + R_j$ and the second term is the sum of the diffusivities of the two colliding clusters (according to the Stokes-Einstein formula, the mobility of a cluster is inversely proportional to its radius).

Assuming that the cluster diffusivity depends on i as i^γ (8.1) one expects for the kernel describing diffusion-limited CCA of fractals of dimension D

$$K_{ij} \sim \left(i^{1/D} + j^{1/D}\right)^{d-2} \left(i^\gamma + j^\gamma\right) . \tag{8.31}$$

The coefficients K_{ij} can be determined from simulations by counting the number of collisions in unit time between clusters of masses i and j and dividing this number by $n_i(t)n_j(t)$. Figure 8.13 shows the results for the effective K_{1s} for $\gamma = 0$. The large fluctuations are due to the small number of clusters of the given size at the given stage of the aggregation process. The mean-field approximation is verified by the fact that the collision kernel is apparently independent of time except for small times (Ziff *et al* 1985).

In the most relevant physical situations the collision kernel is a homogeneous function of its variables (see e.g., Botet and Jullien 1984, Ernst 1985)

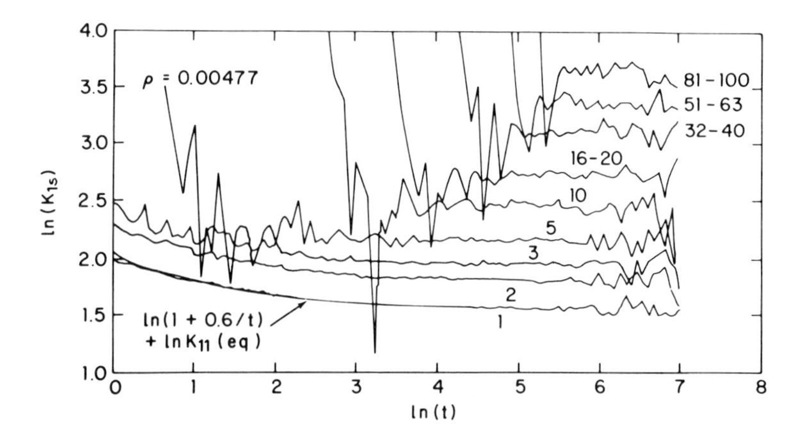

Figure 8.13. Effective collision rates K_{1s} for $\gamma = 0$ determined from simulations of diffusion-limited cluster-cluster aggregation on a simple cubic lattice (Ziff *et al* 1985).

$$K_{\lambda i \lambda j} = \lambda^{2\omega} K_{ij} \,. \tag{8.32}$$

Obviously, (8.31) satisfies this condition with

$$2\omega = (d-2)/D + \gamma \,. \tag{8.33}$$

It is possible to relate ω to the dynamic exponent z describing the time development of the typical cluster size which is defined as the second moment of $n_s(t)$ (8.15). The dominant contribution to its increase arises from aggregates of comparable size. In this approximation, the sums in the rate equation are reduced to a single term involving the reaction of two clusters of half of the typical size. Calculating the second moments of the two sides of the resulting equation gives (Leyvraz 1984, Kang *et al* 1986)

$$\frac{dS(t)}{dt} \sim K_{SS} \sim [S(t)]^{2\omega} \,. \tag{8.34}$$

Since $S(t) \sim t^z$ (8.15), one obtains from (8.34)

$$z = \frac{1}{1 - 2\omega} \, . \tag{8.35}$$

Substituting (8.33) into (8.35) we find a relation between the dynamic exponent z, the mobility exponent γ and the fractal dimension D (Kolb 1984)

$$z = \frac{1}{1 - \gamma - (d-2)/D} \, . \tag{8.36}$$

Note that for the Brownian kernel $\gamma = -1/D$, thus for ordinary diffusion-limited CAA in three dimensions the above expression predicts $z = 1$. For $d \geq 2$, (8.36) is in agreement with the simulation and experimental data, although the numerical values show some deviations from (8.36). For example, the simulations give $z \simeq 0.85$ in $d = 3$ for $\gamma = -1$, while $z \simeq 0.7$ from (8.36). The differences in the values provided by the simulations and the predictions of the Smoluchowski equation might be due to an extremely slow crossover to the asymptotic regime. These results are consistent with the conclusion made in the previous section that the upper critical dimension for the kinetics of CCA is $d_c = 2$ (Kang and Redner 1984, Vicsek *et al* 1985).

Reaction-limited CCA of fractal aggregates can be studied by the Smoluchowski approach assuming that the coagulation kernel is of the form (Ball *et al* 1987)

$$K_{ij} \sim i^{2\omega} \qquad \text{if} \qquad i \simeq j \tag{8.37}$$

since the kernel is expected to be proportional to the reaction surface (the number of active sites to be able to form contacts) which was shown to scale with the cluster size (8.6). For clusters of very different masses one assumes

$$K_{ij} \sim i \, j^{2\omega - 1} \qquad \text{if} \qquad i \gg j \, . \tag{8.38}$$

The above expression follows from a picture in which the larger cluster is imagined being made of i/j blobs of mass j. Then, for $D < d$ the smaller cluster can easily penetrate the larger one and the reaction surface is additive over the accessible blobs.

The solution of the Smoluchowski equation corresponding to the above kernel is known (Ernst 1985). i) For $\omega < 1/2$ the cluster-size distribution has a bell shape. ii) If $\omega = 1/2$, $n_s(t)$ decreases as $s^{-\tau}$ with $\tau = 3/2$ up to a cutoff size. iii) For $\omega > 1/2$ the solution can be interpreted in terms of gelation and $n_s(t) \sim s^{-\tau}$ with $\tau = (3+2\omega)/2 > 2$. Note that these solutions qualitatively correspond to the numerical results obtained in the simulations of CCA (see e.g., Fig. 8.7).

There is a mechanism by which the singularity of the solution at $\omega = 1/2$ can stabilize the system through the *self-adjustment of the fractal dimension D* (Ball *et al* 1987). Let us imagine the effect of increasing ω from 1/2. According to the previous paragraph this leads to an increased τ, i.e., to a greater number of smaller clusters which are able to penetrate within the larger ones. This dominant reaction will increase the fractal dimension of the large clusters. In turn, ω will be decreased since the number of active sites is less for more compact objects. However, if ω is decreased from 1/2, $n_s(t)$ becomes nearly monodisperse. Then the clusters of nearly equal size will interpenetrate substantially less than for $\omega = 1/2$, and the resulting clusters will be more open having a smaller D leading to an increase of ω. Therefore, the system can adjust D to force ω to be equal to 1. This qualitative argument is supported by stability analysis and experimental results.

The Smoluchowski equation can be used to obtain relations among the exponents describing the scaling behaviour of *steady-state CCA*. Feeding single particles into the system can be represented by the term $k\delta_{s1}$, while the elimination of clusters containing more than s_r particles means that $n_s(t) = 0$ for $s > s_r$. With these changes (8.30) has the form

$$dn_s(t)/dt = k\delta_{s1} + 1/2 \sum_{i+j=s\leq s_r} K_{ij}n_in_j - n_s \sum_{j=1}^{s_r} K_{js}n_j \,. \tag{8.39}$$

The above rate equation represents a special case of the more general equation (Rácz 1985a)

$$dn_s(t)/dt = k\delta_{s1} - G_s(n_1, n_2, ..., n_r)\,, \tag{8.40}$$

where G_s is a homogeneous function of degree δ

$$G_s(\lambda n_1, \lambda n_2, ..., \lambda n_r) = \lambda^{1/\delta} G_s(n_1, n_2, ..., n_r) \tag{8.41}$$

with $\delta = 1/2$. The scaling analysis of (8.40) for general δ can be carried out by rescaling the time and the cluster-size distribution

$$\tilde{t} = k^{1-\delta} t, \qquad \tilde{n}_s(\tilde{t}) = \frac{n_s(t)}{k^\delta} . \tag{8.42}$$

Using the above variables k is eliminated from (8.40), and this means that the solution for large t can be written in the form

$$n_s(t) = k^\delta \phi_s(k^{1-\delta} t) \tag{8.43}$$

which is exact only if $n_j(0) \sim k^{1/\delta}$. However, one expects that the steady-state properties are insensitive to the initial conditions. Accordingly, for the total number of clusters, $\sum_s n_s(t)$, one has

$$n(t) = k^\delta \phi(k^{1-\delta} t) . \tag{8.44}$$

This relation is in agreement with the scaling form (8.24) suggested by the simulations with (Rácz 1985a)

$$\delta + \Delta = 1 . \tag{8.45}$$

Since for zero feeding rate $n(t) \sim t^{-z}$ (8.16), $\phi(x)$ must behave for $x \ll 1$ as x^{-z} and from (8.44) we obtain another scaling law

$$\delta = z\delta . \tag{8.46}$$

For the original Smoluchowski equation $\delta = \Delta = 1/2$ and $z = 1$.

Finally, we shall use the rate equation approach to obtain a relation among the exponents describing *reversible CCA*. In the stationary limit the left-hand side of the Smoluchowski equation (8.30) vanishes, and with the breakups taken into account it can be written as

$$0 = \frac{1}{2} \sum_{i+j=s} [K_{ij} n_i n_j - F_{ij} n_{i+j}] - \sum_{j=1}^{\infty} [K_{sj} n_s n_j - F_{sj} n_{s+j}] , \tag{8.47}$$

where F_{ij}, the breakup rate (8.27), is assumed to have the scaling property

$$F_{\lambda i \lambda j} = \lambda^{\alpha} F_{ij} . \tag{8.48}$$

This type of scaling is satisfied by most of the physically relevant forms of F_{ij}. To obtain an expression for the exponent y defined by (8.28), we also assume that the rate equation (8.47) is invariant under the scaling transformation $h \to \lambda h$ and $s \to \lambda^y s$. Then using (8.29), (8.32) and (8.48) in (8.47) one gets (Family *et al* 1986)

$$y = \frac{1}{\alpha - 2\omega + 2} . \tag{8.49}$$

This relation is supported by explicit results for fragmentation models satisfying detailed balance (Ernst and van Dongen 1987) and by numerical simulations (Section 8.2.3) of the particle coalescence model in which $y \simeq 0.66$ was obtained for $\omega = 0$ and $\alpha = -1/2$.

8.3. EXPERIMENTS ON CLUSTER-CLUSTER AGGREGATION

The aggregation of clusters of particles can take place in a wide variety of experimental conditions. For example, the first quantitative analysis of the fractal nature of cluster-cluster aggregates was carried out for clusters of metallic smoke particles formed in air. If the particles interact through an attractive force and the aggregates are mobile, the resulting structures normally have fractal properties. The key point is that the bond which is formed between two particles has to be more or less *rigid*. Otherwise, surface diffusion and evaporation lead to considerable restructuring of the aggregates terminating in simple shapes characteristic for equilibrium morphologies. In most of the cases the relative stiffness of bonds is provided by the size of the aggregating particles: atoms or small molecules are usually mobile on the surface of a growing object, while microscopic particles consisting of a large number of atoms tend to stick to each other rigidly. Colloidal suspensions of metallic or other particles are the most typical systems of this kind. Polymer molecules can also form stiff bonds.

Two approaching particles do not stick necessarily, even if there exists a short range attractive force between them, because a repulsive barrier, V_r, in the interaction potential may prevent the particles from forming a bond. The probability of sticking is proportional to $p_s \sim \exp(-V_r/k_BT)$, and depending on the value of V_r, the aggregation is *diffusion-limited* ($p_s \simeq 1$) or *reaction-limited* ($p_s \ll 1$). A common source for the repulsion term is the electric charge accumulated on the surface of the particles. This charge can be compensated by adding appropriate substances, and the diffusion and reaction-limited regimes can be studied in the same colloidal system.

8.3.1. Structure

Gold colloids are particularly suitable for studying Brownian clustering phenomena. This was recognized by Faraday who studied their stability and resistance to aggregation. The main advantages of using gold colloids are the following (Weitz and Oliveria 1984a, 1984b). i) The freshly made suspension typically consists of highly uniform spherical particles with a size distribution characterized by a root-mean-square deviation of about 10%. ii) the particles stick irreversibly since gold metal bonding is likely to occur at the point two spheres touch. iii) the rate of aggregation can easily be controlled by adding pyridine to the system. Finally, iv) the structure of gold particles can be examined by transmission electron-microscope (TEM) techniques, because they give images with high contrast and do not suffer from charging problems when using an electron beam.

The standard recipe to make gold colloids is the reduction of a gold salt $Na(AuCl_4)$ with sodium citrate. In a typical experiment the mean diameter of particles is about 15 nm, and they are separated at the beginning by about 60 particle diameters corresponding to a volume fraction of $\sim 10^{-6}$.

The diffusion constant associated with the particles is approximately $5 \times 10^{-7} \mathrm{cm}^2/\mathrm{sec}$, which results in a diffusion time $\sim$ 10msec on a length scale of the interparticle distance.

In the course of their formation the gold spheres become covered by citrate ions, creating a large negative surface charge. The ions in the solution produce a Debye-Hückel screening length of the same order as the particle diameter. The resulting short range repulsive interaction makes the colloids rather stable against aggregation. However, it is possible to eliminate the negative charge on the surface of the particles by adding neutral pyridine molecules to the solution which being absorbed on the surface of the gold spheres displace the negative citric ions. As a result aggregation times ranging from several minutes (fast or diffusion-limited CCA) to several weeks (slow or reaction-limited CCA) can be realized.

The real space visualization of *three-dimensional gold colloid* aggregates is achieved by preparing TEM grids using samples of the solution at several points in time (Weitz and Oliveria 1984). The TEM grids consist of an approximately 20nm thick carbon film supported by a copper grid. As the fluid evaporates the surface tension pulls the aggregates straight down to the grid, i.e., "flattens" them causing only a small distortion of the true geometric *projection*. Therefore, in this approach it is the two-dimensional projection of the three-dimensional clusters which is used to obtain information about the structure. According to rule a) in Section 2.2, the fractal dimension of an object projected onto a plane is equal to that of the original one if $D < 2$. In the case $D > 2$, the projected structure does not exhibit scale invariance.

During *fast* aggregation the particles stick at the first time they collide. Figure 8.14 shows representative pictures of aggregates taken from a single grid. Although these pictures are two-dimensional projections, they have open, ramified geometry and the number of areas corresponding to overlapping particles is relatively small. Thus we can assume that the aggregates are fractals with a dimension smaller than 2. In fact, the scale invariance of the structure of aggregates is nicely demonstrated by the fact that the properly magnified small clusters have an overall appearance similar to the shape of the largest aggregate. In order to obtain an estimate for the fractal dimension of the aggregates, one can measure the number of particles N in a cluster as a function of the linear size L. For a fractal of dimension D one expects $N \sim L^D$ (2.2). In Fig. 8.15 $\ln N$ is plotted against $\ln L$, and the straight line giving the best fit to the data indicates that the structure of gold colloid aggregates is characterized by a fractal dimension $D \simeq 1.75$ in good agreement with the computer simulations of diffusion-limited CCA yielding essentially the same value.

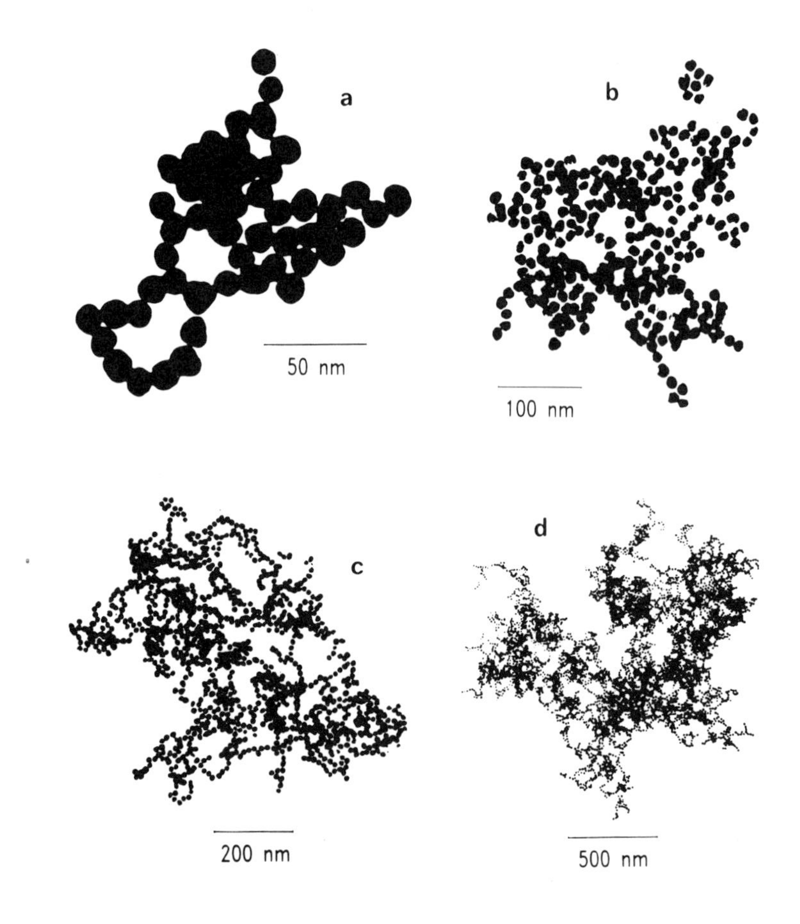

Figure 8.14. Transmission electron micrographs of gold colloid aggregates. These projected images correspond to three-dimensional clusters of various sizes. The largest aggregate consists of 4739 spherical gold particles (Weitz and Oliveria 1984b).

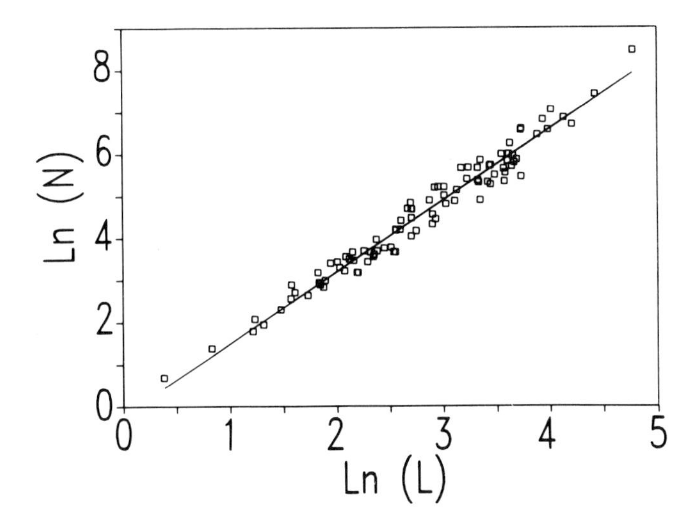

Figure 8.15. Determination of the fractal dimension of gold colloid aggregates by plotting the number of particles in the clusters versus their linear size L. The slope gives $D \simeq 1.75$ (Weitz and Oliveria 1984a).

During reaction-limited or *slow* aggregation the particles form a bond with a small probability (they collide many times before sticking). In addition to gold colloids (Weitz *et al* 1985), this limit can be investigated using silica particles of diameter $\simeq 27$ Å(Schaefer *et al* 1984). Again, the surface charge has to be reduced to provide a sticking probability larger than 0, but much less than 1. This can be accomplished by reducing the pH to 5.5 in the solution containing silica monomers, while increasing the salt concentration to ≥ 0.5 M.

To analyze the structure of the growing silica aggregates (Schaefer *et al* 1984) one can also use light scattering and small angle x-ray scattering (SAXS), as an alternative to transmission electron microscopy. In Section 4.1 it has been shown that scattered intensity from a fractal of dimension D scales with the wave number of the radiation, q, as $I(q) \sim q^{-D}$. This relation is valid for wave numbers corresponding to linear sizes larger than the diameter of the particles and smaller than the linear size of the aggregates.

Obviously, it can be applied to an ensemble of clusters as well, if the size distribution of aggregates is approximately monodisperse (for diffusion-limited dynamics this condition is usually satisfied). Even if the cluster-size distribution is not monodisperse, the scattered intensity at any q is dominated by the contributions from the largest clusters if $\tau < 2$, a condition which is satisfied as well (see Section 8.2.1). An important advantage of the scattering techniques is that they allow *in situ* measurement of the geometrical properties, i.e., there is no need for sample preparation before the application of the method.

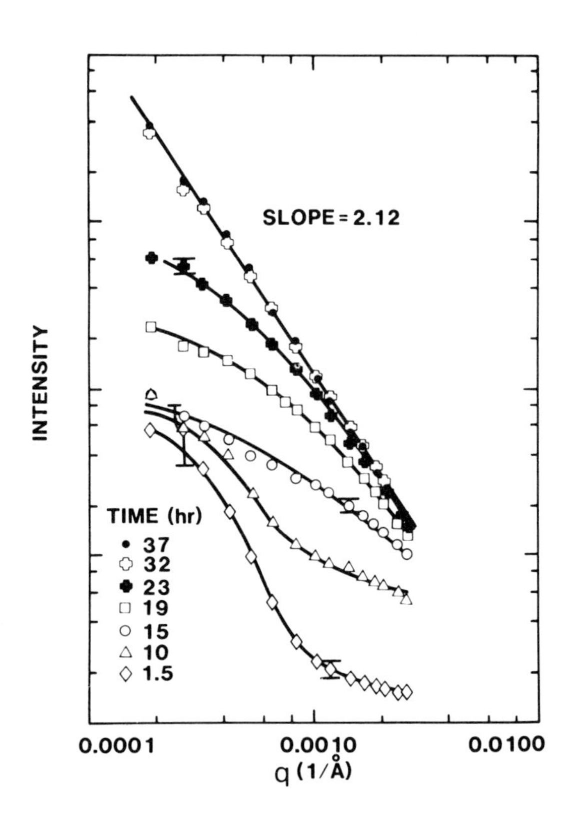

Figure 8.16. Scattered light intensity profiles for various times. The growth of fractal silica aggregates is indicated by the uppermost series of data corresponding to a fractal dimension of $D \simeq 2.12$ (Schaefer *et al* 1984).

Figure 8.16 shows the temporal development of the scattered light intensity. The non-trivial time dependence of the data is explained by the observation that after a relatively short time a few large clusters were visible through a telescope in the scattering volume. These clusters are responsible for the relatively quick increase of the intensity for small q. The error bars decrease as the aggregation proceeds, and the maximum grows two decades in intensity until 37 h after initiation. $I(q)$ behaves as a power law in the range 5000Å$>q^{-1}>$500Å. The slope on this log-log plot corresponds to a fractal dimension $D \simeq 2.1$ which agrees well with the simulation results (for the polydisperse case the reaction-limited CCA model gives $D \simeq 2.1$ as well). The combination of light scattering with SAXS makes it possible to demonstrate that the fractal scaling extends over two decades.

Due to the gravitational force large aggregates growing in three dimensions leave the active volume of the system because of sedimentation. This problem does not arise in *two-dimensional* systems. Carrying out experiments on $2d$ CCA has additional advantages. i) It makes the visualization of the results much easier. This is illustrated by Fig. 8.17, where ordinary photographs taken from aggregates of carbon particles floating on water are shown (Horkai and Bán 1988). ii) Two-dimensional aggregation takes place on surfaces and such processes are interesting from a practical point of view as well. iii) Changing the properties of the interface the aggregation process can be controlled. Finally, the high density limit of CCA can be realized and studied easier.

One of the common ways to study diffusion-limited cluster-cluster aggregation on a surface is to use a fluid–air interface where the charged colloidal particles are trapped by surface tension. For example, a fresh suspension of silica micropheres of diameter 3000 Å can be dispensed onto the flat surface of water with the simultaneous injection of a methanol spreading agent. In this experiment the electrostatic repulsion of particles is screened by adding salt (1.0N CaCl) to the water. The aggregation process can be followed by optical observations and making photomicrographs. The obtained pictures are suitable for digital image processing, and the methods discussed in Section 4.2 can provide estimates for the fractal dimension of the aggregates.

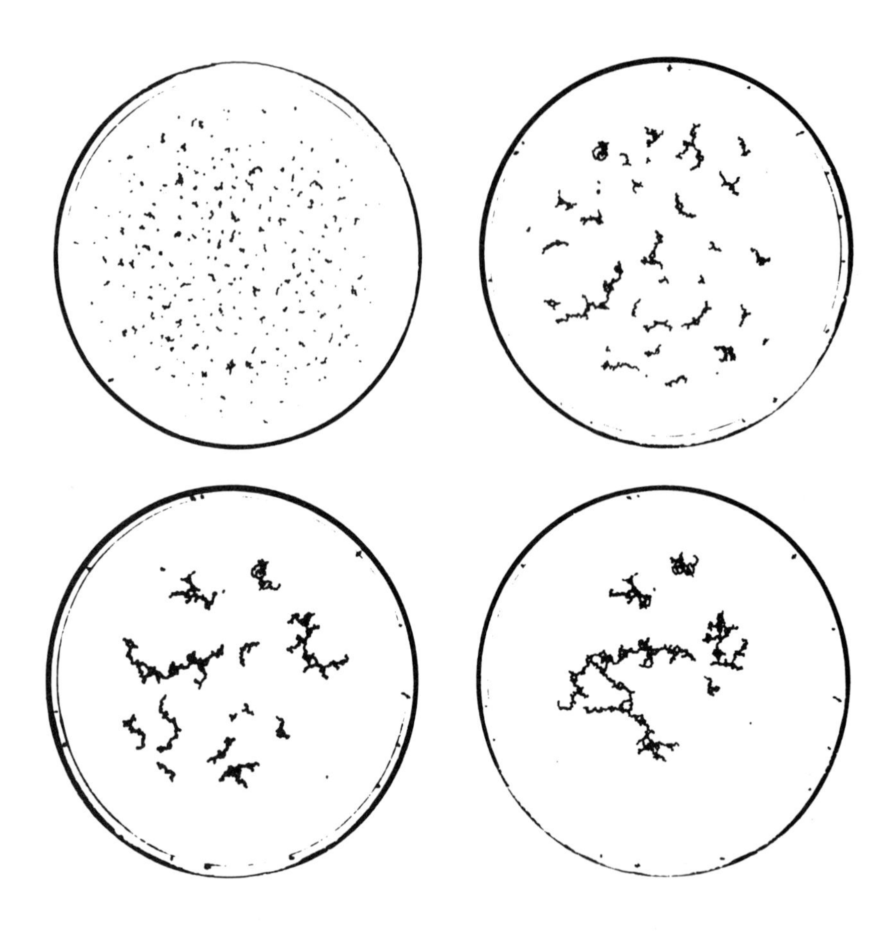

Figure 8.17. Snapshots of carbon particle aggregates illustrating two-dimensional CCA. Since the diameter of the particles used in this experiment was relatively large (approximately 0.5mm), surface tension effects dominated the process and the clusters moved along essentially straight lines rather than Brownian trajectories (Horkai and Bán 1988).

From the log-log plot of the radius of gyration versus the number of particles in the individual aggregates the estimate 1.20 ± 0.15 was obtained for the fractal dimension of the silica clusters (Hurd and Schaefer 1985). This value is considerably smaller than $D \simeq 1.43$ which was calculated for computer simulated cluster-cluster aggregates. The discrepancy can be interpreted using the Debye-Hückel theory for the calculation of the electric field in the vicinity of a small aggregate. The estimates for a particle approaching a dimer show that the barrier for end-on approach is much lower than for particles trying to stick to the side. Thus, a growing cluster has a tendency not to branch when there are repulsive forces present. This effect is expected to decrease the fractal dimension, at least for sizes below the asymptotic limit. An analogous decrease in D was observed in the simulations of chain-chain CCA, where side-branching is entirely prohibited.

The complexity of the situation is illustrated by the fact that there are two more fractal dimension values which can be observed in two-dimensional CCA. Aggregation of polystyrene spheres of diameter 4.7μm confined between two glass plates led to fractal structures with $D \simeq 1.49$ (Skjeltorp 1987). In this system two touching clusters are likely to rotate around the point of contact until at least three particles touch simultaneously. This rearrangement seems to eliminate the above discussed anisotropy of sticking probability and the result is in a better agreement with $D \simeq 1.48$ obtained from the simulations of the corresponding CCA model. Finally, a fractal dimension $D \simeq 1.7$ was observed in the experiments on the two-dimensional aggregation of polyvynil toluene (Armstrong *et al* 1986). To interpret this value we recall that investigations of the diffusion-limited CCA model with relatively high densities showed that structures with a fractal dimension about 1.7 are formed in the system close to its gelation point, i.e., when a cluster spanning the whole cell is formed.

8.3.2. Dynamics

Some of the predictions of the dynamic scaling theory for the cluster-size distribution, n_s, described in Section 8.2.1 can be checked by carrying out quasielastic light scattering experiments (Weitz *et al* 1984, Feder *et al* 1984). Such measurements are made while the aggregation goes on, thus the time dependence of the growth process can be conveniently monitored. According to the standard theory of dynamic light scattering (Berne and Pecoria 1976) the first cumulant of the autocorrelation function of the scattered light is given by

$$K_1 = \frac{1}{I(q,0)} \int s^2 n_s(t) S_s(q)(D_s q^2 + A) ds\,, \tag{8.50}$$

where q is the scattering vector, D_s and A are respectively the translational and rotational diffusion constants of clusters of mass s, and $I(q,0)$ is the time averaged total scattered intensity. The structure factor of an s-cluster depends on the fractal dimension D in the form $S_s(q) \sim s^{-1}q^{-D}$ (see Section 4.1). For fractal aggregates the dominant contribution to the decay of the intensity autocorrelation function comes from the translational term in which the cluster diffusivity can be estimated as $D_s \sim R_s^{-1} \sim s^{-1/D}$, where R_s is the radius of a cluster consisting of s particles. In this approximation the first cumulant becomes a moment of the cluster-size distribution function

$$K_1 \sim \int s^{1-1/D} n_s(t) ds \, . \tag{8.51}$$

The analysis of similar expressions in Section 8.2.1 showed that for $\tau < 2-1/D$ the above integral gives an estimate for $R^{-1}(t) \sim S^{-1/D}(t) \sim t^{-z/D}$, where $R(t)$ is the average cluster radius at time t and $S(t)$ denotes the mean cluster size. Here the exponents τ and z describe the static and dynamic scaling of the cluster-size distribution which scales as $n_s \sim s^{-\tau}$ up to a cutoff at sizes proportional to t^z. For diffusion-limited CCA in $d = 3$ it was also shown that $z \simeq 1$, therefore, we expect

$$(K_1)^{-1} \sim R(t) \sim t^{1/D} \, . \tag{8.52}$$

Gold colloids represent suitable systems for studying aggregation kinetics as well (Weitz *et al* 1984). In Fig. 8.18, $\ln R(t)$ determined from measuring K_1 for gold colloids is plotted versus the time variable t. The straight line fitted to the data has a slope of 0.56 which corresponds to a fractal dimension $D \simeq 1.79$ being close to the value obtained from simulations and independent experiments. The agreement supports the dynamic scaling assumption (8.18) for $n_s(t)$.

For reaction-limited aggregation the dynamics of growth is qualitatively different (Weitz *et al* 1985). Instead of scaling with time according to an exponent that is less than 1, the behaviour of the mean radius is better described by an exponential increase $R \sim e^{Ct}$, where C depends on the experimental conditions. If the repulsion between the gold particles is partially compensated, the dynamics is initially exponential (slow) which crosses

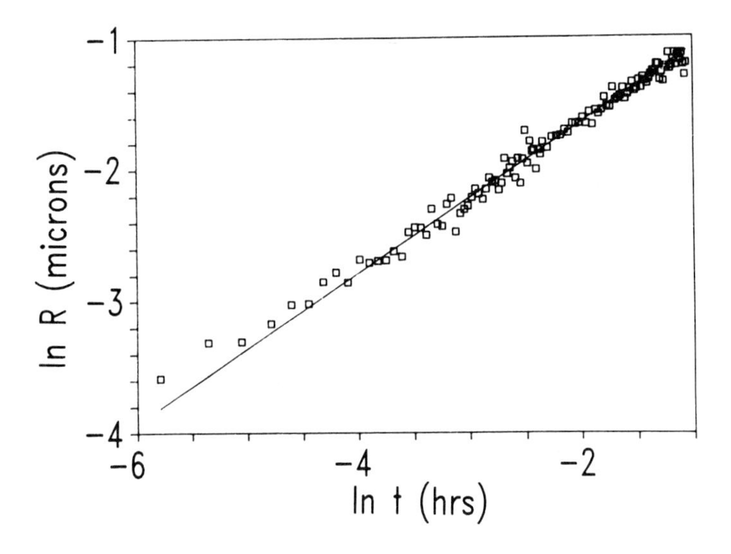

Figure 8.18. Scaling of the mean cluster radius of gold colloid aggregates with time. The slope indicates a fractal dimension $1/0.56 \simeq 1.79$ (Weitz *et al* 1984).

over to exhibit the behaviour characteristic for diffusion-limited aggregation. These regimes are demonstrated in Fig. 8.19. The initially exponential increase of $R(t)$ and its subsequent crossover to a power law growth is consistent with the simulation results shown in Fig. 8.9, where the total number of clusters $N(t)$ is plotted for a small constant sticking probability. For $\tau < 1$ one has $R(t) \sim [N(t)]^{-1/D}$ and indeed, $N(t)$ in Fig. 8.9 behaves as expected from this relation and Fig. 8.19.

A direct test of the dynamic scaling for $n_s(t)$ can be carried out by analysing TEM images of the clusters of gold particles on TEM grids prepared at several times during the aggregation process (Weitz and Lin 1986). The number of particles in the aggregates is counted, and the data are compiled in histograms. Two kinds of plots have been obtained: i) $n_s(t)$ exhibits a reasonably well-defined peak as a function of s for diffusion-limited CCA, while ii) it decays as a power law for reaction-limited aggregation. In addition, the cluster-size distributions can be scaled onto a single curve in both cases, which indicates dynamic scaling of the form (8.18).

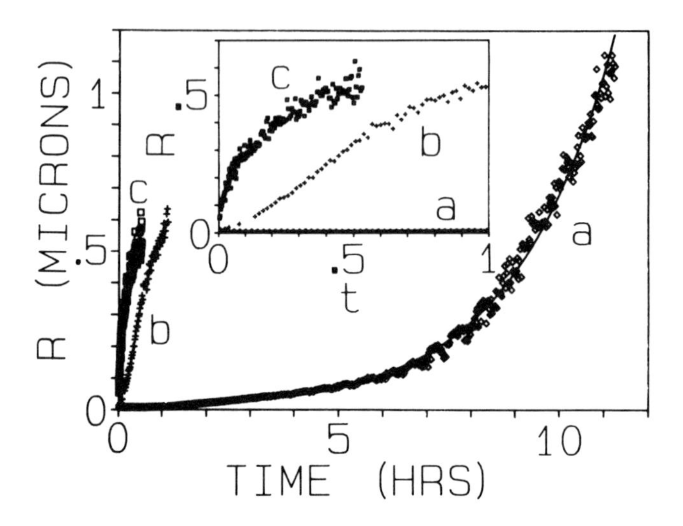

Figure 8.19. Increase of the characteristic cluster size during the aggregation of gold particles. Three kinds of kinetics can be observed: (curve a) reaction-limited, (curve b) crossover, and (curve c) diffusion-limited. The inset shows the initial behaviour on an expanded scale. The reaction-limited kinetics can be well approximated with an exponential (Weitz *et al* 1985).

Dynamic scaling can also be observed in other aggregating systems. For example, in dilute solutions of polyfunctional monomer molecules, an ensemble of branched polymers is formed as smaller molecules are linked through new chemical bonds. A standard technique to monitor the process of polymerization is infrared spectroscopy which gives the number of bonds in the system, instead of the number of monomers. Because of this property of the method the formalism elaborated in 8.2.1 has to be modified (Djordjevic *et al* 1986). To account for the fact that only the bonds are detected experimentally, the system should be characterized by $n_b(t)$, which is the number of clusters with b chemical bonds in them. Since the total number of bonds is not a conserved quantity (it grows permanently), the scaling form for

$$n_b(t) \sim b^{-\theta} f(b/t^z) \tag{8.53}$$

is expected to hold with $\theta = 1.5$ instead of $\theta = 2$, where the latter relation was shown (Section 8.2.1) to be a consequence of mass conservation.

REFERENCES (PART II)

Aharony, A., 1985 in *Scaling Phenomena in Disordered Systems* edited by R. Pynn and A. Skjeltorp (Plenum Press, New York) p. 289

Alexander, S. and Orbach, R., 1982 *J. Physique Lett.* **43**, 2625

Alexandrowitz, Z., 1980 *Phys. Lett.* **80A**, 284

Amit, D. J., Parisi, G. and Peliti, L., 1983 *Phys. Rev.* **B27**, 1635

Amitrano, C., Coniglio, A. and di Liberto, F., 1986 *Phys. Rev. Lett.* **57**, 1016

Armstrong, A. J., Mocklet, R. C. and O'Sullivan, W. J., 1986 *J. Phys.* **A19**, L123

Balberg, I., 1988 *Phys. Rev.* **B37**, 2391

Ball, R. C. and Witten, T. A., 1984a *Phys. Rev.* **A29**, 2966

Ball, R. C. and Witten, T. A., 1984b *J. Stat. Phys.* **36**, 873

Ball, R. C., Brady, R. M., Rossi, G. and Thompson, B. R., 1985 *Phys. Rev. Lett.* **55**, 1406

Ball, R. C., 1986 *Physica* **140A**, 62

Ball, R. C., Weitz, D. A., Witten, T. A. and Leyvraz, F., 1987 *Phys. Rev. Lett.* **58**, 274

Bansil, R., Herrmann, H. J. and Stauffer, D., 1984 *Macromolecules* **17**, 998

Barrow, J. D., 1981 *J. Phys.* **A14**, 729

Berne, B. J., and Pecora, R., 1976 *Dynamic Light Scattering* (Wiley, New York)

Binder, K., 1967 *Ann. Phys. (N.Y.)* **98**, 390

Botet, R., Jullien, R. and Kolb, M., 1984 *J. Phys.* **A17**, L75

Botet, R. and Jullien, R., 1984 *J. Phys.* **A17**, 2517

Botet, R. and Jullien, R., 1985 *Phys. Rev. Lett.* **55**, 1943

Brown, W. D. and Ball, R. C., 1985 *J. Phys.* **A18**, L517

Bunde, A., Herrmann, H., J., Margolina, A. and Stanley, H. E., 1985a *Phys. Rev. Lett.* **55**, 653

Bunde, A., Herrmann, H. J. and Stanley, H. E., 1985b *J. Phys.* **A18**, L523

Cardy, J. L. and Grassberger, P., 1985 *J. Phys.* **A18**, L267

Chandler, R., Koplik, J., Lerman, K. and Willemsen, J., 1982 *J. Fluid Mech.* **119**, 249

Chhabra, A., Matthews-Morgan, D., Landau, D. P. and Herrmann, H. J., 1984 in *Kinetics of Aggregation and Gelation* edited by F. Family and D. P. Landau (North-Holland, Amsterdam)

Chhabra, A., Landau, D. P. and Herrmann, H. J., 1986 in *Fractals in Physics* edited by L. Pietronero and E. Tossati (North-Holland, Amsterdam)

Dhar, D., 1985 in *On Growth and Form* edited by H. E. Stanley and N. Ostrowsky (Martinus Nijhoff, Dordrecht) p. 288

Debierre, J. M. and Turban, L., 1986 *J. Phys.* **A19**, L131

Debierre, J. M. and Turban, L., 1987 *J. Phys.* **A20**, L259

Djordjevic, Z. B., Djordjevic, Z. V. and Djordjevic, M. B., 1986 *Phys. Rev.* **A33**, 3614

Eden, M., 1961 in *Proc. 4-th Berkeley Symp. on Math. Statistics and Probability*, Vol. 4., Ed. F. Neyman (Berkeley, University of California Press)

Edwards, S. F. and Wilkinson, D. R., 1982 *Proc. Royal Soc. London* **A381**, 17

Ernst, M. H., 1985 in *Fundamental Problems in Statistical Mechanics VI* edited by E. G. D. Cohen (Elsevier, Amsterdam) p. 329

Ernst, M. H. and van Dongen, P. G. J., 1987 *Phys. Rev.* **A36**, 435

Family, F., Vicsek, T. and Meakin, P., 1985a *Phys. Rev. Lett.* **55**, 641

Family, F. and Vicsek, T., 1985 *J. Phys.* **A18**, L75

Family, F., Meakin, P. and Vicsek, T., 1985b *J. Chem. Phys.* **83**, 4144

Family, F., 1986 *J. Phys.* **A19**, L441

Family, F., Meakin, P. and Deutch, J. M., 1986 *Phys. Rev. Lett.* **57**, 727

Family, F. and Hentschel, 1987 *Faraday Discuss. Chem. Soc.* **83** paper 6

Family, F., Platt, D. and Vicsek, T., 1987 *J. Phys.* **A20**, L1177

Feder, J., Jossang, T. and Rosenquist, E., 1984 *Phys. Rev. Lett.* **53**, 1403

Fisher, M. E., 1967 *Rep. Prog. Phys.* **30**, 615

Flory, P., 1971 *Principles of Polymer Chemistry* (Cornell Univ. Press, Ithaca)

Forrest, S. R. and Witten, T. A., 1979 *J. Phys.* **A12**, L109

Freche, P., Stauffer, D. and Stanley, H. E., 1985 *J. Phys.* **A18**, L1163

Friedlander, S. K., 1977 *Smoke, Dust and Haze: Fundamentals of Aerosol Behaviour* (Wiley, New York)

Garik, P., Richter, R., Hautman, J. and Ramanlal, P., 1985 *Phys. Rev.* **A32**, 3156

Gefen, Y., and Aharony, A. and Alexander, S., 1983 *Phys. Rev. Lett.* **50**, 77

de Gennes, P. G., 1979 *Scaling Concepts in Polymer Physics* (Cornell Univ. Press, Ithaca)

Gould, H., Family, F. and Stanley, H. E., 1983 *Phys. Rev. Lett.* **50**, 686

Grassberger, P., 1983 *Math. Biosci.* **62**, 157

Grassberger, P., 1985 *J. Phys.* **A18**, L215

Guyer, R. A., 1984 *Phys. Rev.* **A30**, 1112

Halsey, T. C. and Meakin, P., 1985 *Phys. Rev.* **A32**, 2546

Halsey, T. C., Meakin, P. and Procaccia I., 1986 *Phys. Rev. Lett.* **56**, 854

Halsey, T. C., 1987 *Phys. Rev. Lett.* **59**, 2067

Havlin, S. and Ben-Avraham, D., 1987 *Advances in Physics* **36**, 695

Hayakawa, Y., Sato, S. and Matsushita, M., 1987 *Phys. Rev.* **A36**, 1963

Herrmann, H. J., Landau, D. P. and Stauffer, D., 1982 *Phys. Rev. Lett.* **49**, 412

Herrmann, H. J. and Stanley, H. E., 1984 *Phys. Rev. Lett.* **53**, 1121

Herrmann, H. J. and Stanley, H. E., 1985 Z. Physik **60**, 165

Herrmann, H. J., 1986 *Phys. Rep.* **36**, 153

Honda, K., Toyoki, H. and Matsushita, M., 1986 *J. Phys. Soc. Jpn.* **55**, 707

Horkai, F. and Bán, S., 1988 unpublished

Hurd, A. J., and Schaefer, D. W., 1985 *Phys. Rev. Lett.* **54**, 1043

Joag, P. S., Limaye, A. V. and Amritkar R. E., 1987 *Phys. Rev.* **A36**, 3395

Jullien, R. and Botet, R., 1985 *J. Phys.* **A18**, 2279

Jullien, R. and Kolb, M., 1984 *J. Phys.* **17**, L639

Jullien, R. and Meakin, P., 1987 *Europhys. Lett.* **4**, 1385

Jullien, R., 1984 *J. Phys.* **A17**, L771

Kang, K. and Redner, S., 1984 *Phys. Rev.* **A30**, 2833

Kang, K., Redner, S., Meakin, P. and Leyvraz, F., 1986 *Phys. Rev.* **A33**, 1171

Kapitulnik, A., Aharony, A., Deutscher, G. and Stauffer, D., 1983 *J. Phys.* **A16**, L269

Kardar, M., Parisi, G. and Zhang, Y. C., 1986 Phys. Rev. Lett **56**, 889

Kardar, M. and Zhang, Y. C., 1987 *Phys. Rev. Lett.* **58**, 2087
Kerstein, A. R. and Bug, A. L., 1984 unpublished
Kertész J. and Wolf, D. E., 1988 *J. Phys.* **A21**, 747
Kolb, M., Botet, R. and Jullien, R., 1983 *Phys. Rev. Lett.* **51**, 1123
Kolb, M., 1984 *Phys. Rev. Lett.* **53**, 1653
Kolb, M., 1985 *J. Physique Lett.* **46**, L631
Kolb, M. and Herrmann, H. J., 1987 *Phys. Rev. Lett.* **59**, 454
Kremer, K. and Lyklema, J. W., 1985a *Phys. Rev. Lett.* **55**, 2091
Kremer, K. and Lyklema, J. W., 1985b *Phys. Rev. Lett.* **54**, 267
Krug, J., 1987 *Phys. Rev.* **A36**, 5465
Leamy, H. J., Gilmer, G. M. and Dirks, A. G., 1980 in *Current Topics in Mat. Sci.* **6**, (North-Holland, Amsterdam) p. 309
Leath, P. L., 1976 *Phys. Rev.* **B14**, 5046
Lenormand, R. and Bories, S., 1980 *C. R. Acad. Sci. (Paris)* **291**, 279
Leyvraz, F., 1984 *Phys. Rev.* **A29**, 854
Leyvraz, F., 1985 *J. Phys.* **A18**, L941
Liang, S. and Kadanoff, L. P., 1985 *Phys. Rev.* **A31**, 2628
Lyklema, J. W. and Kremer, K., 1984 *J. Phys.* **A17**, L691
Lyklema, J. W., 1985 *J. Phys.* **A18**, L617
Lyklema, J. W. and Evertsz, C., 1986 in *Fractals in Physics* edited by L. Pietronero and E. Tossati (North-Holland, Amsterdam) p. 87
Madelaine, G. J., Perrin, M. L. and Itox, M., 1979 *J. Aerosol Sci.* **12**, 202
Majid, I., Jan, N., Coniglio, A. and Stanley, H. E., 1984 *Phys. Rev. Lett.* **52**, 1257
Makarov, N. G., 1985 *Proc. London Math. Soc.* **51**, 369
Manneville, P. and de Seze, L., 1981 in *Numerical Methods in the Study of Critical Phenomena* edited by I. Della Dora, J. Demongeot and B. Lacolle (Springer, New York)
Matsushita, M., Hayakawa, Y. and Sawada, Y., 1985 *Phys. Rev.* **A32**, 3814
Matsushita, M., Honda, K., Toyoki, H., Hayakawa, Y. and Kondo, H., 1986 *J. Phys. Soc. Jpn.* **55**, 2618
McKane, A. J. and Moore, M. A., 1988 *Phys. Rev. Lett.* **60**, 527
McMurry, P. H., *J. Colloid Interface Sci.* **78**, 513
Meakin, P., 1983a *Phys. Rev.* **A26**, 1495
Meakin, P., 1983b *Phys. Rev.* **A27**, 2616

Meakin, P., 1983c *Phys. Rev. Lett.* **51**, 1119

Meakin, P., 1984 *Phys. Rev.* **B30**, 4207

Meakin, P., 1985a *J. Phys.* **A18**, L661

Meakin, P., 1985b *J. Colloid Interface Sci.* **105**, 240

Meakin, P., 1985c in *On Growth and Form* edited by H. E. Stanley and N. Ostrowsky (Martinus Nijhoff, Dordrecht) p. 111

Meakin, P. and Vicsek, T., 1985 *Phys. Rev.* **A32**, 685

Meakin, P., Leyvraz, F. and Stanley, H. E., 1985a *Phys. Rev.* **A31**, 1195

Meakin, P., Vicsek, T. and Family, F., 1985b *Phys. Rev.* **B31**, 564

Meakin, P., Coniglio, A., Stanley, H. E. and Witten, T. A., 1986a *Phys. Rev.* **A34**, 3325

Meakin, P., Ramanlal, P., Sander, L. M. and Ball, R.C., 1986b *Phys. Rev.* **A34**, 5091

Meakin, P., Ball, R. C., Ramanlal, P. and Sander, L. M., 1987 *Phys. Rev.* **A35**, 5233

Meakin, P. and Vicsek, T., 1987 J. Phys. **A20** L171

Meakin, P., 1987a *Phys. Rev.* **A35**, 2234

Meakin, P., 1987b in *Phase Transitions and Critical Phenomena* Vol. 12 edited by C. Domb and J. L. Lebowitz (Academic Press, New York)

Meakin, P. and Havlin, S., 1987 *Phys. Rev.* **A36**, 4428

Meakin, P. and Jullien, R., 1987 *J. Physique* **48**, 1651

Meakin, P., Kertész, J. and Vicsek, T., 1988 *J. Phys.* **A21**, 1271

Medalia, A. I., 1967 *J. Colloid Interface Sci.* **24**, 393

Miyazima, S., Meakin, P. and Family, F., 1987 *Phys. Rev.* **A36**, 1421

Montag, J. L., Family, F., Vicsek, T. and Nakanishi, H., 1985 *Phys. Rev.* **A32**, 2557

Muthukumar, M., 1983 *Phys. Rev. Lett.* **50**, 839

Niemeyer, L., Pietronero, L. and Wiesmann, H. J., 1984 *Phys. Rev. Lett.* **52**, 1033

Nieuwenheuzen, J. M. and Haanstra, H. B., 1966 *Philips Techn. Rev.* **27**, 87

Nagatani, T., 1987a *J. Phys.* **20**, 6603

Nagatani, T., 1987b *Phys. Rev.* **A36**, 5812

Obukhov, S. P. and Peliti, L., 1983 *J. Phys.* **A16** L147

Ohtsuki T. and Keyes, T., 1985 *Phys. Rev.* **A33**, 1223

O'Shaughnessy, B. and Procaccia, I., 1985 *Phys. Rev. Lett.* **54**, 455

Pandey, R. B. and Stauffer, D., 1983 *Phys. Lett.* **95A**, 511

Pietronero, L., 1983 *Phys. Rev.* **B27**, 5887

Plischke, M. and Rácz, Z., 1984 *Phys. Rev. Lett.* **53**, 415

Rácz, Z. and Vicsek, T., 1983 *Phys. Rev. Lett.* **51**, 2382

Rácz, Z., 1985a *Phys. Rev.* **A32**, 1129

Rácz, Z., 1985b *Phys. Rev. Lett.* **55**, 1707

Ramanlal, P. and Sander, L. M., 1985 *Phys. Rev. Lett.* **54**, 1828

Rammal, R. and Toulouse, G., 1983 *J. Physique Lett.* **44**, 13

Rikvold, P. A., 1982 *Phys. Rev.* **A26**, 647

Rudnick, J. and Gaspari, G., 1986 *J. Phys.* **A19**, L191

Saleur, H. and Duplantier, B., 1987 *Phys. Rev. Lett.* **58**, 2325

Sapoval, B., Rosso, M. and Gouyet, J. F., 1985 *J. Physique Lett.* **46**, 149

Schaefer, D. W., Martin, J. E., Wiltzius, P. and Cannel, D. S., 1984 *Phys. Rev. Lett.* **52**, 2371

Skjeltorp, A. T., 1987 *Phys. Rev. Lett.* **58**, 1444

von Smoluchowski, M., 1917 *Z. Phys. Chem.* **92**, 129

Solc, K, 1973 *Macromolecules* **6**, 378

Stanley, H. E., Reynolds, P. J., Redner, S. and Family, F., 1982 in *Real-Space Renormalization* edited by T. Burkhardt and J. M. J. van Leeuwen (Springer, New York)

Stauffer, D., 1985 *Introduction to Percolation Theory* (Taylor and Francis, London)

Sutherland, D. N., 1966 *J. Colloid Interface Sci.* **22**, 300

Sutherland, D. N., 1967 *J. Colloid Interface Sci.* **25**, 373

Szép, J. and Lugosi, M., 1986 *J. Phys.* **A19**, L1109

Tokuyama, M. and Kawasaki, K., 1984 *Phys. Lett.* **100A**, 337

Turkevich, L. A. and Scher, H., 1985 *Phys. Rev. Lett.* **55**, 1026

Turkevich, L. A. and Scher, H., 1986 *Phys. Rev.* **A33**, 786

Vicsek, T., 1983 *J. Phys.* **A16**, L647

Vicsek, T. and Family, F., 1984a *Phys. Rev. Lett.* **52**, 1669

Vicsek, T. and Family, F., 1984b in *Kinetics of Aggregation and Gelation* edited by F. Family and D. P. Landau (North-Holland, Amsterdam) p. 110

Vicsek, T., Meakin, P. and Family, F., 1985 *Phys. Rev.* **A32**, 1122

Vicsek, T., Family, F., Kertész, J. and Platt, D., 1986 *Europhys. Lett.* **2**, 823

Vold, M. J., 1963 *J. Colloid Interface Sci.* **18**, 684

Weinrib, A. and Trugman, S. A., 1985 *Phys. Rev.* **B31**, 2993

Weitz, D. A. and Oliveria, M., 1984a *Phys. Rev. Lett.* **52**, 1433

Weitz, D. A. and Oliveria, M., 1984b in *Kinetics of Aggregation and Gelation* edited by F. Family and D. P. Landau (North-Holland, Amsterdam)

Weitz, D. A., Huang, J. S., Lin, M. Y. and Sung, J., 1984 *Phys. Rev. Lett.* **53**, 1657

Weitz, D. A., Huang, J. S., Lin, M. Y. and Sung, J., 1985 *Phys. Rev. Lett.* **54**, 1416

Weitz, D. A. and Lin, M. Y., 1986 *Phys. Rev. Lett.* **57**, 2037

White, W. H., 1982 *J. Colloid Interface Sci.* **87**, 204

Wiesmann, H. J. and Pietronero, L., 1986 in *Fractals in Physics* edited by L. Pietronero and E. Tossati (North-Holland, Amsterdam) p. 151

Wilkinson, D. and Willemsen, J., 1983 *J. Phys.* **A16**, 3365

Willemsen, J., 1984 *Phys. Rev. Lett.* **52** 2197

Witten, T. A. and Sander, L. M., 1981 *Phys. Rev. Lett.* **47**, 1400

Witten, T. A. and Sander, L. M., 1983 *Phys. Rev.* **B27**, 5686

Witten, T. A., 1985 in *Physics of Finely Divided Matter* edited by N. Boccara and M. Daoud (Springer, New York) p. 212

Wolf, D. E., 1987 *J. Phys.* **A20**, 1251

Wolf, D. E. and Kertész, J., 1987a *J. Phys.* **A20**, L257

Wolf, D. E. and Kertész, J., 1987b *Europhys. Lett.* **4**, 561

Zabolitzky, J. G. and Stauffer, D., 1986 *Phys. Rev.* **A34**, 1523

Ziff, R. M., Cummings, P. T. and Stell, G., 1984 *J. Phys.* **A17**, 3009

Ziff, R. M. and McGrady, E. D., 1985 *J. Phys.* **A18**, 3027

Ziff, R. M., McGrady, E. D. and Meakin, P., 1985 *J. Chem. Phys.* **82**, 5269

Ziff, R. M., 1986 *Phys. Rev. Lett.* **56**, 545

Part III

FRACTAL PATTERN FORMATION

Chapter 9

COMPUTER SIMULATIONS

The formation of complex patterns by moving unstable interfaces is a common phenomenon in many fields of science and technology. During *pattern formation* in real systems the *surface tension* of the boundary between the growing and the surrounding phases plays an important role. This is in contrast to the case of cluster growth models discussed in Part II, where such effects were not taken into account. Thus we shall use the term pattern formation for growth processes in which surface tension is essential.

As will be discussed in the present part, in addition to surface tension and its *anisotropy*, such further parameters as the amount of *fluctuations* and the *driving force* can influence the geometry of the resulting interfacial patterns. Depending on the values of these parameters a great variety of patterns are found experimentally (see Fig. 1.1). In many cases the morphology of the growing phase is very complex and can be described in terms of fractal geometry.

We shall be concerned with systems in which the motion of the phase boundary is controlled by a field-like quantity which satisfies the Laplace equation. The most typical examples for diffusion-limited or Laplacian growth include solidification, when a crystalline phase is growing in an undercooled melt, the development of viscous fingers which can be observed when a less viscous fluid is injected into a more viscous one, and electrodeposition, where the ions diffusing in an electrolite give rise to beautiful patterns being deposited onto the cathode.

There are three major approaches which can be used to investigate the structure and development of Laplacian patterns:
i) *Stability analysis* of the original equations and their simplified versions allow one to study such questions as the selected velocity and the tip radius of an advancing dendrite or finger. The crucial role of the surface tension and its anisotropy was first pointed out in these investigations. Those who are interested in the theoretical aspects of pattern formation can find excellent treatment of the problem in a number of recent review articles (Langer 1980, Bensimon *et al* 1986, Kessler *et al* 1987, 1988). Here we shall concentrate on the next two approaches.
ii) If one is interested in the description of complex geometrical patterns, it is effective to use such numerical methods as *computer simulations* of aggregation models or other, more standard algorithms. These will be discussed in the present chapter.
iii) *Experiments* on Laplacian growth represent the third approach to the study of interfacial pattern formation. Such investigations are usually inexpensive and in many cases they can be carried out without great efforts. During the past few years the related experiments (to be reviewed in Chapter 10) have made an important contribution to our understanding of diffusion-limited growth.

9.1. EQUATIONS

As discussed above, under certain approximations the same equations can be used for the description of a wide range of pattern forming phenomena. We shall write these equations for the dimensionless field-like variable $u(\vec{x}, t)$, which may denote temperature (solidification), pressure (viscous fingering) electric potential (electrodeposition, dielectric breakdown) or concentration (isothermal solidification, electrodeposition) (Langer 1980, Bensimon *et al* 1986, Kessler *et al* 1987, 1988). In various approximations the interface in diffusion-limited processes is determined by the Laplace equation

$$\nabla^2 u(\vec{x}, t) = 0 \tag{9.1}$$

with appropriate boundary conditions. For example, for solidification equation (9.1) follows from the approximation that the velocity of the interface is small compared with the characteristic time needed for u to relax to a stationary distribution corresponding to the given shape of the changing interface (then the left hand side of the diffusion equation $\partial u/\partial t = C\nabla^2 u$, which expresses heat conservation, can be neglected). In the case of viscous fingering in a porous medium or in two dimensions, (9.1) corresponds to the assumption of incompressibility of the fluids. Obviously, the distribution of the electric potential in electrodeposition experiments also satisfies (9.1) in regions where there are no sources of charge present.

The normal velocity of the interface is given by the first boundary condition

$$v_n = -c\hat{\mathbf{n}} \cdot \nabla u\,, \tag{9.2}$$

where c is a constant and $\hat{\mathbf{n}}$ denotes the unit vector normal to the interface. Thus the local interfacial velocity is proportional to the gradient of the field. As an example, one can mention solidification, where (9.2) is a consequence of heat conservation: the latent heat produced at the interface (which is proportional to v_n) should be equal to the heat flux away from the surface (which is proportional to the temperature gradient). In (9.2) it is assumed that $\nabla u = 0$ in the growing phase.

The growth is induced by the fact that the value of the field far from the interface is a constant

$$u_\infty = Const. \tag{9.3}$$

different from the equilibrium value of u at the interface Γ. If one uses the dimensionless form of the equations, this equilibrium value is equal to zero. Thus, unstable growth (negative gradient at the interface) takes place for $u_\infty < 0$. In the following we shall assume this. Finally, the boundary condition given below prescribes u on the growing interface

$$u_\Gamma = -d_0\kappa - \beta v_n^\eta \,, \tag{9.4}$$

where the so called capillary length d_0 is proportional to the surface tension, κ denotes the local curvature of the interface (with $\kappa > 0$ for a sphere), β is the kinetic coefficient and η is an exponent depending on the physical process considered. The local curvature can be calculated using the expression $\kappa = 1/R_1 + ... + 1/R_{d-1}$, where R_i are the local principal radii of curvature of the surface.

A few additional remarks may be useful in explaining the form of the boundary condition (9.4).
i) It is clear from (9.4) that u is made dimensionless in such a way that $u = 0$ for a resting planar interface.
ii) With $\beta = 0$, (9.4) reduces to the *Gibbs-Thomson* relation which is valid assuming local thermodynamic equilibrium. It can be understood on a qualitative basis. Let us use the language of solidification problems. At equilibrium the same number of particles are leaving the solid as are becoming part of it. If the surface has a part with positive curvature (bump), the situation changes since the molecules in this region can leave the surface easier (because the molecules on the surface have fewer neighbours with attractive interactions bonding them to the growing phase). Assume that the sample is at a temperature (melting temperature) at which the planar interface is in equilibrium. Then a bump will melt back which is equivalent to the statement that the melting temperature at the bump is decreased. For places with negative curvature the situation is reversed. This is expressed by the first term of the right-hand side of (9.4).
iii) The *kinetic coefficient* β has various origins depending on the given growth process (see e.g., Park and Homsy (1985) for viscous fingering), but in each case it reflects the fact that the moving interface represents a departure from equilibrium.

Although the Laplace equation is linear, because of the above boundary conditions the mathematical problem posed by (9.1–9.4) is *non-linear* and except for a few simple cases it can not be solved analytically. This non-linearity is manifested in the *Mullins-Sekerka instability* (Mullins and Sekerka 1963) which takes place whenever one part of the interface advances locally faster than the surrounding region. The gradient of the field around a protrusion becomes larger in analogy with the electric field which is known to become very large close to the tip of a charged needle (see Fig. 9.1). The increased gradient leads to a faster growth of the interface (because of (9.2)) which, in turn, results in a further increase of the gradient.

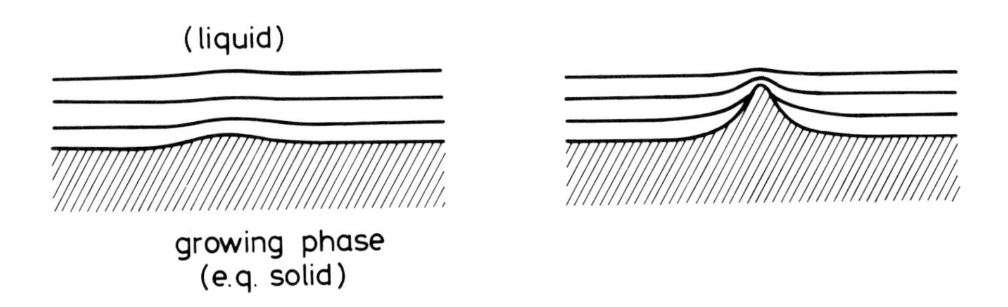

Figure 9.1. Schematic illustration of the Mullins-Sekerka instability. The function u (e.g., the temperature) is the same along the lines drawn close to the interface.

Due to the instability small protruding perturbations of the interface grow exponentially. For short wavelengths, however, the surface tension stabilizes the interface and this mechanism introduces a characteristic length (Mullins and Sekerka 1963). This can be shown by linearizing the differential equation and boundary conditions (9.2) and (9.4) in a perturbation about the steady state solution corresponding to a planar interface moving with a constant velocity v. Assuming that the perturbation has the form $\delta(x,t) = \delta_0 e^{\omega t + ikx}$ with $\delta_0 \ll 1$ the following expression can be obtained for the *dispersion relation*

$$\omega(k) \simeq kv(1 - ld_0k^2)\,, \tag{9.5}$$

where $l = 2C/v$ is the diffusion length with C being the diffusion constant. Because of the above form for $\delta(x,t)$, ω represents the amplification rate of the perturbation, and its sign determines stability. For $\omega > 0$ the perturbations grow exponentially in time (instability), while they quickly die out if $\omega < 0$. Figure 9.2 schematically shows the behaviour of ω as a function of the wave number k. It can be seen from this figure that in a region between $k = 0$ and $k_{max} = \sqrt{ld_0}$ the amplification rate is positive. The upper cutoff is due to the surface tension represented by d_0. If $d_0 = 0$, the growth rate increases indefinitely also for arbitrarily short wavelengths, and the problem is ill-posed from the physical point of view. The fastest growth occurs for

$$\lambda_c = 2\pi/k_c = \sqrt{3ld_0} \tag{9.6}$$

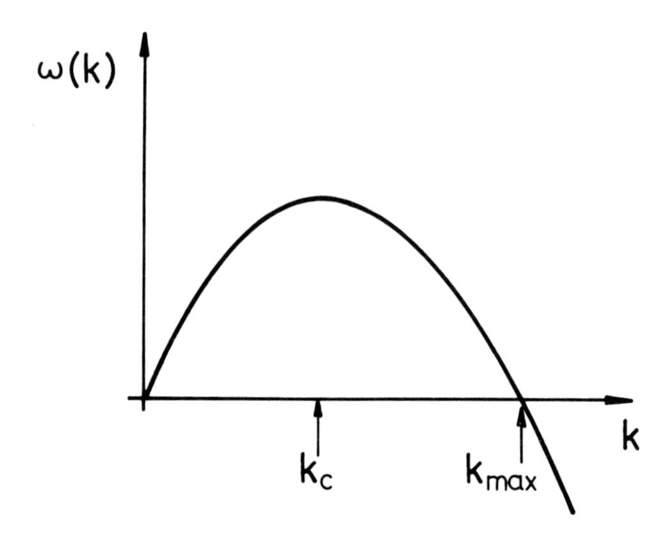

Figure 9.2. Schematic representation of the dispersion relation (9.5). Deformations with a characteristic wave number k for which $\omega > 0$ grow in an unstable manner, while the region $\omega < 0$ corresponds to a stable regime.

which is expected to be close to the characteristic wavelength of the pattern emerging from the competition of the stabilizing effect of the surface tension and the destabilizing effect of the Mullins-Sekerka instability. The capillary length d_0 is a microscopic quantity, typically of the order of Ångstroms. However, the diffusion length l is usually macroscopic, varying in a wide range depending on the given growth process. Thus, Laplacian growth may take place on very different length scales.

It is the unstable, non-linear nature of Eqs. (9.1–9.4) which is behind the sensitive dependence of the solutions on a number of factors influencing the growth of interfaces. Correspondingly, there exists a great variety of possible patterns which can be classified based upon their growth mechanism and geometrical properties (Vicsek 1987, Vicsek and Kertész 1988). Schematic drawings of the *major types of patterns* are shown in Fig. 9.3. In the first set the patterns are divided into two groups depending on the stability of the most advanced parts of the interface, called tips. Pictures Ia and Ib are intended to demonstrate that a tip can be either unstable and go through repeated tip splittings, or stable, and lead to dendritic growth. Unstable tips result in disordered structures with no apparent symmetry. If the tips are stable, in the sense that small perturbations around their stationary shape decay relatively quickly, the obtained patterns have a symmetry of varying degree. Picture Ib shows the three most commonly occurring possibilities.

In the other set (IIa and IIb) it is the geometry of the overall pattern which is qualitatively different for the two types of structures. In some cases the interface bounds a region which is homogeneous on a length scale comparable to the size of the object, while under different conditions the growing pattern has an open branching structure and a corresponding fractal dimension. According to our present understanding all combinations of the above two sets occur in nature.

On the basis of the above discussion it is natural to raise the following questions:

i) What are the relevant parameters determining whether a given growth process leads to a fractal or to another type of complex pattern?

ii) What are the conditions under which a given parameter has a dominating effect on the shape of the interface?

In Part III we shall concentrate on describing numerical and experimental works aimed at answering these questions.

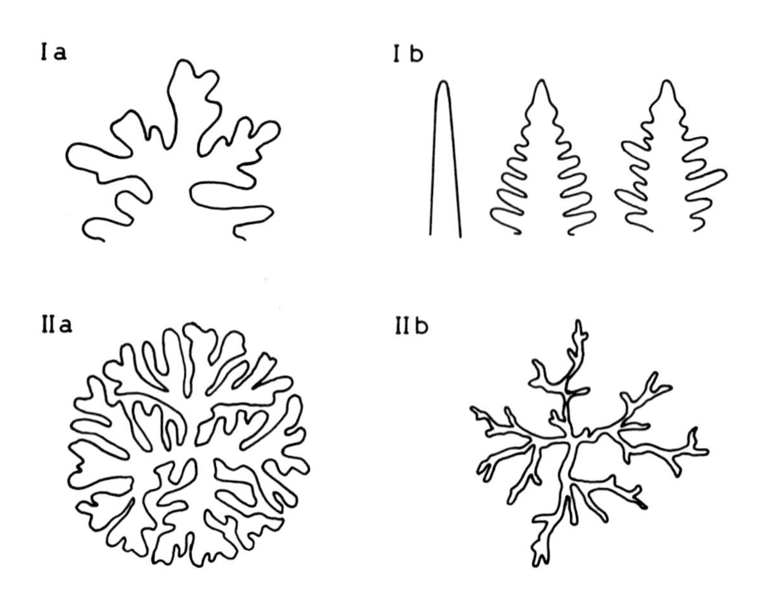

Figure 9.3. Schematic pictures of the major types of patterns which typically occur during unstable interfacial growth.

Experience with pattern forming systems indicated that anisotropy, randomness and driving force $(u_\Gamma - u_\infty)$ can play an essential role during unstable interfacial growth. Fortunately, computer simulations and model experiments allow the investigation of the effects of these parameters on the formation of complex patterns. Of course, the question whether the Laplace equation with no noise and no anisotropy leads to fractal patterns represents one of the most interesting problems. However, at this point the interested reader will be disappointed: there exists no definite answer to this question yet.

9.2. MODELS RELATED TO DIFFUSION-LIMITED AGGREGATION

Aggregation models represent a recent approach to the problem of pattern formation. When using this method, the structure develops on a regular lattice as individual particles are added to a growing cluster. Simulating aggregation is an alternative to the more accurate numerical techniques (Section 9.4) which are based on digitizing along the interface only. Because of this simplification some of the fine details are lost, however, using aggregation models more complex structures can be studied and the fluctuations enter the calculations in a natural way.

Many of the methods related to cluster growth are based on the diffusion-limited aggregation (DLA) model (Chapter 6). It can be easily shown that DLA clusters and the solutions of Eqs. (9.1–9.4) should be closely connected (Witten and Sander 1983). Let $u(\vec{x}, t)$ be the probability that a randomly walking particle released far from the interface will be at point $\vec{x}$ after having made t steps on the lattice. The probability of finding the particle at $\vec{x}$ equals the average of probabilities of finding it at the neighbouring sites at the previous time step

$$u(\vec{x}, t+1) = \frac{1}{z} \sum_{\vec{a}} u(\vec{x} + \vec{a}, t) , \tag{9.7}$$

where $\vec{a}$ runs over the z neighbours of site $\vec{x}$. Deducting a term $u(\vec{x}, t)$ from both sides of (9.7) we see that it represents the discrete version of the diffusion equation $\partial u / \partial t = C \nabla^2 u$. Since in DLA the particles are released one by one the changes in time are slow and we recover the Laplace equation (9.1). The probability that the perimeter site $\vec{x}$ gains a particle at time $t+1$ can be expressed analogously to (9.2)

$$v_n(\vec{x}, t+1) = \frac{1}{z} \sum_{\vec{a}} u(\vec{x} + \vec{a}, t) . \tag{9.8}$$

Here v_n was used to express the fact that the velocity of the interface is given by the probability of gaining a particle at $\vec{x}$. Furthermore, the right-hand side of (9.8) corresponds to the gradient of u, because $u = 0$ at the sites adjacent to the growing cluster. In this way (9.8) is equivalent to the discretized version of the boundary condition (9.2). It should be pointed out that the above relations are true for the probability distribution, and in an actual simulation a single diffusing particle may visit sites with a frequency quite different from that predicted by the average behaviour. Finally, (9.3) is fulfilled by providing a steady flux of the particles released from distant points.

However, the boundary condition (9.4) is not satisfied in DLA. Instead of being a smooth function of the local curvature, the value of u at the surface is equal to zero, as was mentioned above ($d_0 = 0$ because the surface tension is assumed to be zero. On the other hand, the lattice constant introduces a lower typical length scale, a property charcteristic for interfaces with non-zero surface tension). In fact, there exists no well defined interface curvature for diffusion-limited aggregation clusters: Fig. 6.1 shows that the typical curvature is trivially given by the lattice constant, or it is equal to zero. This is the first problem which has to be treated by the aggregation models constructed to simulate pattern formation. In addition, the structure of DLA clusters is largely determined by the fluctuations represented by the individual random walks of the incoming particles. To simulate experimental situations with varying degree of randomness one has to be able to control the amount of fluctuations present in the aggregation process. The algorithms resolving these problems will be described in the next two subsections.

9.2.1. Effects of surface tension

To solve the Laplace equation it is sufficient to determine the value of the field u at the boundary of the two phases. Then the solution can be written as the sum over the geometry dependent Green's function $G(\vec{x}, \vec{s})$ weighted by the boundary value $u_\Gamma(\vec{s})$

$$u(\vec{x}) = \sum_{\vec{s} \in \Gamma} G(\vec{x}, \vec{s}) u_\Gamma(\vec{s}) , \tag{9.9}$$

where the summation runs over all points on the interface and $G(\vec{x}, \vec{s})$ is the electric field generated by a point charge source located at $\vec{s}$ on a grounded conductor.

It can be shown that $G(\vec{x}, \vec{s})$ is proportional to the probability that a random walk starting from $\vec{s}$ visits $\vec{x}$ before hitting the interface again (Kadanoff 1985, Szép *et al* 1985). This quantity can be estimated by counting the number of times the point $\vec{x}$ is visited by random walkers which are released from the surface point $\vec{s}$ and terminate whenever they hit any occupied site. Thus we need two types of walks: i) particles released far from the interface represent the constant boundary condition $u_\infty = Const$ (9.3). ii) The second type of walks leave the surface point $\vec{s}$ with a probability $u_\Gamma(\vec{s}) = d_0\kappa$ corresponding to the boundary condition (9.4), and end at another surface point. These walks transfer flux from one part of the interface to the other, changing its shape.

During the simulations (Kadanoff 1985, Szép *et al* 1985, Liang 1986) one first determines the bonds connecting the surface sites of the growing cluster with their nearest neigbouring empty sites. The total number n_t of net crossings of random walkers through these bonds is recorded (adding to n_t 1 for incoming and deducting from n_t 1 for departing particles). Then the interface is moved forward if n_t reaches a previously fixed number m and moved backward if n_t becomes smaller than $-m$. Here m controls the fluctuations: for $m \gg 1$ the noise due to the randomness of the walks is almost averaged out (see next section). There is a practical difficulty in using (9.4) for the determination of the probability of releasing a particle from a surface point since κ can have both positive and negative values, while the probability $u_\Gamma(\vec{s})$ has to be always positive. This problem is resolved by letting the particles carry a "flux" f and using, instead of (9.4),

$$f u_\Gamma(\vec{s}) = d_0\kappa \tag{9.10}$$

and let κ and f have the same sign ($f = 1$ or $f = -1$).

There are several possibilities for the estimation of the surface curvature κ at site $\vec{s}$. A simple procedure is based on counting the number of particles, N_L, belonging to the aggregate and being within a distance L from $\vec{s}$ (Vicsek 1984). Assuming that the characteristic changes in the shape of the surface occur on a scale larger than L, for example, the quantity

$$\kappa = \frac{3}{L}\left(\frac{N_L}{L^2} + \frac{\pi}{2}\right) \tag{9.11}$$

provides a reasonable estimate for κ on a square lattice.

The above method can be used to simulate pattern formation under various conditions. Figure 9.4 demonstrates that it leads to patterns (Liang 1986) matching the experimental results obtained in the studies of viscous fingering in a longitudinal Hele-Shaw cell (air injected into glycerin placed between two close parallel glass plates). Very different shapes can be generated by changing m and the ration of the number of walkers released from the interface and far from it. For large m the generated structures resemble dendritic growth (Szép *et al* 1985).

Figure 9.4. Experimental (upper picture, Park and Homsy 1985) and simulated (Liang 1986) viscous fingers obtained in the longitudinal geometry.

Pattern formation in diffusion-limited aggregation can be studied by a considerably simpler model as well. Perhaps the easiest way to take into account the effects of surface tension is to introduce a sticking probability $p_s(\kappa)$ depending on the local surface curvature (Vicsek 1984). A plausible choice is

$$p_s(\kappa) = A\kappa + B\,, \tag{9.12}$$

where A and B are constants. The analogy with the boundary condition (9.4) is provided by the fact that the sticking probability is proportional to the local growth velocity. If p_s calculated from (9.12) is less than 0 it is set to a small threshold value, and if $p_s > 1$, it is assumed to be equal to 1. For short times there is a direct correspondence between this method and the algorithm described at the beginning of this section (Sarkar 1985). The analogy can be understood on a qualitative basis: A particle which does not stick to the surface can be regarded as starting a walk from the given point of the interface. In the present model one has to use an additional rule to obtain well defined surfaces. According to this rule a particle previously allowed to stick is relaxed to its final position which is one of the nearest neighbour sites with the largest number of nearest neighbours (with the lowest potential energy).

This curvature dependent sticking probability model can be used to demonstrate the role of surface tension in the development of diffusion-limited patterns in the presence of noise. Figure 9.5 shows a series of simulations carried out in the strip geometry with an increasing value of the parameter A corresponding to the surface tension. These pictures illustrate that for a fixed size of the system there is a crossover from a fractal-type structure to a less disordered, quasi-regular geometry. Although not shown in Fig. 9.5, it has to be noted that none of the structures is stable, and for longer "times" (more particles added) the competition among the fingers results in patterns similar to either Fig. 9.5a or to a single finger (the latter is observed for larger A).

The patterns displayed in Fig. 9.5 were generated on a square lattice whose anisotropy is known to affect the results when aggregation on a single seed particle is simulated. To study *viscous fingering* in the radial Hele-Shaw cell (Section 10.1.1) under *isotropic* conditions one has to use the off-lattice version of DLA with a sticking probability given by (9.12) (Meakin *et al* 1987). A typical pattern generated using this approach is displayed in Fig. 9.6, where the black and white layers indicate the successive stages of the

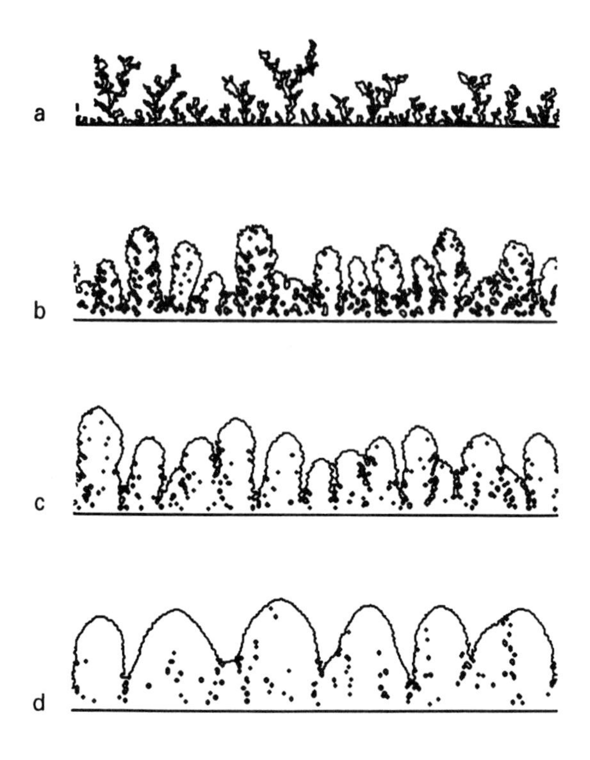

Figure 9.5. Interface of the diffusion-limited deposits with curvature-dependent sticking probability and relaxation. The following values were used for the parameters in Eq. (9.12): (a) A=0, B=1; (b) A=3, B=0.5; (c) A=6, B=0.5 and (d) A=12, B=0.5 (Vicsek 1984).

growth. The initially circular shape becomes unstable when its radius exceeds the radius of curvature characteristic for the given value of A. In DLA tip splitting occurs as a one time event due to the microscopic fluctuations only, while in Fig. 9.6 the growing tips seem to split gradually.

On the basis of Figs. 9.5 and 9.6 we conclude that for a fixed system size, decreasing the surface tension leads to structures with a more pronounced fractal behaviour. This is in agreement with the experimental observation of fractal viscous fingers in a system with zero interfacial tension (Daccord *et al* 1986). One possible explanation is that noise plays an increasingly relevant role as the surface tension is decreased. Similarly, in Fig. 9.5 a crossover from a quasi-regular to a disordered pattern can be seen as a function of the size of the aggregates. Thus, these simulations suggest that

Figure 9.6. Various stages in the growth of an off-lattice cluster generated using a curvature-dependent sticking probability. This figure illustrates the crossover from a compact to a fractal structure as the aggregate grows larger (Meakin *et al* 1987).

in the presence of noise (even if it is small) the asymptotic shape of Laplacian patterns growing out from a centre has a fractal structure presumably analogous to that of DLA (Vicsek 1985, Meakin *et al* 1987).

9.2.2. Noise reduction in DLA

We have seen above that the surface tension does not change the asymptotic behaviour of patterns if i) the simulations are carried out off-lattice (isotropy), and ii) the fluctuations can not be neglected. In this section we shall concentrate on the *interplay of anisotropy and noise* during the growth of complex diffusion-limited patterns. This problem can be investigated by the noise reduced version of on-lattice DLA described below (Tang 1985, Szép *et al* 1985, Kertész and Vicsek 1986). For simplicity, we shall not treat the effects of surface tension in the form discussed in the previous section since the finite size of the particles can be regarded as representing a small finite surface tension.

There is a natural way to decrease the fluctuations in DLA (Kertész and Vicsek 1986). Instead of adding a particle to the aggregate immediately after it hits a growth site, one keeps a record of how many times each of the perimeter sites (empty sites adjacent to the cluster) becomes a termination point for a randomly walking particle. After a perimeter site has been contacted m times it is filled and the new perimeter sites are identified. The scores (number of contacts) associated with these sites are set to zero. The scores associated with all of the other surface sites remain at their values before this event. Clearly, this procedure decreases the noise with growing m, because probing the surface with many walks provides a better estimate of the expectation value of the growth rate at the given point than a single walk. In the limit $m \to \infty$, application of the method yields the solution of the lattice version of the noiseless Laplace equation (9.1) with the boundary conditions (9.2) and (9.3).

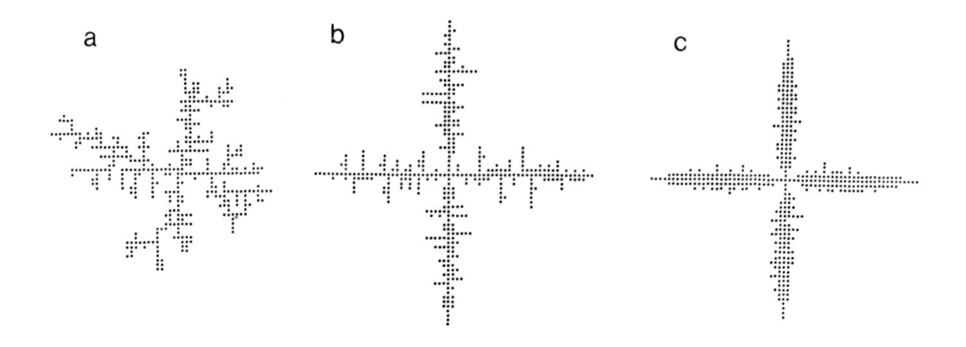

Figure 9.7. Clusters consisting of 400 particles generated on the square lattice using the noise-reduced diffusion-limited aggregation algorithm. (a) m=2, random fractal; (b) m=20, dendritic growth, and (c) m=400, noisy needle crystal (Kertész and Vicsek 1986).

Figure 9.7 shows three representative clusters generated on a square lattice with various values of the noise reduction parameter m. The overall appearance of the aggregates indicates that as a function of decreasing noise (increasing m) two types of morphological changes occur in these small scale simulations. At about $m = 5$ the random, tip-splitting structure typical for DLA clusters crosses over into a dendritic pattern with well defined but irregularly spaced side branches having stable tips. Further increasing m results in the growth of a structure consisting of four needles growing out from the centre along the lattice axes.

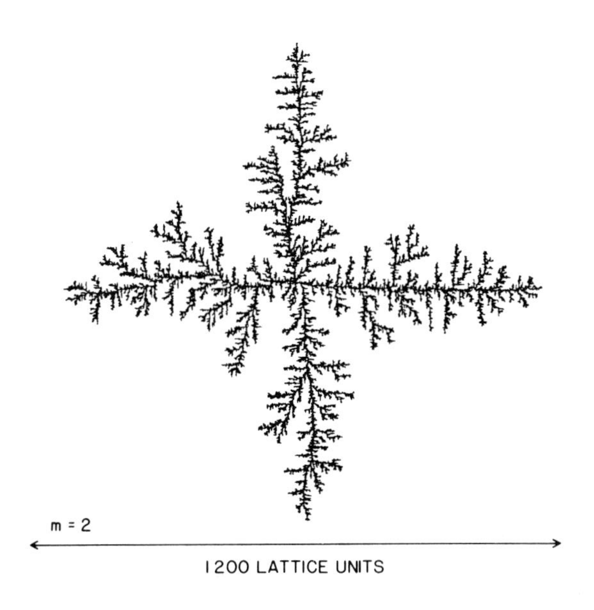

Figure 9.8. Noise-reduced diffusion-limited aggregate (m=2, N= 50,000) generated on the square lattice (Kertész *et al* 1986).

The above described morphological changes taking place in the sequence *random fractal* → *dendritic* → *needle* are results of the competition between the anisotropy provided by the underlying square lattice and the fluctuations due to the random walks. An analogous result is obtained for fixed m and increasing cluster size (Kertész *et al* 1986). Figure 9.8 shows that aggregates with $m = 2$ have asymptotically a cross-like shape, while for much smaller sizes their envelope is approximately circular. As was discussed in Section 6.1.2 the overall shape of extremely large diffusion-limited aggregates grown on a square lattice is approaching a cross. Thus, the method of noise reduction seems to reveal the *asymptotic behaviour* of DLA clusters using considerably smaller number of particles. In conclusion, in a far-from-equilibrium growth process the structure of the interface can change non-trivially as a *function of size.* In the present case this is due to the fact that on a long run lattice anisotropy dominates over the disorder due to fluctuations.

The clusters generated on lattices using noise reduced DLA are expected to reflect some of the relevant features of large Laplacian patterns with anisotropic surface tension growing in the presence of fluctuations. These aggregates usually have the overall shape of a $2n$-fold star, where n is the number of axes of the underlying lattice. A convenient approach to the characterization of such clusters is to define two exponents $\nu_\parallel$ and $\nu_\perp$ through the expressions (6.13) and (6.14) describing respectively the scaling of the average length l and width w of the arms of the clusters as a function of the number of particles in them. Large scale simulations for $m > 1$ indicate that in the asymptotic limit both exponents are somewhat larger, but close to $\nu_\parallel \simeq \nu_\perp \simeq 2/3$ (Meakin 1987). Correspondingly, the fractal dimension of the noise-reduced DLA clusters is about $D \simeq 1.5$.

The situation is quite delicate, as is illustrated by Fig. 9.9, where the logarithm of the ratio $R = l/w$ is plotted versus the logarithm of the number of particles, N, in the clusters. For intermediate cluster sizes the data show scaling of R with N of the form $R \sim N^{\nu_\parallel - \nu_\perp}$, and the $R \simeq Const$ behaviour is manifested only above a certain size. This result suggest that $\nu_\parallel > \nu_\perp$ obtained for very large ordinary diffusion-limited aggregates is not necessarily valid in the asymptotic limit (it describes a transient situation).

9.3. GENERALIZATIONS OF THE DIELECTRIC BREAKDOWN MODEL

In its simplest form the dielectric breakdown model (DBM) discussed in Section 6.3 exhibits a fractal scaling analogous to that of DLA. However, various modifications of DBM result in qualitative deviations between the two models. We briefly recall that in DBM the Laplace equation is solved on a lattice to obtain the growth probabilities associated with the perimeter sites, instead of releasing random walkers as is done in DLA. Then a perimeter site is filled

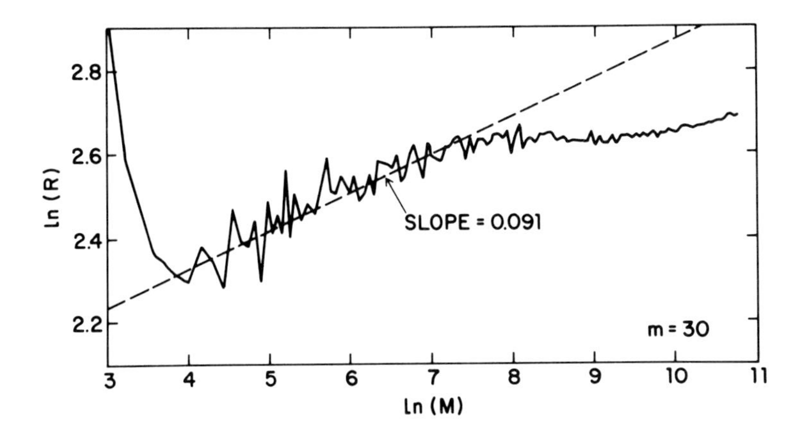

Figure 9.9. Dependence of the ratio l/w on the size of the aggregates grown on the square lattice with a noise reduction parameter $m = 30$. l and w denote respectively the length and the width of the main arms of the clusters (Meakin 1987).

with a probability proportional to some power (η) of the calculated growth probability. This procedure makes the method quite versatile, e.g., it can be used to grow deterministic patterns as well (Section 6.4). Furthermore, it has a parameter η which provides an additional opportunity to generate a family of qualitatively different patterns corresponding to a non-linear dependence of the growth velocity on the local gradient.

There is a difference between the boundary conditions which are used when DLA and DBM clusters are generated. The condition that a randomly walking particle sticks to the aggregate when it arrives at a site adjacent to the cluster (DLA) is equivalent to keeping the probability equal to zero at the *perimeter* sites. In the case of DBM one sets the field equal to zero on the *surface* sites (sites already filled). This seemingly small difference becomes relevant when noise-reduced versions of DLA and DBM are considered. To decrease the fluctuations in DBM one fills a new perimeter site only after it has been chosen m times. Again, $m \to \infty$ corresponds to deterministic growth.

Application of the above method with $m = 20$ on the triangular lattice (Nittmann and Stanley 1986) results in tip splitting patterns very similar to that shown in Fig. 9.6. On the other hand, noise-reduced DLA clusters on the same lattice have stable tips, and look like random snowflakes since they are sixfold analogues of the patterns displayed in Fig. 9.7. Obviously, DBM is less susceptible to lattice anisotropy than DLA. It has to be emphasized that using DBM with $m > 1$ one obtains disordered structures with branches having a well defined, m-dependent thickness, without taking into account the surface tension. Although this finding is somewhat surprising, a similar behaviour is found in the experiments on viscous fingering with miscible fluids, where interfaces with characteristic curvatures can be observed (Paterson 1985). Thus the noise-reduced DBM may offer help to understand some of the properties of moving interfaces with zero surface tension.

Figures 9.4–9.7 demonstrate that aggregation models can be useful in studying the growth of realistic geometrical structures of various kinds, including viscous fingers. One of the most appealing challenges is, however, to understand the development of such intricate objects as snowflakes and other related dendritic crystals. In the remaining part of this section we shall concentrate on models producing patterns relevant to the growth of complex dendritic structures.

Snowflakes exhibit a number of very characteristic features. They (a) are quasi two-dimensional, (b) have sixfold symmetry and (c) can have a large number of entirely different shapes. To grow snow crystals in the computer one can use variations of the DBM on the triangular lattice, which provides the two-dimensional nature and the sixfold symmetry of the patterns simultaneously. Thus the properties (a) and (b) are built into the simulations by this restriction. However, at this point our goal is not to explain (a) or the fact that snowflakes have six arms (these properties are due to the anisotropy of the surface tension). Instead, we shall be more concerned with the statement (c).

Figure 9.10 shows patterns which are quite similar to the real snowflakes also displayed in the figure. The simulated clusters were obtained using the following model (Nittmann and Stanley 1987). i) The growth

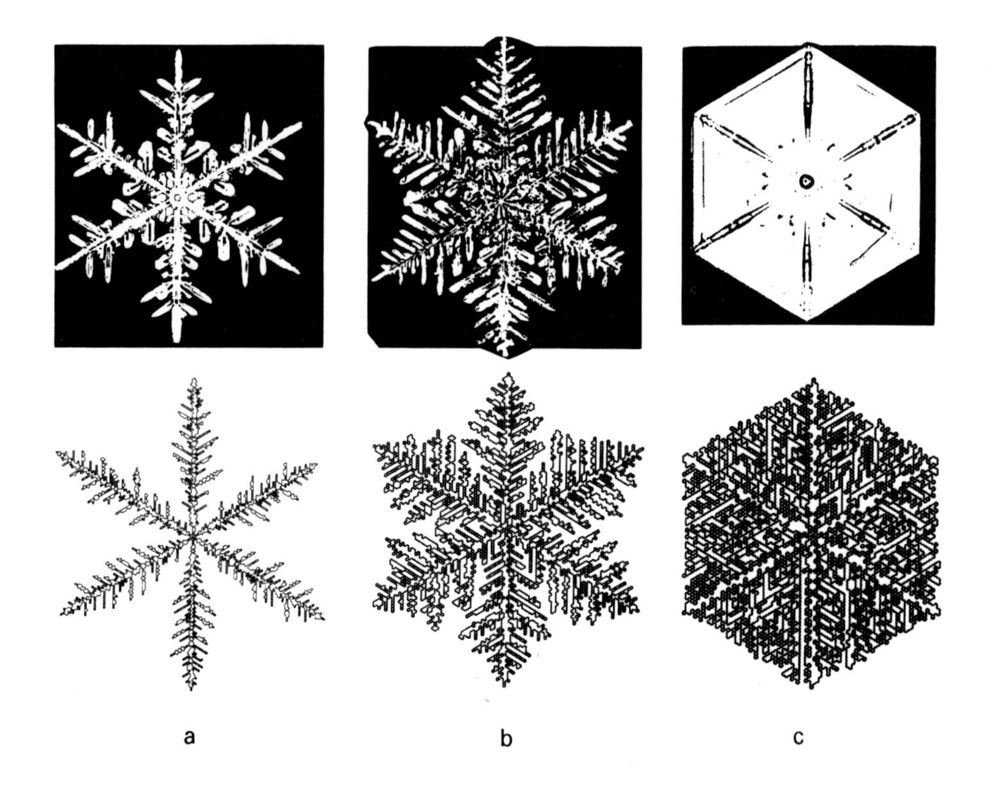

Figure 9.10. Real (top row, reproduced from Bentley and Humpreys (1962)) and simulated (bottom row) snowflakes. The simulated patterns were generated on the triangular lattice up to 4000 particles with $m = 200$. The value of the exponent η for patterns (a-c) was equal to 0.05, 0.5 and 1.0 (Nittmann and Stanley 1987).

probability is determined by solving the Laplace equation on the triangular lattice with a boundary condition corresponding to DLA, i.e., setting the field u equal to zero at the perimeter sites. ii) Perimeter sites are picked randomly, with a probability proportional to $(\nabla u)^\eta$. iii) A perimeter site is filled if it has been chosen m times. As the non-linearity parameter η is increased, qualitatively different patterns are obtained. Although there exists no direct physical interpretation of the parameter η, one expects that the complicated processes taking place on the surface of a snowflake may give rise to a non-linear response to the local gradient of the field (temperature). In addition, one can simulate the effects of surface tension which leads to the thickening of arms and the disappearance of small holes.

Since snowflakes are *almost perfectly symmetric*, one expects that they can be studied effectively by *deterministic growth models* (Section 6.4). It is natural to assume that the low level of randomness in the shape of snowflakes is due to the fact that the fluctuations in the conditions affecting the growth process take place on a length scale larger than the size of a snowflake. To simulate the development of dendritic patterns with the spatial fluctuations neglected, the following method can be used.

The process starts with a seed particle placed on a triangular lattice. At each time step the value of the field u is calculated by solving the lattice version of the Laplace equation (9.1). The value of the field is assumed to be zero on a circle having a radius a couple of times larger than the size of the cluster. For the surface sites u is prescribed by (9.4), in which κ is calculated by a method described in the previous section and the kinetic term is neglected. The gradients at the surface are normalized onto the unit interval and those perimeter sites for which the gradient is larger than q are filled.

To approximate the boundary condition (9.2) one assumes that q varies in time as (Family *et al* 1987)

$$q(t) = a + b(t \bmod [c]) \tag{9.13}$$

which is a piecewise function depending on the parameters a, b and c. This expression ensures that the growth velocity is proportional to the local gradient in a discretized manner (if the local gradient is small, its value exceeds $q(t)$ less frequently and this results in a slower growth). By varying a and b the effects caused by surface diffusion and changing undercooling can be simulated.

The above method is capable of producing most of the symmetric patterns observed in the related experiments. The various cases include faceted growth and structures corresponding to needle and fractal crystals. If the parameter a is changing in time, the combinations of these patterns are obtained within a single cluster. In Fig. 9.11 a few examples are shown, where the sixfold symmetry is provided by the underlying triangular lattice.

There is a striking similarity between the simulated and the real snowflakes (also displayed).

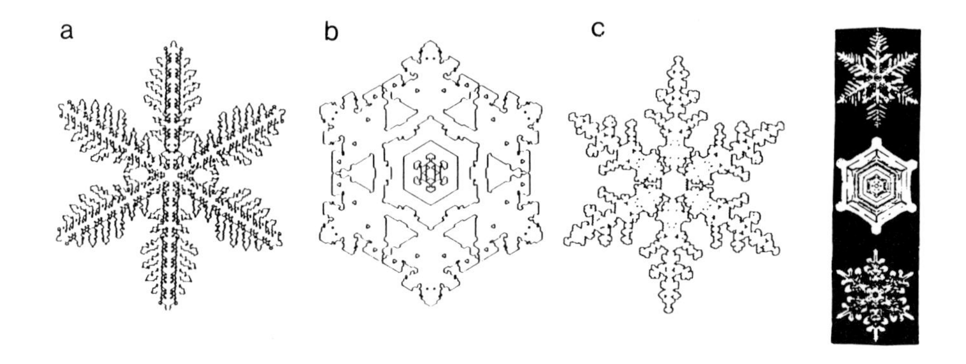

Figure 9.11. Three examples for patterns generated by the deterministic growth model on a triangular lattice. (b) and (c) were obtained by changing the parameters a and b of equation (9.13) randomly during the growth (Family *et al* 1987). The inset shows a few typical snowflakes reproduced from Bentley and Humpreys (1962).

Perhaps the most important conclusion which can be drawn from the study of the above methods is related to the seemingly unlimited number of symmetric structures produced in dendritic growth. These models demonstrate that the great variety of patterns appearing as a result of the same solidification process is likely to be induced by the *temporal changes* in the environmental conditions during crystallization. Indeed, snowflakes fall through regions of air with more or less randomly changing temperature and vapour pressure, and it is the time dependent interplay of such factors as undercooling, surface tension and surface diffusion which leads to the observed rich behaviour.

9.4. BOUNDARY INTEGRAL METHODS

Simulation techniques involving aggregation of particles are bound to describe the behaviour of the discrete version of the Laplace equation. Correspondingly, the obtained patterns are either influenced by the fluctuations present in the algorithms or are determined by the lattice anisotropy. However, in many experimental situations (for example in viscous fingering) the motion of the interface is expected to be described by the continuum version of (9.1) without any significant amount of external noise.

The complex interfacial patterns in such cases seem to emerge as a result of the initial surface geometry and the subsequent proliferation of tip-splitting instabilities. This observation suggests an analogy with deterministic chaos: slightly different initial configurations may lead to very different disordered patterns because the instabilities amplify the smallest deviations.

To study this aspect of fractal pattern formation one is led to solving the Laplace equation without using an underlying lattice. This can be achieved by the so called *boundary integral method* in which for $d = 2$ the free interface is approximated as a piecewise linear with uniform monopole sources along each leg. A given interface uniquely determines the source distribution which, in turn, uniquely determines the velocity distribution along the interface.

The above idea is based on converting Eqs. (9.1–9.4) to an integro-differential equation (Kessler *et al* 1984, Gregoria and Schwartz 1986). Before doing so, for convenience we confine ourselves to the two-dimensional case and replace the boundary condition (9.3) and (9.4) with the essentially equivalent conditions $u(R_0) = 0$ and $u_\Gamma = 1 - \kappa(\vec{x}_s)$, where R_0 is the radius of a circle much larger than the growing object and the kinetic term in (9.4) is neglected. Furthermore, using the language of solidification, we can say that the boundary condition (9.2) corresponds to the fact that an element of the interface at position $\vec{x}_s$ represents a source of temperature of an amount proportional to $v_n(\vec{x}_s)$. Then the temperature at any other point is the background temperature (which is now assumed to be equal to zero) plus the superposition of sources at all interface points propagated by the diffusion

Green's function. Therefore,

$$u(\vec{x}) = \int dx'_s G(\vec{x}, \vec{x}'_s) v_n(\vec{x}'_s), \tag{9.14}$$

where the Green's function for the two-dimensional Laplace equation is

$$G(\vec{x}, \vec{y}) = \ln(\vec{x} - \vec{y})^2 - \ln(R_0^2 \vec{y}/y^2 - \vec{x})^2 - \ln(R_0^2/y^2). \tag{9.15}$$

The equivalence of (9.14) to (9.1) and (9.2) can be shown by noting that for points not belonging to the interface the Laplace equation is satisfied, because

$$\nabla^2 G(\vec{x}, \vec{y}) = -\delta(\vec{x} - \vec{y}), \tag{9.16}$$

where $\delta(\vec{x})$ denotes the Dirac delta function. The boundary condition (9.2) is also satisfied since

$$-c\frac{\partial u}{\partial n}|_\Gamma = -\int\int \frac{\partial G}{\partial n}|_\Gamma \cdot v_n(\vec{x}') = v_n(\vec{x}). \tag{9.17}$$

The above expression is based on a standard result of potential theory which can be obtained using Green's theorem. According to this result the discontinuity in the normal derivative of G is a delta function. The final equation is obtained by calculating u at the interface and satisfying the boundary condition $u_\Gamma = 1 - \kappa(\vec{x}_s)$. The result is

$$1 + c\int dx'_s \kappa(\vec{x}_s)\frac{\partial}{\partial n'} G(\vec{x}_s, \vec{x}'_s) = \int dx'_s G(\vec{x}_s, \vec{x}'_s) v_n(\vec{x}'_s), \tag{9.18}$$

where the integral on the left-hand side is the potential due to a dipole layer of strength $-c\kappa$, which provides a discontinuous jump in the field from 1 in the interior to $1-\kappa$ on the interface.

Next one parametrizes the interface by using the variables $\theta(s)$ and s_T, where $\theta(s)$ is the angle between the normal to the curve and a fixed direction in space, and s_T is the total arc length. The angle $\theta(s)$ is defined as a function of the distance (arc length) s along the interface. The equations of motion for these quantities are (Kessler *et al* 1984)

$$\partial\theta(s)/\partial t = -\partial v_n/\partial s \tag{9.19}$$

$$\partial s_T/\partial t = \int_0^{s_T} ds \kappa(s) v_n(s) \,. \tag{9.20}$$

The following procedure is used to determine the development of the interface. At any fixed time the discretized $\theta(s_i)$ is given, thus we can construct the curve $\vec{x}_{s_i}$ by quadrature. The integral equation (9.18) is rewritten as a discrete matrix equation for $v_n(\vec{x}_{s_i})$. After having solved the corresponding system of linear algebraic equation for $v_n(\vec{x}_{s_i})$, (9.19) and (9.20) are used to step the interface forward in time using a predictor-corrector method.

The main limitation of the above procedure is represented by the number of grid points M required for discretizing the arc length. There is a condition depending on the surface tension which sets an upper limit for distance between the grid points, thus M increases with the total arc length. Since M is the number of columns of the matrix to be set up in Eq. (9.18), in a medium-scale calculation with a fourfold symmetric object (Sander *et al* 1985) one reaches the computational limit when M is about 4000. Figure 9.12 shows a typical interface at this limit demonstrating the difficulty of obtaining a structure close to the complexity of a DLA type fractal pattern using the boundary integral approach. Furthermore, the displayed pattern seems to be space filling rather than becoming increasingly sparse (the latter property would indicate the fractal nature of the interface).

The boundary integral method is particularly suitable for the investigation of the effects caused by a small amount of controllable *anisotropy* (Kessler *et al* 1984). Let us assume that the angular dependence of the surface tension in the boundary condition (9.4) is of the form

$$u_\Gamma = -d_0[1 + \epsilon \cos(n\theta)]\kappa \,, \tag{9.21}$$

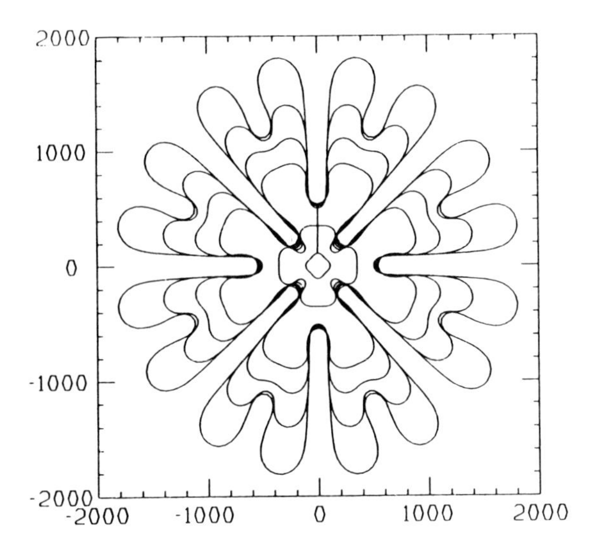

Figure 9.12. Snapshots of the interface growing out from a fourfold seed. This figure was obtained by the boundary integral method. The axes are in the units of the capillary length (Sander *et al* 1985).

where ϵ is a small parameter and $n = 4$ for a pattern with four preferred growth directions. According to the numerical simulations with various ϵ, the anisotropy of the surface tension has to be larger than a well defined threshold value (close to 0.1 for $n = 4$) in order to give rise to dendritic growth with stable tips. However, the calculated patterns are again of relatively simple structure because of the above mentioned computational limitations.

One possibility to obtain more complex interfaces is to change the curvature dependence in the boundary condition (9.4) arbitrarily to κ^K, where K is an odd positive integer (Sander *et al* 1985). Obviously, for large K small curvatures have no effect, which corresponds to an effective decrease of the surface tension. Consequently, the typical curvature increases to a K-dependent value, and more complex patterns using the same number of grid points can be obtained. The zero surface tension limit can be simulated by $K \to \infty$. In the diffusion-limited aggregation model there is no surface tension except for an upper cutoff for the surface curvature provided by the lattice spacing. A calculation with $K \gg 1$ is expected to simulate the deterministic version of DLA, but it should be noted that the real physical growth is determined by the boundary condition with $K = 1$.

Figure 9.13a shows a representative pattern obtained for $K = 5$. (In these calculations the term $\kappa(\vec{x}_s)$ in (9.18) has to be replaced with $\kappa^K(\vec{x}_s)$.) The fractal dimension of the object bounded by the interface can be estimated by determining the scaling of its area with its radius of gyration R_g. The resulting plot (Fig. 9.13b) suggests that one can associate a fractal dimension $D \simeq 1.72$ with the object shown in Fig 9.13a, although it does not possess a high degree of complexity.

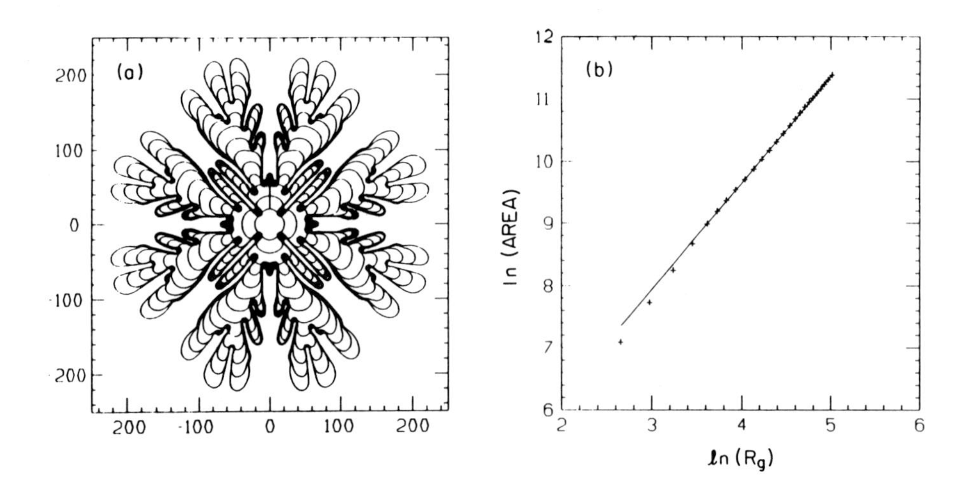

Figure 9.13. (a) Growth of a fourfold interface obtained for $K = 5$. (b) From the scaling of the area of the pattern in Fig. 13a with its radius the estimate $D \simeq 1.72$ can be obtained for the effective fractal dimension (Sander *et al* 1985).

As we have seen, the boundary integral method directly provides a set of velocities $v_n(\vec{x}_{s_j}) = v_j$ corresponding to the motion of the jth grid point on the interface. Let us introduce $p_j = v_j / \sum_j v_j$, the set of normalized velocities, which has a relation to the growth probability measure defined e.g., for DLA. Obviously, the local velocity of the interface determines the distance by which the interface advances during the next time step, and this distance is linearly proportional to the growth probability in a cluster growth model (both are given by the gradient of the associated fields). This analogy can be used to compare the *multifractal* properties of DLA clusters and the continuous interfaces generated by the boundary integral method with a boundary condition similar to (9.4), but depending on the surface curvature according to κ^K (Ramanlal and Sander 1987). Defining $p_i(\epsilon)$ as the average normalized velocity in the i-th region of size ϵ one can use the formalism described in Chapter 3 to calculate the spectrum of fractal dimensionalities $f(\alpha)$.

To determine $f(\alpha)$, the generalized fractal dimensions D_q are evaluated numerically for the final stage of growth in Fig. 9.13a, by plotting $\ln \sum p_i^q(\epsilon)$ as a function of $\ln \epsilon$ and measuring the slope in the region $16 < \epsilon < 100$. The functions α_q, f_q and $f(\alpha)$ are subsequently calculated using (3.11) and (3.13–3.15).

The result for $f(\alpha)$ is in reasonable agreement with the multifractal spectrum obtained for DLA clusters (see Section 6.1.4). (There is a region about the numerically singular point $q = 1$ where no data are given.) One can use two methods to subtract estimates from $f(\alpha)$ for the fractal dimension D. First, in a large system its maximum, f_{max}, is equal to D. In the present case $f_{max} \simeq 1.627$. Second, one obtains another value for the fractal dimension from the relation (6.31) $D = 1 + \alpha_{min} \simeq 1.684$, where α_{min} is the smallest measured value of α. Since with growing size f_{max} is increasing while α_{min} is decreasing, one is led to the conclusion that the fractal dimension of the continuous interfaces used to simulate continuum DLA has to be in the range $1.627 < D < 1.684$. Further careful analysis of the results suggests a value close to 1.65. This estimate is different from $D \simeq 1.71$ obtained from large scale simulations of the diffusion-limited aggregation model. The relevance of this discrepancy is not clear because of the relatively small size of the investigated patterns.

In conclusion, investigations based on the boundary integral method demonstrate that i) noise is not a necessary ingredient of an algorithm producing DLA-type patterns since they can be grown using deterministic equations, ii) this result is supported by multifractal analysis and iii) the fractal dimension of off-lattice DLA clusters in the asymptotic limit may be approximately equal to 1.65. However, studies of larger systems are needed for making more definite statements.

Chapter 10

EXPERIMENTS ON LAPLACIAN GROWTH

The large number of relevant new results obtained by various theoretical approaches and computer simulations has stimulated an increased interest in the experimental systems exhibiting fractal pattern formation. It has turned out that there are many publications and unpublished results on the desks of scientists which, in the light of the recent theoretical progress, may represent good starting points for further investigations. In particular, the formation of random branching structures in thin solid films was observed some time ago in several laboratories, but it was the theoretical framework of fractal geometry which revived the interest in the related experiments.

The majority of the latest experiments, however, have been designed to produce data for additional theoretical research and to check the validity of various predictions. The most suitable systems to carry out this kind of investigation are those in which the behaviour of the growing interface is determined by a relatively small number of well defined parameters. Viscous fingering in the nearly two-dimensional Hele-Shaw cell is a good example for such processes. The related results will be discussed at the beginning of this chapter. The most studied further phenomena leading to complex interfacial patterns are solidification and electrodeposition which will be treated in separate sections. Finally, a few more types of experiments will be reviewed.

As we shall see, in the above mentioned experiments the motion of the interface is dominated by a quantity which in various approximations satisfies the Laplace equation. They also share the property of being relevant from the point of view of *applications*. For example, viscous fingering plays an essential role in the process of secondary oil recovery, where water is pumped into the ground through a well to force the oil to flow closer to the neighbouring wells. Furthermore, the formation of dendritic structures during solidification determines the final internal texture of many alloys, in this way strongly influencing their mechanical and other properties.

10.1. VISCOUS FINGERING

The phenomenon of viscous fingering takes place when a less viscous fluid is injected into a more viscous one under circumstances leading to a fingered interface. In general, the motion of the two fluids and the interface between them is described by the Navier-Stokes equation, which is a non-linear equation containing terms depending on such factors as the gravitation, heat diffusion or shear viscosity. Here we shall discuss experiments where the conditions are such that most of the terms in the Navier-Stokes equation can be neglected and the resulting mathematical problem corresponds to Laplacian growth (Bensimon *et al* 1986).

If the fluids are embedded in a porous medium, it is the term related to viscous forces which dominates the flow. In such media the fluid flows through narrow channels and its velocity is limited by the walls of the channels. A similar effect occurs in the case of quasi-two-dimensional flows without the presence of a porous medium, where instead of narrow channels the flow is confined to a thin layer between two closely placed plates.

10.1.1. The Hele-Shaw cell

At the end of the last century Hele-Shaw (1898) introduced a simple system to study the flow of water around various objects for low Reynolds numbers. The cell he designed consists of two transparent plates of linear size w separated by a relatively small distance b (typical sizes are in the region $w \sim 30$ cm and $b \sim 1$ mm). The viscous fluids are placed between the plates and pressure can be applied either at one of the edges (longitudinal version, Fig. 10.1) or at the centre of the upper plate (radial version) of the cell. In Appendix B more details are given, together with a few useful suggestions for those who are interested in building a simple Hele-Shaw cell. Clearly, the Hele-Shaw cell in its above described form is suitable for the investigation of two-dimensional flows, and correspondingly, most of the results concerning fractal viscous fingering have been obtained for $d = 2$.

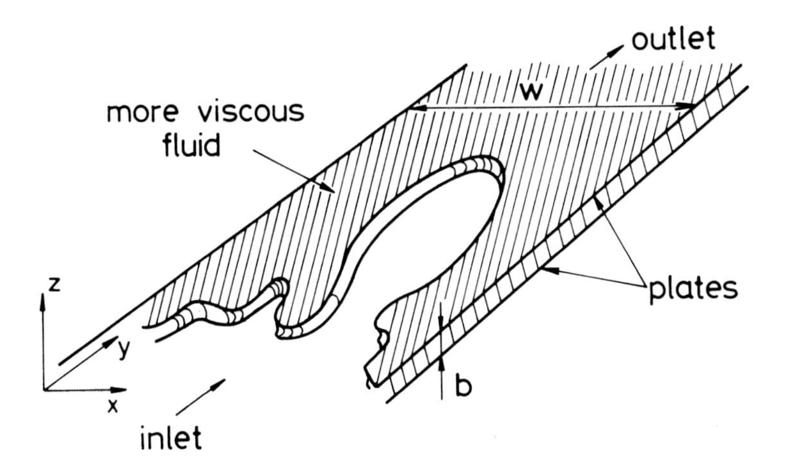

Figure 10.1. Schematic representation of the longitudinal Hele-Shaw cell. Unstable, finger-like patterns are observed if the less viscous fluid is injected into the more viscous one.

The relation of viscous fingering to Laplacian growth can be shown by assuming that the plates are horizontal, and the flow in the x, y plane has a velocity profile $v(z) = [v_x^2(z) + v_y^2(z)]^{1/2}$ which is approximately parabolic in the direction z perpendicular to the plates

$$v(z) = a(b^2/4 - z^2)\,. \tag{10.1}$$

Furthermore, we assume that $v_z = 0$ and $\partial v_x/\partial x = \partial v_y/\partial y = 0$. For the average velocity one has

$$\bar{v} = \frac{1}{b}\int_{-b/2}^{b/2} v(z)dz = \frac{ab^2}{6}\,. \tag{10.2}$$

If the gravitational effects can be neglected the Navier-Stokes equation has the form

$$\nabla p = \mu\nabla^2\vec{v} + \rho\partial\vec{v}/\partial t\,, \tag{10.3}$$

where μ is the viscosity of the fluid, ρ is its density, and p denotes the pressure. For small b the first term of the right-hand side in (10.3) dominates, because it is proportional to $1/b^2$. Inserting (10.1) into (10.3) (where the second term of the right-hand side is neglected) and using (10.2) we get

$$\vec{v} = -\frac{b^2}{12\mu_i}\nabla p\,, \tag{10.4}$$

The above equation represents the so called Darcy's law expressing the fact that for small b the average velocity is proportional to the local force. Assuming that the fluids are incompressible one arrives at the Laplace equation $\nabla^2 p = 0$ (9.1) for the pressure distribution p, from the condition that the divergence of the velocity vanishes. The boundary condition (9.2) is essentially equivalent to (10.4) evaluated at the interface.

When writing down the boundary condition for the pressure jump Δp at the interface (corresponding to (9.4)), one has to take into account the specific geometry of a Hele-Shaw cell. This expression contains three terms: i) There is a relatively large pressure jump due to the parabolic profile mentioned above. The associated surface curvature is typically about

$2/b$. This curvature measured in a direction perpendicular to the plates is approximately independent of the actual position of the interface and it does not have a relevant effect on the motion. ii) The contribution of the surface curvature κ observed in the $x-y$ plane is proportional to the surface tension γ and to κ. Finally, iii) due to the wetting of the cell's walls by the displaced fluid, Δp is increased by a term proportional to $v_n^{2/3}$ (Park and Homsy 1985). Thus, for viscous fingering (9.4) has the form

$$\Delta p \simeq \frac{2\gamma}{b} + \gamma\kappa + \frac{3.8\,(\mu v_n)^{2/3}\gamma^{1/3}}{b}\,. \tag{10.5}$$

In a longitudinal cell of width w (Fig. 10.1) there is a single dimensionless quantity which can be used as a control parameter of the problem. In such a system

$$d_0 = \frac{b^2}{w^2}\frac{\gamma}{\mu V} \tag{10.6}$$

is an analogue of the capillary number in (9.4). Here V is the velocity of the injected fluid far from the interface. The experiments to be discussed below have demonstrated that on increasing d_0 from a small (or negative) value various regimes take place concerning both the dynamics and geometry of viscous fingering.

The traditional experiment is carried out using two inmiscible, Newtonian fluids with a *high viscosity ratio*. For example, air can be used to displace glycerine or oil. Injecting an inviscid fluid into a longitudinal Hele-Shaw cell containing a viscous one results in an initial transient behaviour followed by the development of a single finger which propagates along the channel in a stationary way. According to the related experiments (Saffman and Taylor 1958, Tabeling and Libchaber 1986), numerical simulations (Gregoria and Schwartz, Sarkar and Jasnow 1987) and theoretical approaches (Bensimon *et al* 1986), the width of the finger λ for intermediate values of d_0 is close to $w/2$, where w is the width of the cell. In fact, theory predicts $\lambda = w/2$ in the $d_0 \to 0$ limit, while finger widths somewhat smaller than $w/2$ can be observed experimentally and in numerical solutions of the corresponding equations. For large V the single finger shows rather unstable behaviour (see Fig. 9.4).

In the *radial* Hele-Shaw cell (Paterson 1981, 1985) the stabilizing effect of the side walls is absent and *no steady-state fingers* can develop in the cell. Consequently, the Mullins-Sekerka instability (Saffman and Taylor 1958, Mullins and Sekerka 1963) leads to the growth of disordered interfaces shown in Fig. 10.2. Interestingly, the question whether such viscous fingering patterns become fractals in the large size limit has not been satisfactorily answered yet. For intermediate injection rates (Fig. 10.2a) relatively simple structures can be obtained with an effective radius of gyration exponent corresponding to a fractal dimension close to 1.8 (Rauseo *et al* 1987).

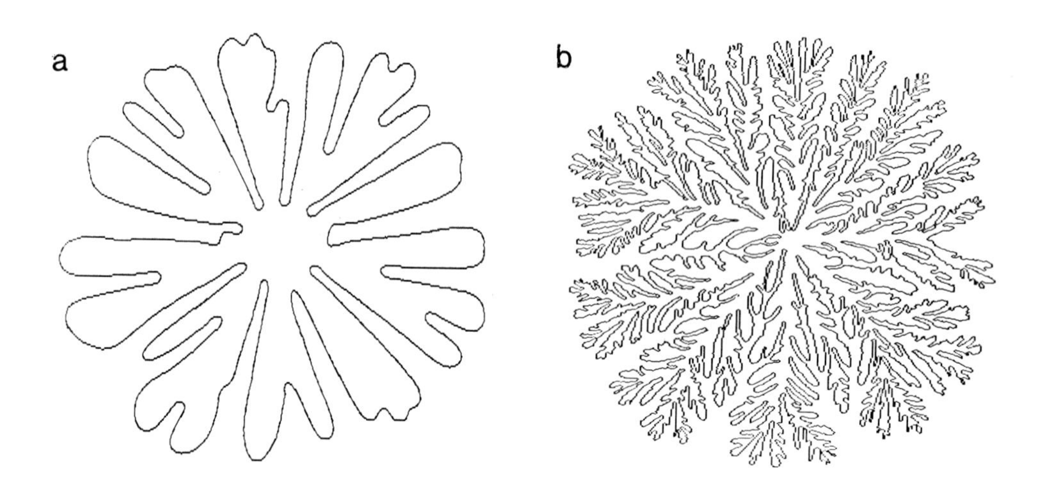

Figure 10.2. Typical viscous fingering patterns obtained in the radial Hele-Shaw cell with glycerine for (a) intermediate and (b) large pressure of the injected air.

When *large injection rates* are applied (Fig. 10.2b) the growing interface seems to branch randomly without forming holes of increasing size, and the structure can be regarded as *homogeneous* (non-fractal) on a length scale comparable to its diameter. In this case the surface has a geometry analogous to the Peano curve discussed in Section 2.3.1. Similar interfacial patterns have been observed in other types of experiments as well, thus it is useful to introduce a common name for these homogeneous structures. An interface has *dense radial structure* (DRS) if it i) grows outward from a centre, ii) is not a fractal, and iii) has a well defined spherical (circular) envelope.

10.1.2. Fractal viscous fingering

Figure 10.2 demonstrates that fractal viscous fingering patterns are not expected to grow in the traditional Hele-Shaw cell. However, the situation is completely changed if one introduces some kind of randomness into the experiment by placing a porous medium between the two plates. The investigation of viscous fingering in a random environment can be accomplished by using such model media as layers of small spheres (Maloy *et al* 1985) or carefully manufactured networks of channels having stochastically varying diameters (Chen and Wilkinson 1985, Lenormand and Zarcone 1985, Lenormand 1986).

For a given *random medium* very different regimes of fractal growth can be observed as a function of the applied flow rate or the wetting properties of the injected fluid. The following micromodel (e.g., Lenormand 1986) has been successfully used to investigate various crossovers in the shape of the interface between two fluids moving in a network. The main part of the model is a transparent resin plate consisting of channels and pores cast on a photographically etched mould. The flow is restricted to the channels by appropriately closing the resin plate. In a typical experiment the depth of the channels following the bonds of a square lattice is about 1mm, while their width varies around an average value according to a given distribution. The less viscous fluid is injected into the cell through a hole at the centre of the upper plate.

The two basic geometries are presented in Fig. 10.3 for the case of air displacing very viscous oil wetting the walls of the model. If the injection rate is extremely slow (many hours per experiment), the capillary forces dominate the motion of the interface. These forces may prevent the air from entering very narrow channels, while the interface advances faster at places where the channels have large cross-sections. In this limit the viscous forces and the effects of pressure distribution can be neglected. The above conditions correspond to the *invasion percolation* model with trapping (Section 5.2), and accordingly, the shape of the region filled with air is reminescent of percolation clusters (the perimeter of a large percolation cluster is shown in Fig. 5.10). The analogy is supported by an agreement between the fractal dimension $D \simeq 1.82$ calculated for invasion percolation with trapping and the value $1.80 < D < 1.83$ determined experimentally.

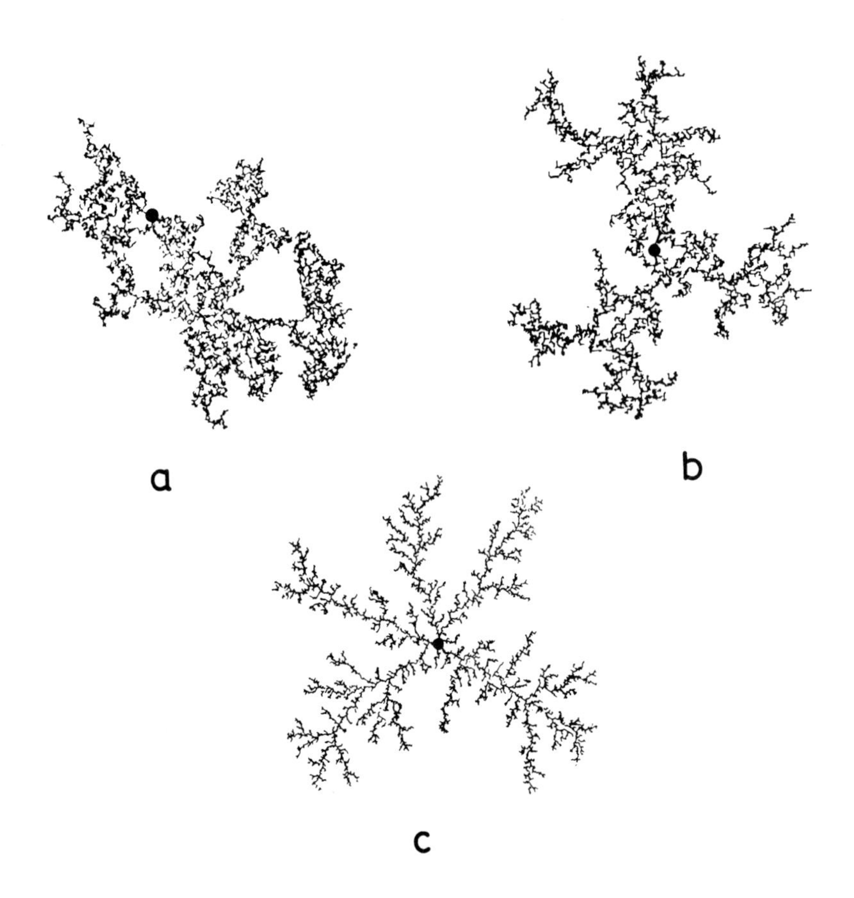

Figure 10.3. The black structures correspond to the channels in which the injected air displaced oil in a radial micromodel consisting of 250,000 capillary tubes having randomly distributed radii. (a) Very slow injection rate: invasion percolation regime, (b) crossover, and (c) high injection rate: in this regime the pattern is similar to DLA clusters (Lenormand 1986).

If the flow rate is considerably increased (few seconds per experiment) the capillary forces become negligible compared with the viscous ones. In this case, it is the non-local distribution of pressure within the more viscous fluid which dominates the flow. The situation is very similar to that existing in an ordinary Hele-Shaw cell, except that the varying channel widths represent local fluctuations in the parameter d_0. As a result fractal viscous fingering patterns are observed (Fig. 10.3c) having a geometry almost indistinguishable from diffusion-limited aggregates (Chen and Wilkinson 1985, Lenormand 1986). Decreasing the level of randomness in the channel widths leads to structures analogous to noise-reduced DLA clusters (Kertész and Vicsek 1986).

Porous media can also be modelled by placing a layer of randomly distributed glass spheres of diameter about 1mm between the plates of a Hele-Shaw cell (Maloy *et al* 1985). In these experiments DLA type patterns have been observed as well. The corresponding fractal dimension can be calculated by digitizing the pictures taken from the growing structure. The application of methods described in Section 4.2 gives an estimate $D \simeq 1.62$ somewhat lower than $D \simeq 1.71$ characteristic for two-dimensional DLA clusters. In such experiments the width of an individual finger is comparable to the holes between the glass spheres. If the injected fluid preferentially wets the medium, the structure of the interface remains qualitatively the same, but the finger width becomes much larger than the pore size (Stokes *et al* 1986).

Cells packed with non-consolidated, crushed glass have also been used to study viscous fingering in three dimensions (Clèment *et al* 1985) which represents a more important case from the practical point of view. The observed interfaces are complex, however, it is quite difficult to obtain an accurate estimate for the fractal dimension from the experimental data. An elegant three-dimensional experiment related to viscous fingering will be discussed in Section 10.4.

One interesting aspect of the micromodels is related to a special distribution of channel widths. Assuming that a fraction of the channels is completely blocked, while the rest of them have the same width, one arrives at networks of channels corresponding to *percolation* clusters (Oxaal *et al*

1987). Let us assume that the concentration of conducting channels is equal to p. Then at the percolation threshold, p_c, there exists a fractal network of open channels in the system and the model can be used to investigate the phenomenon of viscous fingering on a fractal. The incompressibility of the trapped fluid regions and the presence of singly connected paths gives rise to a number of non-trivial effects in such experimental systems. Most importantly, the fractal dimension of the viscous fingering patterns is considerably reduced compared with its value on networks, where each of the channels has a finite conductivity of varying degree. Since a finger which connects two sites on the backbone must pass through all the singly connected bonds on the link between these two sites, the overlap of patterns obtained for independent injections is relatively high. The resulting fractal dimension is $D \simeq 1.3$.

Fractal viscous fingering can be observed in systems *without any randomness imposed* by stochastic boundary conditions. Using a *smectic A liquid crystal* as the more viscous fluid in a radial Hele-Shaw cell it is possible to study a fractal $\rightarrow$ non-fractal crossover in the morphology of viscous fingers as the role of the inherent fluctuations decreases (Horváth *et al* 1987). Figure 10.4 shows the patterns which are obtained if gaseous nitrogen of varying pressure is injected into an 8CB Licrystal (BHD) at a temperature of 24°C corresponding to the smectic phase. The experiments with liquid crystals are usually carried out in a cell which is smaller than the standard versions. The linear size w is typically in the region of 7-10 cm, while the distance between the plates is about $b \simeq 40\mu$m.

If the applied pressure is small, randomly branching interfacial patterns are observed whose open structure is similar to that of DLA clusters. The fractal dimension associated with the bubble shown in Fig. 10.4a is close to 1.6. However, this open geometry crosses over into a homogeneous DRS (Fig. 10.4c) when the pressure of the injected N_2 is increased.

The above results can be interpreted on the basis of the specific internal structure of smectics. If the plates are not prepared to provide an ordering of the molecules on a length scale comparable to w, the director (the local orientation of the elongated liquid crystal molecules) randomly changes within

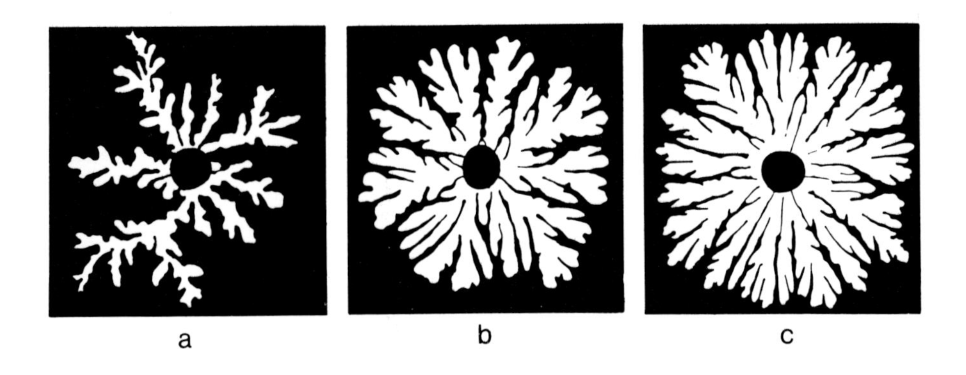

Figure 10.4. Increasing the pressure of air injected into a layer of smectic A liquid crystal results in a crossover in the global behaviour of the patterns. (a) For low pressures ($p = 30$mmHg) the interface has a structure analogous to that of DLA clusters, while (c) the patterns become homogeneous (non-fractal) for large pressures ($p = 70$mmHg) (Horváth *et al* 1987).

the sample. As a result of the *orientational disorder* the screening effects become dominant. The mechanism for the relevance of fluctuations is provided by the fact that the shear viscosity in an anisotropic liquid is strongly orientation dependent, thus the local ordering of the molecules indicates "easy" or "hard" local flow directions. The application of large pressures is likely to destroy the state in which a well defined director can be associated with the different parts of the sample. In this case the behaviour is expected to be analogous to that observed for isotropic fluids. Indeed, Fig. 10.2b and Fig. 10.4c are in qualitative agreement.

In addition to the above discussed examples, the structure of viscous fingering patterns has fractal properties (Fig. 1e) analogous to DLA if two miscible, non-Newtonian (or shear-thinning) fluids are used in the experiments (Nittmann *et al* 1985, Daccord *et al* 1986). In such a system the surface tension is practically zero and the flow velocity is determined by the expression

$$\vec{v} \sim (\nabla p)^k , \tag{10.7}$$

where the exponent k is in a typical experiment, larger than 1. For Newtonian fluids $k = 1$ (10.4). Correspondingly, the pressure distribution satisfies the equation $\nabla(|\nabla p|^{k-1}\nabla p) = 0$, which is also different from the Laplace equation (9.1). Because of the $(\nabla p)^k$ term in (10.7), for $k > 1$ the preferential growth of the tips is more pronounced in the non-Newtonian case than for Newtonian growth.

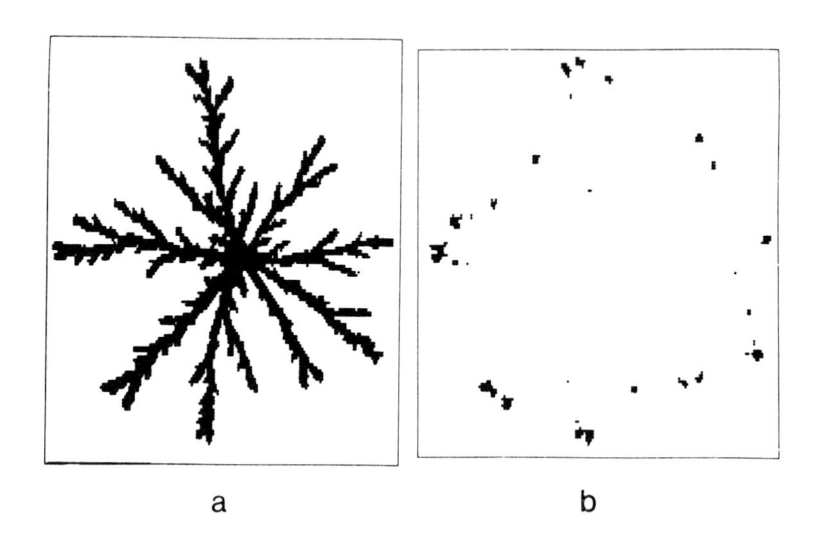

Figure 10.5. (a) Digitized image of a fractal viscous fingering pattern obtained by injecting water into a non-Newtonian, miscible liquid. (b) The region of fastest growth is visualized by subtracting the images of the same finger photographed at slightly different times (Daccord *et al* 1986).

The related experiment can be carried out by injecting dyed water into a layer of aqueous solution of polysaccharide. The viscosity of the polymer solution depends on the actual velocity, but it is usually $10^2 - 10^4$ times larger than that of the pure water. Figure 10.5 shows the digitized image of a representative pattern together with a picture demonstrating the preferential growth of tips. The measured fractal dimension $D \simeq 1.7$ is close to the DLA value in this case as well. Analogous patterns are grown if water is injected into clay slurry (van Damme *et al* 1986), which is a system of practical importance.

Figure 10.5 also indicates a possible method to calculate the *multifractal spectra* describing the distribution of experimental growth velocities (Nittmann *et al* 1987). These velocities can be estimated by simply subtracting subsequent images of the same finger taken at slightly different times. Then an analysis along the lines discussed in Section 3.2 and Section 9.4 can be applied to determine the spectrum of fractal dimensions $f(\alpha)$. Because of the limited available resolution it is not possible to obtain quantitatively accurate results using this technique; however, the qualitative behaviour of the calculated quantities has been found to be of DLA type.

10.1.3. Viscous fingering with anisotropy

The crucial role of the anisotropy of surface tension in the formation of interfacial patterns has recently been demonstrated by a number of theoretical approaches and computer simulations (Sections 9.2.2 and 9.3). These studies have also indicated that diverse sources of anisotropy may lead to similar results. Experiments on viscous fingering in appropriately modified versions of the radial Hele-Shaw cell represent a valuable tool for the investigation of the combined effects of anisotropy and the driving force (pressure difference) in real systems. In particular, the conditions leading to the stabilization or destabilization of the tips of fingers are of special interest.

Anisotropy can be introduced into the experiments on viscous fingering in isotropic liquids by *engraving a mesh* on the surface of the glass plates of the cell (Ben-Jacob *et al* 1985). For a given set of grooves there are two parameters which can be monitored in the course of an experiment. Increasing the distance b between the two plates generally corresponds to decreasing the effective anisotropy, while in order to investigate the role of driving force one changes the pressure of the less viscous fluid. In the basic version of such experiments air is injected into an ordinary viscous liquid (e.g., glycerine). The grooves commonly have a depth, a width and an edge-to-edge distance about 0.5-1.0 mm. It has been demonstrated that a small separate bubble located at and moving together with the tip of a growing finger introduces an effective anisotropy as well (Couder *et al* 1986a, 1986b) leading to interfaces analogous to one of the structures shown in Fig. 9.3

Figure 10.6. These patterns are observed in a radial Hele-Shaw cell as a function of increasing pressure if one of the plates has a triangular lattice engraved on it.

(middle picture of Ib).

The types of patterns which are observed if the grooves are engraved according to the geometry of a triangular lattice are shown in Fig. 10.6. In this set of experiments b is fixed and the *morphological changes* are obtained as a function of the increasing pressure p. For low values of p faceted growth (a) is found. When increasing the pressure the interface becomes unstable against tip splitting (b), and an entirely disordered structure develops in the cell. A needle crystal type pattern having stable tips and a sixfold symmetry (c) is observed, if p is further increased. Finally, for the largest pressures applied, the radius of curvature of the tips is reduced and side branches appear on the main dendrites (d). The sixfold symmetry due to the underlying mesh is approximately conserved in this case, in analogy with the structure shown in Fig. 9.10. Correspondingly, this pattern is likely to have a non-trivial radius of gyration exponent. However, such measurements have not been carried out.

A rich variety of morphological phases can also be obtained in a simpler system, where a set of parallel grooves is etched on one of the plates (Horvàth *et al* 1987). A systematic series of experiments with this geometry allows the construction of a *morphological phase diagram* of the non-equilibrium patterns observed in the system. The results are summarized in Fig. 10.7, where the phases are indicated by their typical patterns.

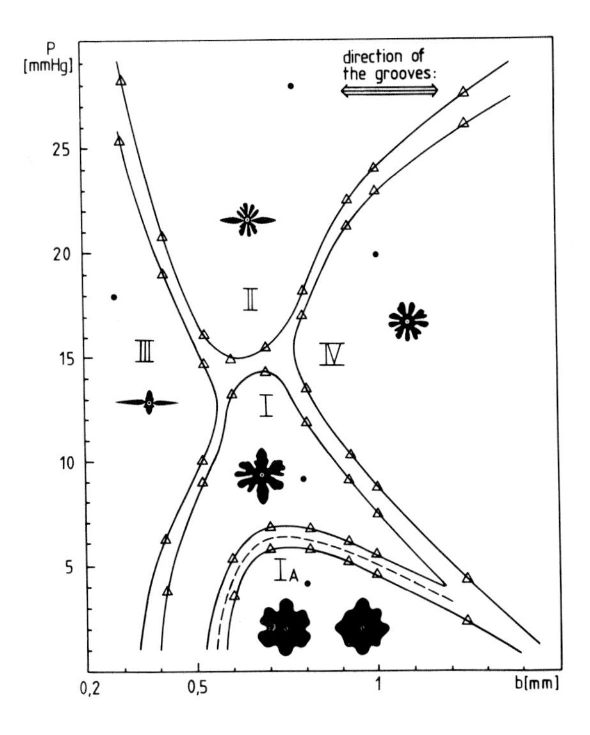

Figure 10.7. Morphological phase diagram of the interfacial patterns obtained in the radial Hele-Shaw cell with a uniaxially engraved plate. Varying the pressure (p) and the distance of the plates (b) relatively sharp transitions can be observed between the different phases denoted by characteristic patterns (Horváth *et al* 1987).

This figure demonstrates the complexity of the phenomenon of viscous fingering with anisotropy as a function of the driving force. Depending on the value of the parameters b and p, the direction of the stable growth changes and virtually all of the possible geometries can be achieved. The *crossover* from one kind of pattern to another one is quite *sharp* taking place in a narrow pressure interval. For low pressures the growth can be stable in the "hard" direction (perpendicular to the grooves) and unstable (splitting tips) along the grooves. This situation is reversed as the pressure is increased. Another typical case is when stable tips grow in both directions.

The experiments with an imposed anisotropy suggest the following picture of the interplay of anisotropy and pressure difference. It is plausible to assume that the mechanism by which a tip growing in the direction perpendicular to the grooves is coupled to the channels is different from that dominating in the case of parallel growth. The response of these mechanisms to the changes in the pressure is expected to be different; the effective anisotropy in the perpendicular direction decreases, while in the parallel direction it increases as larger pressures are applied.

The presence of the direction dependent couplings explains the nontrivial behaviour of the effective anisotropy as a function of b and p and this is the reason leading to the complex phase diagrams of the type shown in Fig. 10.7. It is important to point out that the above situation is typical, because in the real systems exhibiting pattern formation the anisotropy has several sources. In fact, this can be taken into account by the two terms in the boundary condition (9.4), where the equilibrium and kinetic contributions are represented separately. For fast growth the anisotropy in the kinetic coefficient β dominates the growth, but in the slow regime it is not expected to play an important role. If the angular dependence of the static surface tension γ is different from that of β, the *competition* between the anisotropies gives rise to a rich behaviour.

In the experiments discussed above the anisotropy was imposed externally by engraving a macroscopic grid on the plates. Another alternative is to use a liquid which is *inherently anisotropic on a microscopic scale* (Buka *et al* 1986). This can be achieved by placing a liquid crystal having an angular dependent shear viscosity into the cell. The anisotropic nature of liquid crystals is manifested only if there exists a well defined local director, i.e., the elongated molecules are directed approximately uniformly in the given region. In the experiments with *nematic liquid crystals* the flow of the liquid itself provides the necessary orientation. As soon as the fluid pushed by the injected air starts flowing away from the central hole, the molecules become radially oriented (this can be checked by crossed polarizers). Nevertheless, for low injection rates the growing interface goes through repeated tip splittings and one obtains a pattern characteristic for Hele-Shaw cells with isotropic liquids (see Fig. 10.2a).

However, increasing the flow rate results in the qualitatively different behaviour shown in Fig. 10.8. The tips become stable and the observed structure looks like a snowflake type dendritic crystal. A further increase of the pressure induces another morphological transition: the tip splitting mechanism is restored and the resulting patterns (Fig. 1d) seem to be space filling (in analogy with Fig. 10.2b). The above described tip splitting → dendrites → tip splitting *reentrant transition* can be observed as a function of the decreasing temperature as well (Buka *et al* 1987). A possible qualitative interpretation of the above transitions is based on the assumption that for slow flow the molecules are not orientated enough to produce the necessary anisotropy to stabilize the tips. On the other hand, for large pressures the director at the interface is likely to become turbulent and the local ordering of the molecules is lost.

Carrying out experiments with a smectic A liquid crystal can be used to demonstrate the effects of *inherent uniaxial anisotropy* (Horváth *et al* 1987). The local director in smectics is not dependent on the applied pressure as much as in the case of nematics, therefore, it is possible to investigate the influence of a preliminary alignment of the molecules. During such experiments the temperature of the sample is increased to a value corresponding to the isotropic phase and then an external magnetic field parallel to the plates is applied. The field aligns the molecules along its direction and as the liquid crystal is cooling down this orientation freezes in. For a given pressure of the injected air the overall shape of the bubbles changes from nearly circular to an elongated one as the magnitude of the magnetic field is increased.

Figure 10.8. Viscous fingering pattern obtained by injecting air into a thin layer of a nematic liquid crystal (7A Licrystal, BDH) at room temperature. The radial alignment of the molecules and the anisotropic shear viscosity results in stabilization of the tips (Buka *et al* 1986).

In the case of preoriented samples the following crossover can be observed. For very low injection rates the interface is elongated in a direction perpendicular to the director (Fig. 10.9a). This is understandable, since in smectic A liquid crystals the molecules are ordered into separate layers perpendicular to the director and these layers can slip on each other relatively easily. However, when the pressure is increased, the structure of the patterns changes qualitatively. It becomes more ramified and elongated parallel to the field (Fig. 10.9b) resembling DLA clusters obtained by using an anisotropic sticking probability (Fig. 6.7). A possible explanation for this behaviour is that the larger pressure gradient generates dislocations of the layers and these dislocations move easier in a direction parallel to the field (perpendicular to the layers).

10.2. CRYSTALLIZATION

Non-equilibrium solidification processes are known to lead to complex geo-

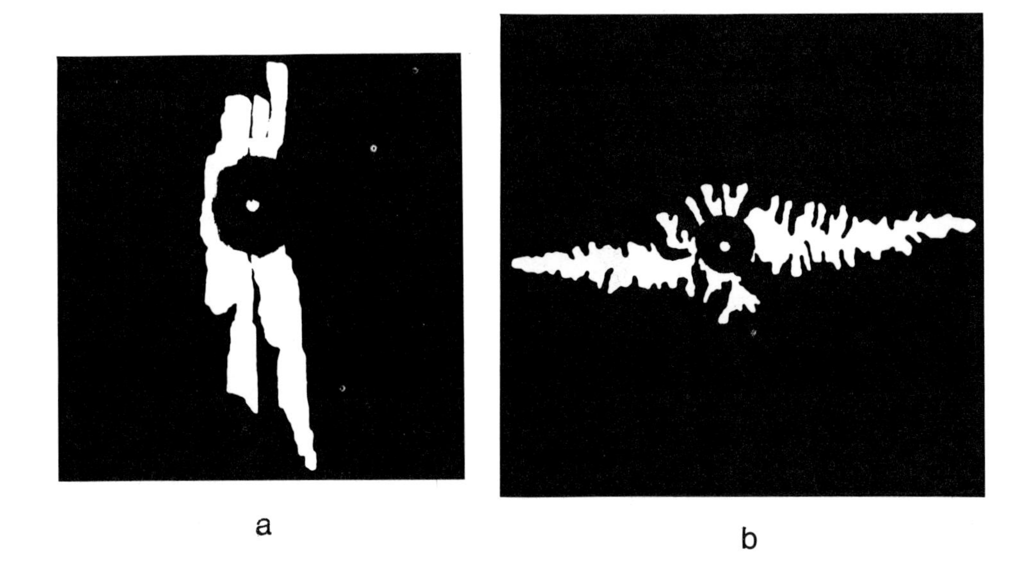

Figure 10.9. If the molecules of the smectic liquid crystal are preoriented by an external magnetic field, the interface becomes elongated. (a) For small pressure the pattern is elongated in a direction perpendicular to the field; (b) at $p \simeq 35$mmHg a crossover takes place and the fingers grow easier parallel to the field (Horváth *et al* 1987).

metrical patterns. There are two basic types of dendritic crystallization: i) in a pure undercooled liquid and, ii) in an isothermal liquid mixture. Here we used the term crystallization (instead of solidification) to express the fact that crystals can grow in heat treated amorphous (i.e., already solid) materials as well. In the undercooled case it is the distribution of temperature which represents the rate limiting quantity, while during isothermal crytallization the concentration of the diffusing atoms dominates the phenomenon. In the following the term diffusion will be used for the transport of both heat and mass.

The equations describing crystallization (Langer 1980) in the limit of large diffusion length, $l = 2C/v_n$, and large diffusivity in the crystalline phase are the same as (9.1-9.4), where C is the diffusion constant and v_n is the normal velocity of the interface. In many cases these assumptions are not valid, and instead of (9.1) it is more appropriate to describe the process by the *diffusion equation* expressing the conservation of heat or mass

$$C_i \nabla^2 u = \partial u / \partial t \,, \tag{10.8}$$

where u denotes either temperature or $\tilde{\mu}$, and C_i is the diffusion constant in substance i. Here $\tilde{\mu}$ is the difference between the chemical potential μ and its equilibrium value for two phase coexistence at the temperature at which the crystallization takes place. If the diffusivity in the crystallized phase is not much larger than in the surrounding one, the boundary condition (representing a continuity equation at the surface) (9.2) should be modified to have the form

$$v_n = [c_{cryst}(\nabla u)_{cryst} - c_{surr}(\nabla u)_{surr}]\hat{\mathbf{n}} \,, \tag{10.9}$$

where the subscripts indicate quantities in the crystalline and the surrounding phases (the latter can be either liquid or amorphous). It is important to be aware of the fact that unlike viscous fingering the approximations leading to the set of equations (9.1-9.4) are frequently not satisfied for crystallization. In some cases this is the reason why the observed complex patterns are not fractals.

Because of the regular microscopic structure of the growing phase, the development of single crystals is typically dominated by the anisotropy of the surface tension. Accordingly, the patterns observed in these experiments are more or less symmetric. To obtain randomly branching fractal structures one can study crystallization in an *amorphous thin film* (Radnóczy *et al* 1987). In this system, the effects caused by the anisotropic surface tension are expected to be small because the new phase is polycrystalline with preferred growth directions randomly distributed.

Figure 10.10a shows a highly ramified polycrystalline branch grown in an amorphous $GeSe_2$ thin film. Such pictures can be obtained by heat treating the amorphous sample prepared by vacuum evaporation and taking transmission electron micrographs of the patterns. Crystallization is observed at 220°C, which is well below the $GeSe_2$ glass transition temperature of

265°C. Digitizing the image an estimate for the fractal dimension of the pattern shown in Fig. 10.10a can be given. One counts the number of dark pixels $n(r)$ at a radius r from the centre of mass and assumes the scaling $n(r) \sim r^{D-1}$. From the related log-log plot shown in Fig. 10.10b the value $D \simeq 1.69$ is obtained (Radnóczy et al 1987) which is very close to the fractal dimension of diffusion-limited aggregates generated in two dimensions.

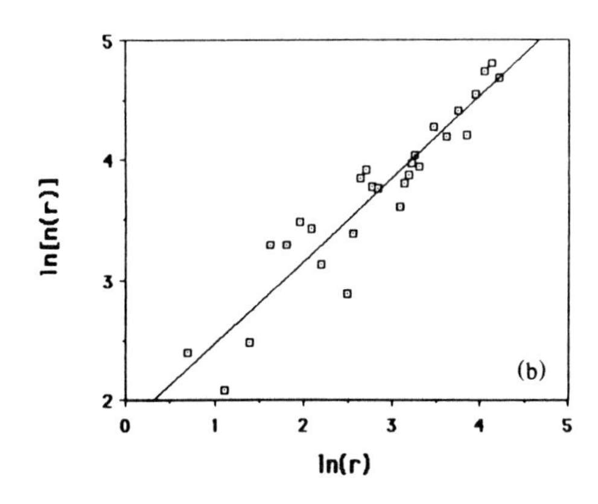

Figure 10.10. Part of the tree-like crystalline phase growing in an amorphous $GeSe_2$ thin film heat treated at 220°C (left). The plot on the right gives an estimate $D \simeq 1.69$ for the fractal dimension of the pattern shown in part (a) (Radnóczy *et al* 1987).

Fractal crystallization in the present example is likely to be due the mechanism analogous to that occurring in the isothermal solidification of a liquid mixture. This mechanism represents one of the ways how the Mullins-Sekerka instability is manifested (Section 9.1). Let us assume that the main components are not present with their exact stochiometric concentrations, e.g., $c_{Se} > c_{Ge}/2$. Then, if the temperature is higher than the amorphous-solid transition temperature, crystallization of $GeSe_2$ starts at places which are somewhat warmer or contain more nucleation centres. The crystallizing phase expels excess Se so that the amorphous region becomes further enriched in Se. This excess of Se must diffuse away before further crystallization can occur. Concentration gradients are greatest at the most advanced parts of the interface, and the corresponding instability drives the system.

To understand the phenomenon better, a few further remarks are in order. An estimation for the velocity of the interface and for the diffusion constant of Se in $GeSe_2$ at 220°C suggests that the diffusion length l in the bulk is very small compared with the size of the pattern. Thus, a fractal interface can not be formed through bulk diffusion of atoms. However, surface diffusion coefficients are commonly measured to be 10^8-10^{10} times larger than those of the bulk. Advancement of the crystallization front through *surface diffusion* would account for the scale invariance of the observed structures. In a solid-solid phase transition a number of additional factors are expected to affect the results. The specific volume of the crystalline phase is smaller than that of the amorphous state and because of this, long-range elastic forces are created during the growth which may also play an important role in the fractal growth.

Interestingly, a qualitatively different behaviour can be observed in a closely related system of vacuum deposited $Al_{0.4}Ge_{0.6}$ amorphous thin films (Ben-Jacob *et al* 1986). After heat treating the sample at a temperature of 230°C *dense radial structures* (DRS) have been found to grow at many sites (Fig. 1a). As revealed by electron diffraction studies the growing phase is polycrystalline Ge, which is surrounded by an Al-rich region being a nearly perfect crystal. One of the possible reasons for the apparent non-fractality of the obtained patterns may be the relative shortness of the diffusion length. The physics of dense radial growth is not completely understood yet and it has recently been subject to intensive theoretical and experimental investigations (Grier *et al* 1987, Alexander *et al* 1988, Goldenfeld 1988).

A number of further experiments on pattern formation in thin films have demonstrated the development of complex interfacial structures. Random dendritic structures were observed to grow in $NbGe_2$ sputter deposited onto silica (Elam *et al* 1985). The skeleton of these objects was found to have a fractal dimension close to 1.7. In some of the cases the obtained structures have been suggested to be fractals, however, in the light of recent results on dense radial growth it is more plausible to assume that the geometry of the observed patterns is analogous to the dense radial structure. The complex crystalline phosphorlipid domains growing in monomolecular layers (Miller *et al* 1986) and the patterns observed in ion irradiated Ni-Mo alloy films (Liu *et al* 1987) seem to fall into this category (schematically shown in Fig. 9.3, IIa). Note, that one can call a physical object fractal if the corresponding non-integer scaling is well satisfied at least for two orders of magnitude (Section 2.1). (Sometimes one makes exceptions when the scaling holds with particularly good accuracy.)

Randomly branching, fractal dendritic growth of single crystals can be produced by *introducing fluctuations externally* (Honjo *et al* 1986). Consider a supersaturated solution of NH_4Cl between two smooth parallel plates separated by a small distance of $\sim 5\mu$m. In this system the anisotropy of the surface tension causes very regular dendritic growth. Strong random perturbations can be imposed by replacing one of the plates with another one having a *rough surface.* This experiment is related to those designed to study viscous fingering in random media. On the other hand, it represents an opposite approach to engraving on the surface of the Hele-Shaw cell a regular lattice (Sections 10.1.2 and 10.1.3). There, anisotropy had to be introduced because the fluid had an isotropic surface tension, while in the present case one intends to get rid of the anisotropy by scratching the plate. It should be noted that for very large undercoolings some of the dendrites can grow in a direction different from the crystallographic one even in a cell without randomness. In such cases spontaneous splittings of the tips have been found.

Figure 10.11a shows a typical pattern obtained with a plate having a characteristic length of roughness approximately equal to 7.5 μm. The development of the structure is indicated by superposing the digitized pictures corresponding to subsequent stages of the process. As in DLA, those are only the tips which advance significantly. The width of the fingers is roughly the same as the mean length of the roughness. The interface is very similar to the small scale viscous fingering patterns observed in Hele-Shaw cells with randomness, and is also reminiscent of the computer simulation shown in Fig.

9.13. The dependence of the crystal's area on the radius of gyration is displayed in Fig. 10.11b. Although fractal scaling with an exponent $D \simeq 1.67$ is found on a length scale somewhat less than a decade, the data points follow a straight line with a surprising accuracy. This suggests the continuation of similar behaviour to considerably larger sizes.

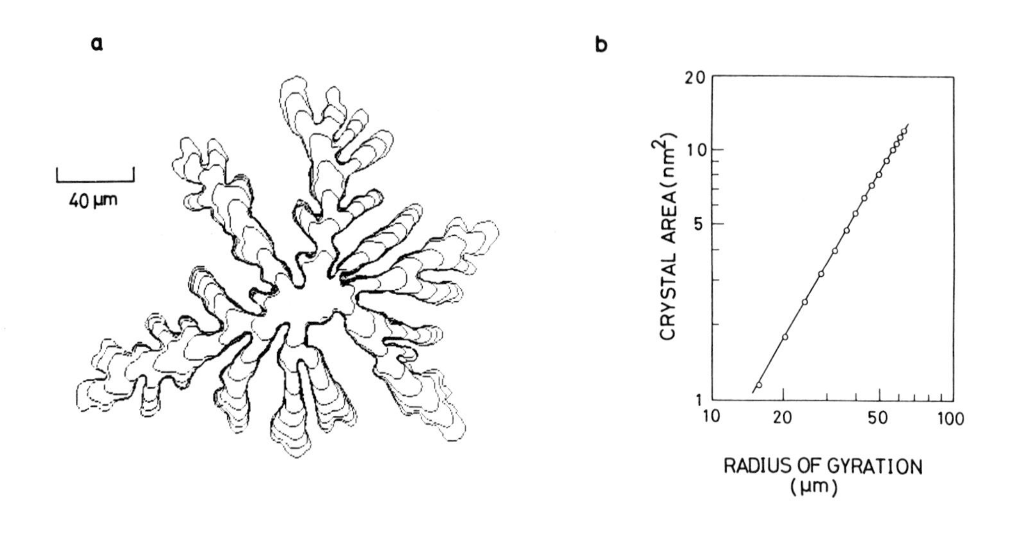

Figure 10.11. (a) Superimposed picture of a growing NH_4Cl crystal taken at 20-sec intervals. The growth is confined between glass plates separated by 5μm. (b) Estimating the fractal dimension of the pattern (a) from the dependence of the crystal's area on its radius of gyration ($D \simeq 1.67$) (Honjo *et al* 1986).

The overlapping patterns shown in Fig. 10.11a provide a suitable basis for the *multifractal analysis* of the growth velocities (Ohta and Honjo 1988). Let us define the growth probability distribution as a set of normalized velocities $p_j = v_j / \sum_j v_j$, where v_j is the normal velocity of the interface at the jth pixel point. Then the $f(\alpha)$ spectrum can be determined using the method discussed in Chapter 3, Section 6.1.4 and at the end of Section 9.4.

There are several ways to determine the set of v_j-s. One possibility is to calculate the velocity of the interface directly by measuring the length of the interval which is perpendicular to the surface and is bounded by two successive perimeter positions. The boundary condition (9.2) can be used for an indirect determination of the interfacial velocity through the knowledge of the temperature gradients at the surface. These gradients are calculated by numerically solving the Laplace equation. This is the same procedure which was used to study the multifractal properties of the growth site distribution of off-lattice diffusion-limited aggregates.

The results for $n(p)$ and $f(\alpha)$ are displayed in Fig. 10.12, where $n(p)dp$ is the number of places with a growth probability between p and $p + dp$. Figure 10.12a demonstrates that the direct determination of the growth velocities is rather limited by the resolution. Thus, this method does not allow the calculation of $f(\alpha)$ for $\alpha > 1$, because this is the region in which low growth probabilities give the dominant contribution. The $f(\alpha)$ spectrum determined from the calculated gradients is in good agreement with the related

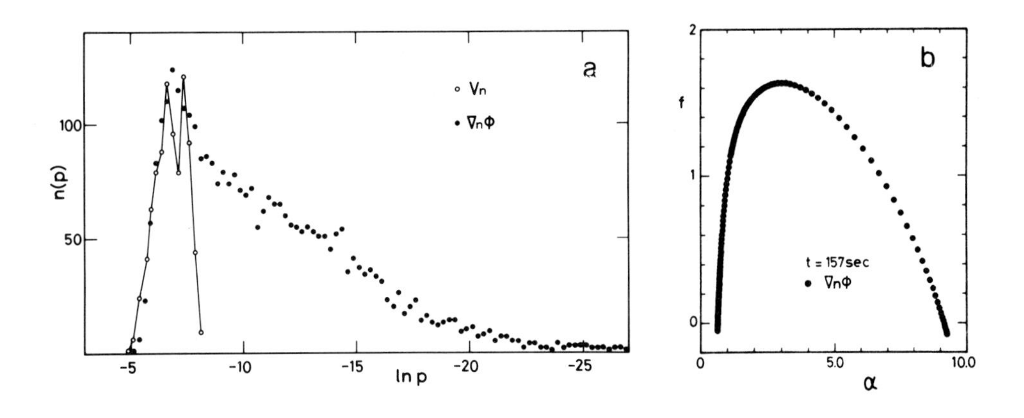

Figure 10.12. (a) The distribution functions, $n(p)$, obtained from determining the velocity of the interface of the experimental pattern shown in Fig. 10.11a. The velocities were both measured directly (V_n) and calculated from the corresponding temperature gradients ($\nabla_n \phi$). (b) The spectrum of fractal dimensionalities ($f(\alpha)$) determined from the calculated data for the temperature gradients (Ohta and Honjo 1988).

results obtained for DLA clusters. The limiting cases $D_0 \simeq 1.63$, $D_1 \simeq 1.13$, $\alpha_{-\infty} \simeq 9.4$ and $\alpha_\infty \simeq 0.6$ are in reasonable accord with the corresponding numerical and the following theoretical results for DLA: $D_0 = 5/3$, $D_1 = 1$ and $\alpha_\infty = 2/3$.

10.3. ELECTROCHEMICAL DEPOSITION

In a typical experiment on electrodeposition two elecrodes are immersed into an ionic solution and one observes the structure made of metal atoms deposited onto the cathode. The anode and the cations are of the same metal and a stationary concentration of the cations is maintained by the dissolution of the anode itself. The most common experimental setup (Matsushita *et al* 1984) is similar to that of the radial Hele-Shaw cell: the electrolite is kept between two close glass plates with a circular anode surrounding the cathode located at the centre. The typical sizes are 10 cm (diameter) and 0.5 mm (distance of the plates), while the applied voltage is in the range of 1 to 20 volts.

The growth of electrodeposits involves a number of simultaneous processes which lead to a quite complicated behaviour of the system as a function of the applied voltage and chemical concentration. As a result, the experimentally observed *morphological phase diagrams* are rich (Grier *et al* 1986, Sawada *et al* 1986) and do not entirely coincide even for the same values of the above parameters. Deviations from a purely Laplacian growth are expected to occur due to the following problems: i) the transport of the ions is affected by convection of the fluid, ii) at high voltages gas evolution takes place because of electrolysis of water, iii) the ion can be driven both by the concentration gradient and the electric field, and iv) the actual voltage drop within the electrolite is unknown (unless reference electrodes are used), because of the voltage drops within the deposit and in the surface layer of the electrodes.

To see the relation of electrochemical deposition to Laplacian growth one first assumes that the effects of convection can be neglected on a length scale much larger than the distance between the plates. Furthermore, let us suppose that only one kind of the ions (having a local concentration c) is electroactive. The total current of this ion $\vec{j}$ (due to both diffusion and drift in the field) is proportional to the gradient of the associated electrochemical potential (e.g., Kessler *et al* 1988)

$$\mu = A(T) + kT \ln c + q\phi\,, \tag{10.10}$$

where $A(T)$ is a temperature dependent constant, k is the Boltzmann factor, q is the electric charge per ion and ϕ is the electric potential. Next we introduce the dimensionless diffusion field $u = \mu/qV$ (with V denoting the applied voltage), and note that in the quasi-stationary limit the divergence of the total current vanishes, i.e., $\nabla\vec{j} \simeq \nabla^2 u = 0$.

As an extension of the above conservation law to the interface one obtains the boundary condition concerning the velocity of the interface $v_n \sim \hat{\mathbf{n}}\,\vec{j} \sim \hat{\mathbf{n}}\nabla u$ in accord with (9.2). It is less straightforward to specify the boundary condition prescribing the value of u on the growing interface. It can be shown that in the limit of small growth rate this last boundary condition becomes analogous to (9.4) (without the kinetic term). In general, however, the condition of local thermodynamic equilibrium is not satisfied in electrodeposition processes and the kinetics of charge transfer at the interface results in a relation for u on the surface more complicated than (9.4). On the other hand, in the far-from-equilibrium limit the interface acts as a perfect absorber and this condition corresponds to the situation characteristic for DLA. It has also been argued that in the case of dense radial growth the potential at the interface is determined by the resistivity of the filaments (in other words by the voltage drop across the deposit).

The above mentioned *complex reaction kinetics* at the interface is likely to be the reason for the *rich behaviour* of electrochemical deposition. In fact, it is only in the process of electrodeposition that the three basic types of patterns (homogeneous, random fractal and dendritic) can be observed without changing the conditions of the experiment qualitatively (Grier *et al* 1986, Sawada *et al* 1986). This is demonstrated by the bottom row of Fig. 1, where zinc metal deposits of various morphologies are displayed. These rather different patterns were obtained as a function of the $ZnSO_4$ concentration and the applied voltage only. For the homogeneous or dense radial structure $D \simeq 2$, while the value $D \simeq 1.65$ obtained for the random fractal patterns (Matsushita *et al* 1984) is in good agreement with the DLA result. Estimates for the fractal dimension of the observed dendritic patterns have not been reported.

The differences in the macroscopic geometrical properties of the observed patterns is in a close relation to their microscopic structure. This can be shown using electron microscopy and x-ray diffraction for examining the deposits. Transmission electron micrographs suggest that during the dendritic and tip splitting modes of growth the structures are different on the microscopic level as well (Grier *et al* 1986). In Fig. 10.13a a small part of a DLA-like, random fractal deposit is displayed. The isotropic rings in the inset (which is a selected area diffraction pattern from such a region) demonstrate that there is no long range ordering of atoms in the tip splitting deposits. On the other hand, micrographs taken from the dendritic tips (Fig. 10.13b) suggest that they are characterized by large, rounded crystal facets. In addition, the corresponding diffraction patterns (inset in Fig. 10.13b) consist of well defined diffraction peaks indicating long range order.

As in the case of diffusion-limited deposition models (Section 6.2) studies of electrochemical deposition onto *linear electrodes* provide further information about the growth. In particular, this arrangement is practical for carrying out three-dimensional experiments (Brady and Ball 1984). More importantly, in this approach an ensemble of separate metal trees is found to grow during the process of deposition (Fig. 10.14) (Matsushita *et al* 1986). This feature of the experiment is helpful to strengthen the analogy between DLA and electrodeposition. To describe the distribution of trees in a quantitative manner, the picture of the deposit is digitized, and the size of a given tree is defined as the number of pixels s belonging to it. Then n_s the number of trees of size s is determined making an average over several forests of approximately the same size.

Assuming that for intermediate s the *tree-size distribution* scales as $n_s \sim s^{-\tau}$ the estimate $\tau \simeq 1.54$ can be obtained for the exponent describing

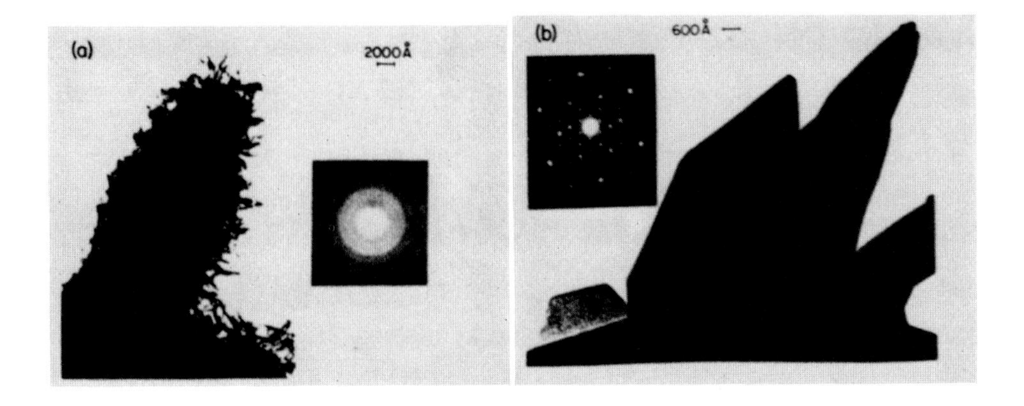

Figure 10.13. Transmission electron micrographs and diffraction patterns of electrodeposited zinc. (a) Picture of a tip observed in the DLA-type regime. The diffuse rings in the diffraction pattern (inset) indicate amorphous microstructure. (b) Dendritic growth with crystalline microstructure (demonstrated by the superlattice diffraction pattern shown in the inset) (Grier *et al* 1986).

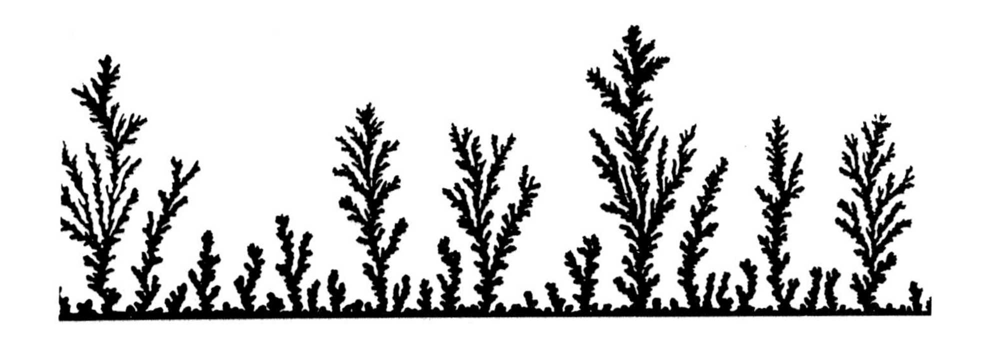

Figure 10.14. Representative picture of a forest of zinc metal trees deposited along a linear carbon cathode. The actual size of this part of the sample is about 6cm (Matsushita *et al* 1985).

the algebraic decay of n_s. This value is in remarkable agreement with the related result for τ determined from large scale simulations of diffusion-limited deposition (see the end of Section 6.3).

Electrochemical *polymerization* of conducting polypyrrole has also been shown to exhibit rich behaviour (Kaufmann *et al* 1987). An important difference between these experiments and electrochemical deposition of metal ions is that the polypyrrol monomer is a neutral species. Polypyrrol is grown by the removal of two electrons from pyrrole monomers and the subsequent polymerization with loss of alpha hydrogens. In addition, the pyrrole chains are oxidized with about one hole, and one counterion, for every three monomer units. Because of the neutrality of the pyrrol monomers the experiment is subject to a smaller number of side effects of electrical origin than the electrodeposition of metal ions. This could be the reason for the observation that in the polymerization experiment an increase of the voltage produces a dendritic $\rightarrow$ tip splitting crossover, while in the case of depositing ions, increasing V results in a tip splitting $\rightarrow$ dendritic morphological transition.

10.4. OTHER RELATED EXPERIMENTS

In this section two more examples for Laplacian growth will be described. i) Dielectric breakdown represents a typical random growth process. It occurs whenever the electric field is strong enough to generate a conducting phase within an insulator. Lightning is the best known version of this phenomenon. ii) The other example is concerned with the formation of dissolution patterns in a porous medium. The nature of these experiments allows one to produce three dimensional objects which can be preserved for later studies of their fractal properties.

i) Fractal *dielectric breakdown* patterns can be studied by inducing a two dimensional radial discharge. To obtain a leader surface discharge (Lichtenberg figure) one can use the following arrangement (Niemeyer *et al* 1984). A starter electrode is brought into contact with an insulating (glass) plate. The other side of the plate is covered by a grounded conducting material, and the whole system is kept in compressed SF_6 gas of pressure

0.3 MPa. Applying a voltage pulse 30kV×1μs at the starter electrode a propagating discharge pattern can be observed. Figure 10.15a shows a typical time-integrated image of such structures.

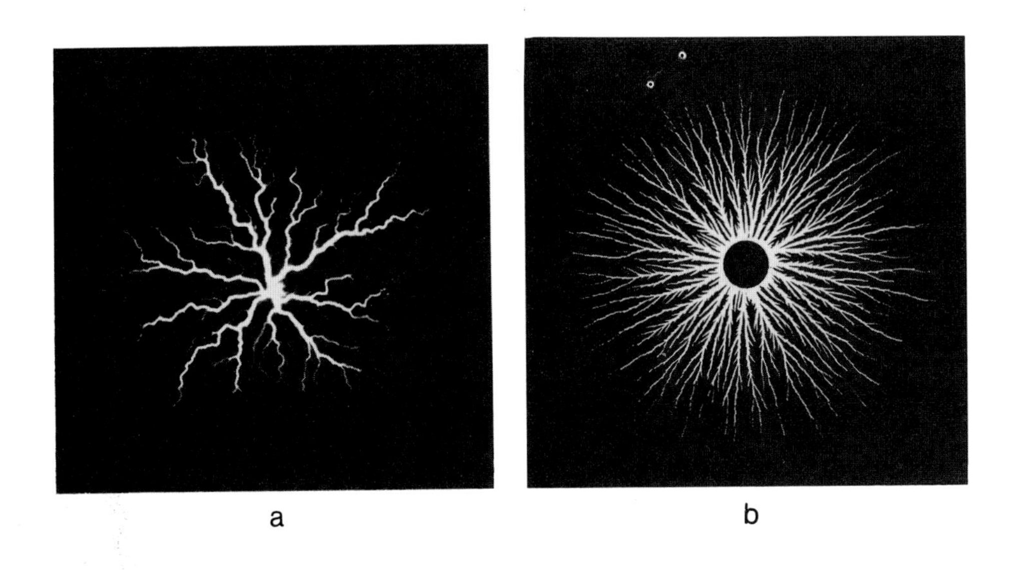

Figure 10.15. Time-integrated images of dielectric breakdown patterns. (a) Fractal structure obtained in a quasi-three-dimensional experiment (Niemeyer *et al* 1984) and (b) a dense branching structure observed under two-dimensional boundary conditions (courtesy of N. Allen).

To estimate the fractal dimension of the pattern displayed in Fig. 10.15a one counts the number of branches $n(r)$ at a given distance from the centre. For a fractal of dimension D this number should scale as $n(r) \sim dN(r)/dr \sim r^{D-1}$, where $N(r)$ is the total length of all branches (or the area measured with a given resolution) inside a circle of radius r. The branches are supposed to be extremely thin, and their thickness does not grow with the size of the object. (The apparent thickening of the branches in Fig. 10.15 is just an optical effect which is due to the number of carriers that have passed through a given branch.) In fact, there are more branches visible on the original negative, because the thinnest ones are lost in the course of reproduction. A careful counting of $n(r)$ suggests a value $D \simeq 1.7$ (Niemeyer *et al* 1984) in agreement with the dielectric breakdown model for $\eta = 1$ (Section 6.3).

The physics of dielectric breakdown is quite complicated. During the growth of a discharge pattern a conducting phase is created as the interface advances. The channels consist of non-equilibrium plasma with mobile electrons produced by a critical electric field $\vec{E}_c$ through electron impact ionization. The phenomenon can be described by the standard equation of electrodynamics: $-\nabla^2\phi = Q$ and $\vec{E} = -\nabla\phi$, where ϕ is the electric potential and $Q = e\sum_{k,\sigma}\sigma n_k(\sigma)$ is the density of total charge with $n_k(\sigma)$ denoting the density of charge carriers of charge σ e. In the insulating phase Q is equal to zero, however, within the discharge pattern charge is created, annihilated and transported according to the expression

$$\partial n_k(\sigma)/\partial t = D_k(\sigma)n_k(\sigma)\,, \tag{10.11}$$

where $D_k(\sigma)$ is a local but non-linear operator depending on $\vec{E}$ and $n_k(\sigma)$. In the dielectric breakdown model (DBM) all the details represented by (10.11) are ignored by simply assuming that $\phi = Constant$ within the pattern. (This is a trivial solution of the Laplace equation for a conducting object.) Thus the charge density is assumed to be different from zero only at the interface, which is consistent with the observations. In addition to the above assumption, in the dielectric breakdown model the apparent stochasticity of the process is taken into account by selecting surface sites for occupation randomly.

In the original simulations of DBM it is assumed that the phenomenon is essentially two-dimensional. Correspondingly, as a boundary condition for the potential (9.3) is used, i.e., ϕ is supposed to be constant along a large circle. However, the experiment is *quasi three-dimensional*, and the conducting material attached to the bottom of the glass plate provides a constant potential on the surface of a disc being at a small distance from the discharge in the third dimension. The simulations with $\eta = 1$ and boundary conditions corresponding to this situation resulted in nearly homogeneous clusters with a dimension close to 2 (Satpathy 1986). On the other hand, the fractal dimension of the simulated patterns agrees well with the experimental value if an exponent $\eta = 4$ is used in the calculation of the growth probabilities.

The complexity of the problem is further emphasized by Fig. 10.15b, where a discharge pattern obtained in a *two-dimensional experiment* (Allen 1986) is shown (with a circular electrode in the same plane with the pattern). In this case the structure seems to be dense ($D \simeq 2$) (Niemeyer *et al* 1986), while DBM gives $D \simeq 1.7$. Since on physical grounds one expects $\eta = 1$, one concludes that to make the DBM become a realistic model for dielectric breakdown a number of such additional effects have to be taken into account as the existence of a threshold field and the internal resistance of the plasma channels.

ii) *Chemical dissolution of a porous medium* involves the flow of a liquid in the medium coupled with a chemical reaction. In this sense the process is similar to viscous fingering in a random environment, with the difference that with the motion of the interface part of the medium is removed (dissolved). The main idea of the experiment (Daccord and Lenormand 1987, Daccord 1987) is based on the phenomenon that plaster (hydrated calcium sulfate) is slightly soluble in pure water.

The related investigations can be carried out in a *three-dimensional* sample with a characteristic linear size ~ 5 cm. The plaster is prepared by mixing 10 parts of pure water with 11 parts of $CaSO_4 \cdot 0.5H_2O$. Initially the sample is saturated with water so that the system is in chemical equilibrium. Then pure water is pumped through one of the faces at a constant rate, displacing the saturated water and subsequently dissolving some of the plaster. The resulting three-dimensional dissolution pattern is displayed in Fig. 10.16 which is obtained by the following method. The channels etched by the water are filled with melted Wood's metal, and after cooling the plaster is completely dissolved.

The structure shown in Fig. 10.16 (and those obtained in the two-dimensional version of the experiment) reminiscent of the geometry of DLA clusters. To interpret this analogy one notes that the process of chemical dissolution shares many features with viscous fingering in a porous medium which, together with DLA, is governed by the Laplace equation. In ordinary

Figure 10.16. Three-dimensional trees obtained at the end of an experiment on chemical dissolution. These structures are made of Wood's metal which was used to fill the channels previously etched by water injected into plaster (Daccord and Lenormand 1987).

viscous fingering there is a sharp increase in the viscosity (mobility) at the interface due to the high viscosity ratio of the fluids, and the motion of the interface is determined by the pressure distribution in the more viscous phase. In the experiments on chemical dissolution the injected reactive fluid and the saturating displaced fluid has the same viscosity. However, at the interface (the reactive front), the permeability jumps from a low value in the porous medium to a much larger value in the etched channels. This corresponds to a considerable change in the effective viscosities. The randomness of the medium is provided by the porous structure of the plaster.

At places with high pressure gradient the flow is faster. In these regions the dissolution of plaster is more effective (its rate is proportional to the amount of incoming pure water), thus the geometry of the network of channels follows the flow of injected water. The effects of injection rate are rather complex in this experiment, but can be taken into account by appropriate rules in the related computer simulations (Daccord 1987).

It is far from trivial to give a reliable estimate for the fractal dimension of the obtained three-dimensional macroscopic objects. One possibility is to cut out quasi two-dimensional sections from the structure and evaluate the digitized image of these cross-sections. This procedure destroys the pattern. An alternative method based on capillary effects has recently been suggested to determine D for the type of objects produced by chemical dissolution.

The principle of the technique (Lenormand *et al* 1987) is to cover the fractal with a layer of *wetting fluid.* The structure is first immersed into the wetting fluid and next slowly lowered into a non-wetting one. Because of capillary effects the wetting fluid remains around the object. The characteristic radius of curvature R of the interface between the two fluids is determined by a balance between capillary and hydrostatic pressures. To evaluate the fractal dimension one needs to measure the volume of the wetting fluid $V(R)$ for various R. This can be achieved by using different pairs of fluids. Then the fractal dimension is obtained using the expression $V(R) \sim R^{3-D}$ which is the same as Eq. (4.9). Application of this method gives an estimate $D \simeq 1.8$ for the dimension of the dissolution patterns obtained by drilling a thin tube into the original sample and injecting the water radially from this central hole.

With the two examples presented in this section we close the discussion of phenomena related to fractal growth governed by the Laplace equation. Of course, there are numerous further growth processes leading to fractal structures. Many of these involve mechanisms which are related to, but are more complex than those reviewed in the present Part. Some of the *biological patterns* (trees, roots, blood vessels) (Mandelbrot 1982) or *networks of cracks* (Louis and Guinea 1987) in solids can be regarded as examples for growing fractals as well.

REFERENCES (PART III)

Alexander, S., Bruinsma, R., Hilfer, R., Deutscher, G. and Lereah Y., 1988 *Phys. Rev. Lett.* **60**, 1514

Ben-Jacob, E., Godbey, Y., Goldenfeld, N. D., Koplik, J., Levine, H., Mueller, T., and Sander, L. M., 1985 *Phys. Rev. Lett.* **55**, 1315

Ben-Jacob, E., Deutscher, G., Garik, P., Goldenfeld, N., and Lereah, Y., 1986 *Phys. Rev. Lett.* **57**, 1903

Bensimon, D., Kadanoff, L. P., Liang, S., Shraiman, B. I., and Tang, L., 1986 *Rev. Mod. Phys.* **58**, 977

Bentley, W. A. and Humpreys, W. J., 1962 *Snow Crystals* (Dover, New York)

Brady, R. M. and Ball, R. C., 1984 *Nature* **309**, 225

Buka, A., Kertész, J. Vicsek, T., 1986 *Nature*, **323**, 424

Buka, A., Palffy-Muhoray, P. and Rácz, Z., 1987 *Phys. Rev.* **A36**, 3984

Chen, J. D. and Wilkinson, D., 1985 *Phys. Rev. Lett.* **55**, 1982

Clément, E., Baudet, C. and Hulin, J. P., 1985 *J. Physique Lett.* **46**, L1163

Couder, Y., Cardoso, O., Dupuy, D., Tavernier, P. and Thom, W., 1986 *Europhys. Lett.* **2**, 437

Couder, Y., Gerard, N. and Rabaud, N., 1986 *Phys. Rev.* **A34**, 5175

Daccord, G., Nittmann, J., and Stanley, H. E., 1986 *Phys.Rev. Lett.* **56**, 336

Daccord, G., 1987 *Phys. Rev. Lett.* **58**, 479

Daccord, G. and Lenormand, R., 1987 *Nature* **325**, 41

DeGregoria, A. J. and Schwartz, L. W., 1986 *J. Fluid. Mech.* **164**, 383

DeGregoria, A. J. and Schwartz, L. M., 1987 *Phys. Rev. Lett.* **58**, 1742

Elam, W. T., Wolf, S. A., Sprague, J., Gubser, D. U., Van Vechten, D., Barz, G. L. and Meakin, P., 1985 *Phys. Rev. Lett.* **54**, 701

Family, F., Platt, D. and Vicsek, T., 1987 *J. Phys.* **A20**, L1177

Goldenfeld, N., 1988 *Phys. Rev.* **A37**, xx

Grier, D. G., Kessler, D. A. and Sander, L. M., 1987 *Phys. Rev. Lett.* **59**, 2315

Grier, D., Ben-Jacob, E., Clarke, R., and Sander, L. M., 1986 *Phys. Rev. Lett.* **56**, 1264

Hele-Shaw, J. S. S., 1898 *Nature* **58**, 34

Honjo, H., Ohta, S. and Matsushita, M., 1986 *J. Phys. Soc. Japan* **55**, 2487

Horváth, V., Vicsek, T. and Kertész, J., 1987 *Phys. Rev.* **A35**, 2353

Horváth, V., Kertész, J. and Vicsek, T., 1987 *Europhys. Lett.* **4**, 1133

Kadanoff, L. P., 1985 *J. Stat. Phys.* **39**, 267

Kaufman, J. H., Nazzal, A. I., Melroy, O. R. and Kapitulnik, A., 1987 *Phys. Rev.* **B35**, 1881

Kertész, J., and Vicsek, T., 1986 J. Phys. **A19**, L257

Kertész, J., Vicsek, T. and Meakin, P., 1986 *Phys. Rev. Lett.* **57**, 3303

Kessler, A., Koplik, J. and Levine, H., 1984 *Phys. Rev.* **A30**, 2820

Kessler, D. A., Koplik, J. and Levine, H., 1987 in *Patterns, Defects and Microstructures in Nonequilibrium Systems* edited by D. Walgraef (Martinus Nijhoff, Dordrecht)

Kessler, D. A., Koplik, J. and Levine, H., 1988 it Advances in Physics to appear

Langer, J. S., 1980 *Rev. Mod. Phys.* **52**, 1

Lenormand, R. and Zarcone, C., 1985 *Phys. Rev. Lett.* **54**, 2226

Lenormand, R., 1986 *Physica* **140A**, 114

Lenormand, R., Soucémarianadin, A., Tourboul, E. and Daccord, G., 1987 *Phys. Rev.* **A36**, 1855

Liang, S., 1986 *Phys. Rev.* **A33**, 2663

Liu, B. X., Huang, L. J., Tao, K., Shang, C. H. and Li, H. D., 1987 *Phys. Rev. Lett.* **59**, 745

Louis, E. and Guinea, F., 1987 *Europhys. Lett.*, **3**, 871

Maloy, K. J., Feder J. and Jossang, J., 1985 *Phys. Rev. Lett.* **55**, 2681

Mandelbrot, B. B., 1982 *The Fractal Geometry of Nature* (Freeman, San Francisco)

Matsushita, M., Sano, M., Hayakawa, Y., Honjo H. and Sawada, Y., 1984 *Phys. Rev. Lett.* **53**, 286

Matsushita, M., Hayakawa, Y. and Sawada, Y., 1985 *Phys. Rev.* **A32**, 3814

Meakin, P., Family, F. and Vicsek, T., 1987 *J. Colloid Interface Sci.* **117**, 394

Miller, A., Knoll, W. and Möhwald, H., 1986 *Phys. Rev. Lett.* **56**, 2633

Mullins, W. W. and Sekerka, R. F., 1963 *J. Appl. Phys.* **34**, 323

Niemeyer, L., Pietronero, L. and Wiesmann, H. J., 1984 *Phys. Rev. Lett.* **52**, 1033

Niemeyer, L., Pietronero, L. and Wiesmann, H. J., 1986 *Phys. Rev. Lett.* **57**, 650

Nittmann, J., Daccord, G., and Stanley, H. E., 1985 *Nature* **314**, 141

Nittmann, J. and Stanley, H. E., 1986 *Nature* **321**, 663

Nittmann, J. and Stanley, H. E., 1987 *J. Phys.* **A20**, L1184

Nittmann, J., Stanley, H. E., Touboul, E. and Daccord, G., 1987 *Phys. Rev. Lett.* **58**, 619

Ohta, S. and Honjo, H., 1988 *Phys. Rev. Lett.* **60**, 611

Oxaal, U., Murat, M., Borger, F., Aharony, A., Feder, J. and Jossang, T., 1987 *Nature* **329**, 32

Park, C. W. and Homsy, G. M., 1985 *Phys. Fluids* **28**, 1583

Paterson, L., 1981 *J. Fluid Mech.* **113**, 513

Paterson, L., 1984 *Phys. Rev. Lett.* **52**, 1621

Paterson, L., 1985 *Phys. Fluids* **28**, 26

Radnóczy, G., Vicsek, T., Sander, L. M. and Grier, D., 1987 *Phys. Rev.* **A35**, 4012

Ramanlal, P. and Sander, L. M., 1987 preprint

Rauseo, S. N., Barnes, P. D. and Maher, J. V., 1987 *Phys. Rev.* **A35**, 1245

Saffman, P. G. and Taylor, G. I., 1958 *Proc. Roy. Soc. London* Ser. **A245**, 312

Sander, L. M., Ramanlal, P. and Ben-Jacob, E., 1985 *Phys. Rev.* **A32**, 3160

Sarkar, S., 1985 *Phys. Rev.* **A32**, 3114

Sarkar, S. and Jasnow, D., 1987 *Phys. Rev.* **A35**, 4900

Satpathy, S., 1986 *Phys. Rev. Lett.* **57**, 649

Sawada, Y., 1986 *Physica* **140A**, 134

Sawada, Y., Dougherty, A., and Gollub, J. P., 1986 *Phys. Rev. Lett.* **56**, 1260

Stokes, J. P., and Weitz, D. A., Gollub, J. P., Dougherty, A., Robbins, M. O., Chaikin, P. M. and Lindsay, H. M., 1986 *Phys. Rev. Lett.* **57**, 1718

Szép, J., Cserti J. and Kertész J., 1985 *J. Phys.* **A18**, L413

Tabeling, P. and Libchaber, A., 1986 *Phys. Rev.* **A33**, 794

Tang, C., 1985 *Phys. Rev.* **A31**, 1977

Van Damme, H., Olbrecht, F., Levitz, P., Gatineau, L. and Laroche, C., 1986 *Nature* **320**, 731

Vicsek, T., 1984 *Phys. Rev. Lett.* **53**, 2281

Vicsek, T., 1985 *Phys. Rev.* **A32**, 3084

Vicsek, T., 1987 *Physica Scripta*, **T19**, 334

Vicsek, T. and Kertész, J., 1988 *Europhys. News* **19**, 24

Witten, T. A. and Sander, L. M., 1983 *Phys. Rev.* **B27**, 5686

Part IV

RECENT DEVELOPMENTS

Chapter 11

CLUSTER MODELS OF SELF-SIMILAR GROWTH

Before starting the discussion of the latest results on fractal growth it should be pointed out that this field has continued to flourish and a large number of important new findings have been published in a variety of journals during the past three years since the manuscript of the first edition of this book has been completed. In addition, a few recent edited books including papers on topics related to our subject have also appeared, including school and conference proceedings (Stanley and Ostrowsky 1988, 1990, Aharony and Feder 1989, Pietronero 1990, Herrmann and Roux 1990, Avnir 1990).

The purpose of Part IV is to review the most recent developments in the field of fractal growth phenomena in order to make the book up to date. The following chapters represent natural extensions of the earlier ones; they are built on the information given previously. Since the material presented in the first three parts can still be used effectively as it is, it did not seem necessary to change those parts of the book, instead, the new results will be given as complementing primarily Chapters 6, 7 and 10. However, developments related to other chapters will be discussed as well.

Studies of *self-similar growth* (Chapter 11) have been mainly concentrating on a number of new aspects of the growth mechanisms and possible applications to various related phenomena. The investigations of diffusion-limited aggregation attempted to obtain a more complete picture of the multifractal distribution of the harmonic measure with a special attention paid to the behaviour of the scaling properties of the smallest growth probabilities. This question has turned out to be intimately related to the details of the asymptotic shape of the aggregates. Studies on fracture have been undertaken as a new application of fractal cluster growth models. Finally, recent theoretical ideas will be discussed in this chapter.

During the last few years the interest in the dynamic scaling behaviour of *self-affine surfaces* (Chapter 12) has been rapidly growing. This is at present perhaps the most active area of fractal research. Several exciting questions have been raised and many of them were successfully treated with a unique combination of theoretical, numerical and experimental approaches. Here also the application to physically relevant cases has started to play an essential role.

One of the new important conceptual developments has been the appearance of interesting *new experimental systems* (Chapter 13) which have been motivated by the interest in getting a better insight into the relevance of simulational and theoretical results to the actual fractal growth phenomena occurring in nature. In particular, the publication of studies on biological growth and viscous flows represents a challenge for appropriate interpretations.

11.1. DIFFUSION-LIMITED AGGREGATION

The structure and the growth of DLA clusters is so rich that the continued interest has led to many recent papers devoted to the further exploration of these aspects of aggregation. Particular attention has been paid to the multifractal nature of the growth probability distribution and the geometry of diffusion-limited aggregates. Finally, DLA has been used to describe pattern formation under conditions governed by the Laplace equation.

11.1.1. Global structure

Since aggregation models are known to exhibit extremely slow crossover behaviour, it is of primary importance to obtain as large clusters as possible when the asymptotic structure of the aggregates is investigated. The anisotropic shape of very large clusters (containing 4×10^6 particles) was discussed in Section 6.1.2. However, before 1989 much less was known about the geometry of off-lattice aggregates of the same size.

Tolman and Meakin (1989) has improved the earlier algorithms and used 2500 hours of CPU time on an IBM 3090 supercomputer to generate many off-lattice and on-lattice DLA clusters in dimensions $d = 2$ to 8. The most important trick of their algorithm (originally suggested by Ball and Brady 1985) is the following. For a method to be efficient the particle should be allowed to make large jumps even if it is in the region between the large, already present branches of the aggregate. (A possible way to treat this case is discussed in Appendix A.) To achieve this one has to know a good underestimate of the largest unoccupied hypersphere (a hypersphere which does not intersect with the cluster) centred on the current position of the particle. This estimate can be obtained by constructing maps of the cluster on different scales (coarse graining). Then the maps on different scales are examined from the point whether a non-intersecting jump can be taken. If yes, the particle moves to a new position. On the other hand, if the move is not allowed, the situation is examined on a lower level with more detailed information (map) about the cluster in the vicinity of the particle. This process continues until the lowest level, corresponding to the actual position of the aggregated particles, is reached and the particle is finally added to the cluster. After the particle has been added to the aggregate, the maps in each level of the hierarchy have to be updated.

An off-lattice DLA cluster generated in $d = 2$ and containing $N = 6 \times 10^6$ particles is shown in Fig. 11.1. The results obtained for such aggregates indicate that the fractal nature of the clusters observed for considerably smaller sizes is robust: the dependence of the effective fractal dimension on its linear size appears to level off at a value somewhere between 1.705 and 1.710 beyond 400 particle diameters (Tolman and Meakin 1989). Extrapolation

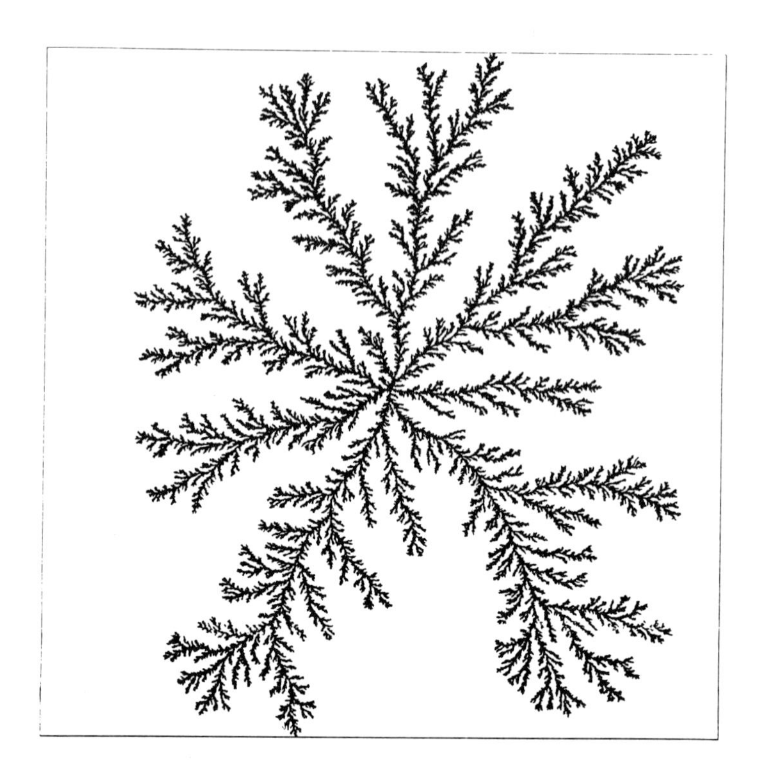

Figure 11.1. A typical off-lattice diffusion-limited aggregate of 6×10^6 particles grown in two dimensions. Comparison with smaller clusters (Figures 6.1 and 6.3) shows some characteristic changes, however, self-similarity and the fractal dimension are not affected (Ossadnik 1991).

suggests $D = 1.715 \pm 0.004$. This is demonstrated in Fig. 11.2. Thus, no peculiar crossovers could be detected as the simulations have been pushed to a much larger scale. These findings have been confirmed by very recent simulations (Ossadnik 1991) of 6×10^6 DLA clusters.

The three-dimensional results are also interesting. First, the best estimate for the fractal dimension of the off-lattice aggregates is so close to the mean-field prediction $(d^2+1)/(d+1) = 5/2 = 2.50$ that the validity of the latter cannot be numerically ruled out. (This observation is true for the dimension of off-lattice aggregates generated up to $d = 8$.) Second, the large clusters grown on the cubic lattice possess the kind of dendritic anisotropy which was previously observed for $d = 2$. Figure 11.3 shows the cross section

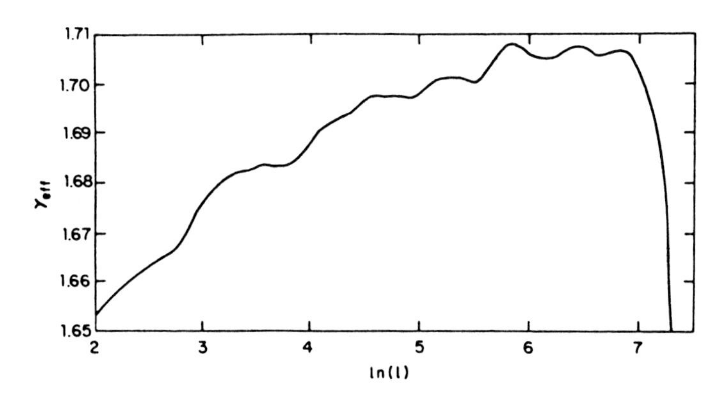

Figure 11.2. Dependence of the effective fractal dimension $D = \gamma_{eff}$ as determined from the expression $M(l) \sim l^{\gamma_{eff}}$ on the linear size of the region l in which the scaling of the mass M was investigated (Tolman and Meakin 1989).

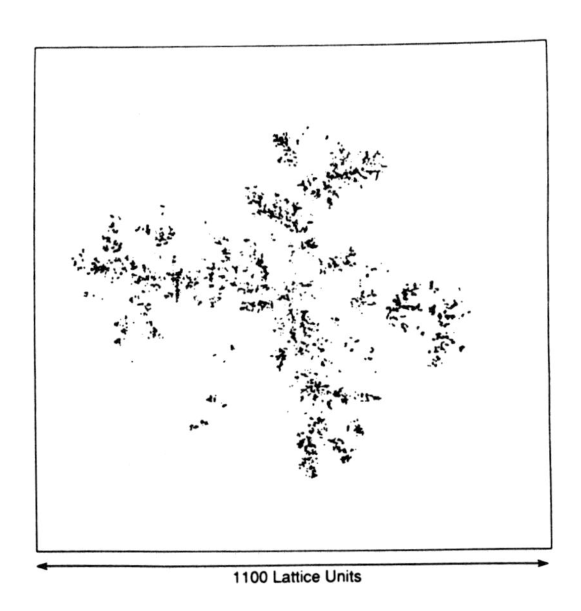

Figure 11.3. Cross section of a large DLA cluster (3×10^6 particles) generated on the simple cubic lattice (Tolman and Meakin 1989).

of a cluster obtained in 3d. The anisotropy is indicated by the overall cross-like shape of this disconnected set of points. According to rule c) of Section 2.2 the dimension of the cross-section is about 2.5-1=1.5.

At this point the reader is encouraged to compare Figures 6.1, 6.3 and 11.1. Although DLA clusters are self-similar, one may notice a gradual change in the overall morphology as the size of the aggregates becomes larger. The role of the fluctuations on the scale comparable to the particle size becomes less relevant as $N \to \infty$. As a result, the main branches become straighter and directed away from the center in a more pronounced manner.

The directedness of the patterns raises the question of self-similarity of such structures. In other words: can directed clusters be self-similar? When the standard recursive methods of generating a deterministic fractal are applied the resulting structure is not directed. If one uses a generator which leads to branching structures the branches are turned with a given angle at each step of the iteration procedure resulting in branches directed in all possible directions, moreover, in the appearance of unphysical spirals.

The above undesirable effects are avoided and the directedness is preserved in the following directed recursive model (Mandelbrot and Vicsek 1989). In the construction to be described below the units of the previous stage are replaced with an appropriately rotated and reflected (mirror image) version of the generating configuration. The first stage ($k = 1$) is the generator: a simple branching structure made of three units (intervals of the same length). At the next stages each of the units obtained at the previous steps are replaced by the $k = 1$ configuration while simultaneously obeying the following rules: a) none of the branches should point in a direction below the horizontal, b) no branches are allowed to overlap or touch each other. Fig. 11.4 (left) shows the $k = 7$th stage, while in the right side of Fig. 11.4b a random version of the model is displayed.

A construction similar to the above has been proposed to characterize the nature of branching in DLA clusters by Hinrichsen et al 1989. They studied the bifurcation ratio r_n between the number of branches and the length ratio r_l for two subsequent branching orders. Here each branch is

Figure 11.4. This figure illustrates the structure of the directed self-similar model which can be used to mimic the geometry of large off-lattice DLA clusters. The left part shows the $k = 7$th stage of the deterministic construction, while right figure shows a random version of the model (Mandelbrot and Vicsek 1989).

represented by a continuous line starting at a tip and ending on another branch of lower order. The highest order branches have no side branches. The next-to-highest-order branches have side branches of the highest order, and so on. From a scaling analysis of a deterministic model with fixed r_n and r_l the expression $D = \ln r_n / \ln(1/r_l)$ was obtained.

Several further interesting attempts have been made to uncover the complex structure of diffusion-limited aggregates. I would like to single out the extensive work on the *wavelet transform* approach to the description of the mass distribution in fractal aggregates (Argoul et al 1989, 1990) and the study of the cluster size distribution in the incremental growth of DLA clusters (Tolman et al 1989).

11.1.2. Growth probability distribution

The evolution of the aggregates is determined by the growth probabilities p associated with the surface sites. At the tips, where p is larger, the advancement of a branch is faster, while deep in the screened fjords p quickly tends to zero as the linear size of the aggregate increases. The growth-site probability distribution (GSPD) has been one of the central quantities of interest in recent years.

As discussed in 6.1.4 the GSPD for DLA clusters is the same as their harmonic measure and can be shown to exhibit multifractal scaling. This means that in an aggregate of linear size L, sites with a local singularity exponent $\alpha = \ln p / \ln(1/L)$ are distributed on a fractal subset of dimension $f(\alpha)$. Early numerical estimates of $f(\alpha)$ indicated that it has a bell shape with a well defined maximum and two characteristic values for α_{min} and α_{max}. The smallest singularity α_{min} corresponds to the fastest growth and was found to be close to 0.7, in good agreement with the related theoretical prediction $D = 1 + \alpha_{min}$.

The situation has been much less clear concerning the dependence of α_{max} or, which is the same, the minimum growth probability p_{min} on the cluster diameter L. In the following we shall systematically go through the possible alternatives and discuss the recent results supporting the various propositions. Mainly the two-dimensional case will be treated, but in higher dimensions analogous problems are expected to arise.

i) The existence of a well defined finite α_{max} is equivalent to the expression

$$p_{min} \sim L^{-\alpha_{max}} . \tag{11.1}$$

In the view of a few very recent studies, however, further possibilities (to be discussed below) leading to the breakdown of the above standard multifractal behaviour should be considered. ii) If the growth sites are extremely strongly screened one can assume that their probabilities to grow tend to zero exponentially with the linear size of the aggregates

$$p_{min} \simeq e^{-aL}, \tag{11.2}$$

where a is some constant. iii) The most recent suggestion is that the scaling of the smallest growth probability is given by the expression

$$p_{min} \simeq e^{-b(\ln L)^c}, \tag{11.3}$$

with b and $c \simeq 2$ constants.

Let us now discuss the geometrical pictures corresponding to the behaviours given by Eqs. (11.1-11.3) and the consequence of Eqs. (11.1-11.3) regarding the multifractal spectrum in the light of the corresponding works. i) p_{min} scales as indicated by (11.1) if, as it has been argued (Harris and Cohen 1990, Barabási and Vicsek 1990), the structure of the aggregates is such that the major fjords always have a wedge or cone shaped structure leading to the most screened sites.

This picture is consistent with the directed recursive model (Mandelbrot and Vicsek 1989) described in the previous section. A close inspection of, e.g., the left picture in Fig. 11.4 shows that the structure is made of two kinds of almost closed loops being open to a different degree. (In the following we shall simply use the word loop for these incompletely closed parts of the structure and the word opening for the entrance of these loops.) This observation makes it easier to understand what is the mechanism which produces the multifractal nature of the harmonic measure (which behaves the same way as the electric field around a charged aggregate). Each time we get deeper into the pattern and enter a loop, the field is decreased (screened) by a given factor λ_1 or λ_2 (depending which type of loop we enter). As a result, the strength of the field is determined by a *multiplicative process* with different weights λ_1 and λ_2 which is well known to lead to fractal measures.

To determine p_{min} we assume that the ragged interface between the fractal and the cone-shaped empty regions can be approximated by a smooth surface like in the probability scaling theory (Section 6.1.3) of DLA. However, in the present case instead of the convex shape with a tip we treat an incision having the shape of a cone. Nevertheless, we shall use the term tip for the place where the smoothed-out interface has a sharp turn. The potential along the above described cones is constant and we expect that its gradient (the electric field playing the role of the measure defined on the fractal) goes to zero as some power of the distance from the tip as in the probability scaling approach. Thus, we can make use of the known solution of the simplified problem shown in Fig. 6.9. This solution is $\nabla\phi(r,\varphi) = \frac{C\pi}{2\varphi} r^{\pi/\varphi - 1}$, where ϕ denotes the electric potential, C is a constant and r is the distance from the tip. Imagine now that the clusters of size L are rescaled into the unit square. Then the size of a particle (sitting at the end of a fjord) is $1/L$. Let us assume that the structure is covered by boxes of linear size $\epsilon = 1/L$. The amount of measure in a box consisting the tip of angle φ is given by

$$p_{tip} \sim (1/L)^{\pi/\varphi} \tag{11.4}$$

which is is equivalent to (11.1). Harris and Cohen (1990) made a detailed analysis of the probability of the possible fjord configurations in DLA clusters. Using the electrostatic analogy and scaling considerations they argued that the occurrence probability of fjords with structures leading to a p_{min} scaling differently from a power law is negligible. In addition, Barabási and Vicsek (1990) and Blumenfeld and Ball (1991) obtained results indirectly supporting (11.1). The quantity which was investigated numerically was $n(x)$, the distribution of the number of fjords with a given ratio $x = w/l$ of their width w (entrance size, or lip) to their length l (distance of the entrance from the closest point of the fjord to the centre). This distribution was found to scale numerically for large scale off-lattice aggregates (Barabási and Vicsek 1990) in a manner which was in a very good agreement with the theoretical formulae of Blumenfeld and Ball for $n(x)$ compatible with the existence of an α_{max}.

ii) The growth probability becomes exponentially small (Eq. 11.2) in clusters consisting of long "tubes". The main feature of these tubes needed for (11.2) is that the ratio $x = w/l$ of their width w to their length l has to go to zero with the size of the clusters as $1/L$. In this case there is a phase transition in the multifractal spectrum at some α_0, where $f(\alpha)$ becomes non-analytic. Since (11.2) corresponds to contributions coming from arbitrary large $\alpha \to \infty$, and the multifractal spectrum has to be a convex function (this is a consequence of the monotonicity conditions for D_q), it follows from (11.2) that the shape of $f(\alpha)$ must be one of the following: a) If we also require that $f(\alpha)$ has to be positive then it first reaches its maximum at some α_0 and starting from that point it remains constant as $\alpha \to \infty$ (left-sided $f(\alpha)$). This behaviour means that in α_0 the function is continuous but most of its higher order derivatives are not. b) $f(\alpha)$ becomes negative for large α values. A phase transition is still expected to take place at the point where the smoothly curving $f(\alpha)$ crosses over into a straight line.

The notion phase transition is motivated by the formal analogy with thermodynamics: χ_q, the qth moment of the GSPD (see Eq. (3.5)) can be rewritten in the form

$$Z(\beta, L) = \sum_i p^\beta \sim L^{-F(\beta)}\,, \tag{11.5}$$

where $Z \to \chi$, $\beta \to q$ and $F \to (q-1)D_q$ play respectively the role of the partition function, the temperature and the free energy. Then the Legendre transform of F corresponds to the entropy $S(\beta)$ which is, in turn, in analogy with $f(\alpha)$. This is why a non-analytic behaviour of $f(\alpha)$ or D_q is associated with phase transitions.

The possibility of a phase transition for diffusion-limited aggregates was first pointed out by Lee and Stanley (1988). In their exact enumeration study they obtained a behaviour corresponding to (11.2) for a complete ensemble of small DLA clusters. Similar conclusions were made in the investigations of aggregate-like Julia sets (Bohr et al 1988) and typical diffusion-limited aggregates (Blumenfeld and Aharony 1989, Havlin et al 1989). Finally, in a recent work Mandelbrot and Evertsz (1991) conclude that $f(\alpha)$ is left-sided (see also Mandelbrot et al 1990) because of the existence of long narrow fjords in their simulations of medium size (about 30,000 particles) simulations of diffusion-limited deposition.

iii) The possibility represented by (11.3) is particularly interesting and has the most direct numerical support. Equation (11.3) results in a phase transition in $f(\alpha)$ as well, although the growth probabilities go to zero much slower than in the case ii). This behaviour was first indicated by the numerical determination of p_{min} for clusters containing 2600 particles (Schwarzer et al 1990). In a more extensive subsequent study (Schwarzer et al 1991) the authors considered clusters up to $N = 20,000$ particles and found that their results for the distribution of growth probabilities $n(\alpha, N)$ could be well described by the expression

$$\ln n(\alpha, N) \sim -\alpha^\gamma / \ln^\delta N\,, \tag{11.6}$$

where α denotes now $\alpha = -\ln p / \ln N$ differing from the above used definition only by a constant prefactor $1/D$, $\gamma = 2 \pm 0.3$ and $\delta = 1.3 \pm 0.3$. The validity of this expression is demonstrated by Fig. 11.5, where $M = N$. Note that $\ln n(\alpha, N)$ is different from $f(\alpha)$ only by an additive constant and that the standard $f(\alpha)$ plot would not have the $\log_{10}^{1.3} M$ term in the denominator of the quantity plotted along the horizontal axis. Without this rescaling the $n(\alpha, N)$ (or $f(\alpha)$) curves would be gradually shifted to the left with a *less steep slope* when clusters consisting of increasing number of particles are evaluated. In the asymptotic limit this shift would result in a horizontal line correspondng to a one-sided $f(\alpha)$. The above results also follow from a hierarchical model proposed by Lee et al (1990) for the structure of fjords.

It is extremely difficult to provide a decisive answer for the question discussed in this section. The medium size simulations in the circular geometry (Schwarzer et al 1990) undoubtedly favour expression (11.3). A recent theoretical approach leading to exact results on superscreening (Ball and Blumenfeld 1991) is more consistent with the existence of a well defined α_{max}. Finally, from a study of the dielectric breakdown model Mandelbrot and Evertsz (1991) concluded that the smallest growth probability decays exponentially as a function of L. However, any method based on the direct calculation of the growth probabilities is not feasible for cluster sizes beyond 20 to 50 thousand particles. In the view of the slow crossovers in the overall behaviour of diffusion-limited aggregates the question of the asymptotic

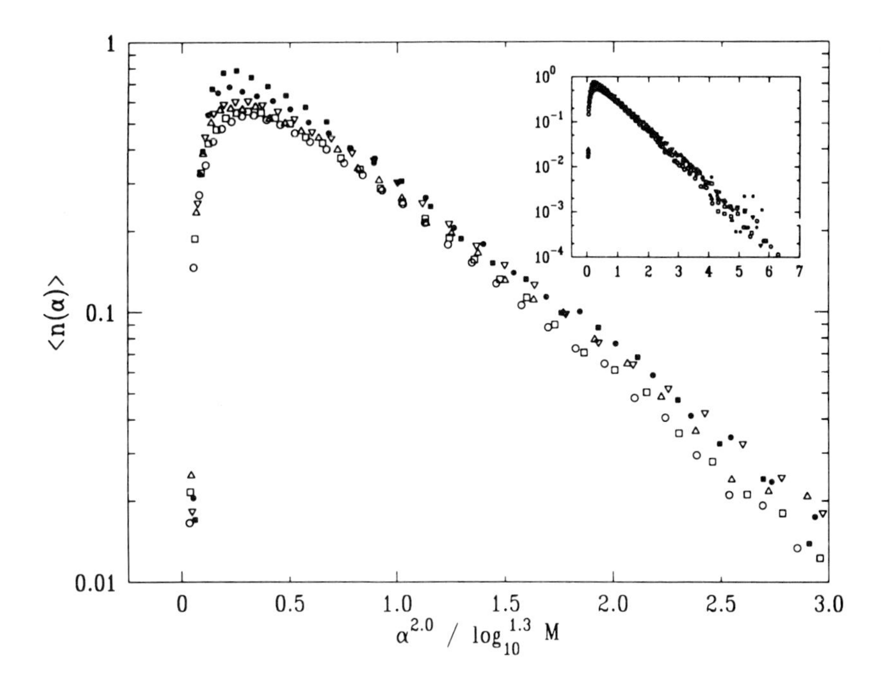

Figure 11.5. Results for the distribution of growth probabilities $n(\alpha, N)$, where $\alpha = -\ln p / \ln M$ $(M = N)$. Note that $\ln n(\alpha, N)$ is different from $f(\alpha)$ only by an additive constant and that the standard $f(\alpha)$ plot would not have the $\log_{10}^{1.3} M$ term in the denominator of the quantity plotted along the horizontal axis. The different symbols are for clusters of different sizes ranging from $M = 753$ to $M = 20000$. The inset shows the same data for an extended range of α values (Schwarzer et al 1991).

behaviour of the phase transition in DLA may not become fully resolved for a long time.

11.1.3. Multifractal geometry

The concept of geometrical multifractality has already been described in Section 3.4. There it was shown that the mass or particle density distribution in clusters in principle can be a multifractal. In the spirit of Appendix C this means that if the cluster is covered by a grid of boxes of linear size l and the number of particles in the ith box is denoted by M_i, then the number of boxes with a given M, $N(M)$, has to scale as follows: Assume that we plot $\ln N(M)$ versus $\ln M/M_0$ for various l. If these histograms fall onto the same universal (size independent) curve after rescaling both coordinates by a factor $\ln(l/L)$, the structure is a mass (geometrical) multifractal. It is required that $l \ll L$ and $a \ll l$, where a is the particle size. This is equivalent to the requirement that the clusters must be extremely large. From the formalism for fractal measures it follows that for multifractals there exists an infinite hierarchy of generalized dimensions D_q which for the mass distribution are defined by

$$\sum_i \left(\frac{M_i}{M_0}\right)^q \sim \left(\frac{l}{L}\right)^{(q-1)D_q}, \tag{11.7}$$

where the normalization factor M_0 is the total mass of the cluster.

For mathematical fractals with no lower cutoff length scale the situation is delicate. In the formalism presented here for growing fractals all masses are measured using the smallest unit that we call particle. When this unit becomes exactly equal to zero (this is the case when the limiting set of finite mathematical fractals is attained) it is less straightforward to provide a physically plausible picture. This situation is somewhat analogous to the well know cases from analysis when the value of an expression for a parameter approaching a critical value does not coincide with the value obtained by inserting this critical value into the expression (e.g., $\lim_{x\to 0}(e^x - 1)/x$). In any case, in the physically relevant situations the condition of the existence of a smallest unit is always satisfied.

It is a natural idea to check whether the mass distribution in DLA clusters can be described by a single exponent or it is a multifractal. The main motivations to think that the latter is the more likely case are the following: i) A "monofractal" distribution would be a rather special case of a more general possible distribution, ii) more importantly, the density of the particles in a cluster is determined by the growth probability distribution which is known to be a multifractal.

A medium scale investigation of D_q for $q > 0$ using the standard box counting approach did not detect any relevant q dependence of the multifractal spectrum (Argoul et al 1988). Another surprising, but questioned (Li et al 1989) conclusion of this study was that the box counting dimension of DLA clusters is smaller (approximately 1.63) than that obtained by other methods. On the other hand, Nagatani (1988) in a small cell renormalization group calculation study obtained a non-trivial D_q spectrum for the mass distribution.

From the condition $a \ll l \ll L$ and the work of Argoul et al (1988) it is clear that non-trivial results can be obtained only by using a method different from box counting for very large clusters. To answer our question one has to determine the generalized dimensions. When one tries to calculate D_q from the expression (11.7) for $q < 0$ the boxes which contain small number of particles (because they barely overlap with the cluster) give an anomalously large contribution to the sum in the left-hand side of Eq. (11.7) and consequently it is not possible to get reliable results for this case.

This problem can be solved by using the generalized sand box method which is based on studying the *average* of the masses $M(R)$ (and their qth moments) within boxes of size R with the centers of the boxes *randomly distributed* on the fractal. The usefulness of this method has been demonstrated for exact fractals by Tél et al (1989). Thus, the question is reduced to how $\langle M^q(R)\rangle$ scales with increasing R, where $\langle ...\rangle$ denotes the average over the centers. To use (11.7) in this context one first notes that the sum in (11.7) corresponds to taking an average of the quantity $(M_i/M_0)^{q-1}$ according to the probability distribution M_i/M_0. When the centers of the boxes are chosen randomly the averaging is made over this distribution and, therefore

$$\left\langle \left(\frac{M(R)}{M_0}\right)^{q-1} \right\rangle \sim \left(\frac{R}{L}\right)^{(q-1)D_q} . \tag{11.8}$$

Since in this method the boxes are centered on the fractal there will be no boxes with too few particles in them and the scaling of the negative moments can also be investigated.

Figure 11.6 shows the results (Vicsek et al 1990a) for D_q obtained by applying the above expression to 7 off-lattice DLA clusters generated in two dimensions each consisting of one million particles. It is clear from this figure that the various moments scale differently, although the actual D_q values are rather similar. However, one does not expect them to be very different, since mass multifractals embedded into two dimensions must satisfy the condition $1 \leq D_q \leq 2$.

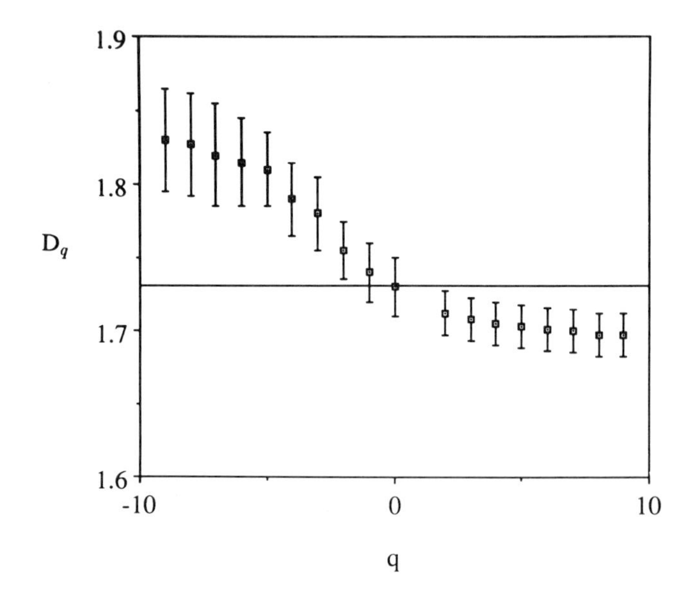

Figure 11.6. The multifractal spectrum D_q describing the scaling of the mass density inside large off-lattice DLA clusters. The effect is relatively small, but significant, because if the analysis leading to this figure is repeated for structures with monofractal geometry, the obtained D_q values are virtually identical (Vicsek et al 1990).

Because of the relatively weak effect it is important to cross examine the results. As a test of the method, the D_q spectrum for two other kinds of models, for which the mass distribution is well understood, was calculated (Vicsek et al 1990). In the first simulation the generalized dimensions for randomly distributed particles was determined. The second simulation was made to test whether an apparent multifractal mass distribution is obtained for an object which is known to be described by a single fractal dimension. The calculations were repeated for the Sierpinski gasket, which is a simple deterministic fractal having a unique fractal dimension. Both tests resulted in D_q spectra which were virtually indistinguishable from a constant behaviour indicating that in contrast to DLA clusters the analysis in the above cases detects a monofractal distribution of mass.

Consequently, the structure of DLA is described by an infinite set of mass exponents. The overall dependence of D_q on q is consistent with the expected behavior, namely, it exhibits a monotonic decay with growing q. An interesting consequence of Fig. 11.6 is that $D_{q=2} \neq D_{q=0}$, where $D_{q=0}$ is the true fractal dimension D of the aggregates. This is significant because the standard methods for the determination of the fractal dimension (such as the radius of gyration method or the calculation of the density-density correlation function) yield $D_{q=2}$.

An interesting proposal related to multifractality of DLA clusters has been made by Coniglio and Zanetti (1990) who have assumed a very general form for the scaling of the density profile of the aggregates. Based on general considerations they arrive at the scaling form

$$g(r, R) = r^{-d+D(r/R)} A(r/R) \tag{11.9}$$

where $g(r, R)$ is the density of particles at a distance r from the origin in a cluster having a radius of gyration R and $A(x) = g(1, x^{-1})$. This multiscaling behaviour predicts a different fractal dimension for each shell corresponding to a given ratio $x = r/R$. The actual shape of $D(x)$ is such that $D(x) \simeq D(0) = D$ up to some characteristic x, and beyond that value gradually decays to a lower value. Numerical simulations of clusters with $N = 10^5$ particles were found to be consistent with multiscaling. A very recent much larger scale simulation (Ossadnik 1991) has indicated that the effect may vanish in the asymptotic limit. Further investigations seem to be necessary to draw the final conclusion.

The calculations of Coniglio and Zanetti (1990) also led to results for the multiscaling in spinodal decomposition. As an interesting connection we mention here that the early stage morphology in spinodal decomposition has been shown by Klein (1990) to be related to mass multifractality. Finally, the concept of multifractal geometry can be applied to a number of other type of systems including the space filling bearings model for turbulence (Herrmann 1990) and the Apollonian gaskets in which a non-trivial D_q spectrum characterizing the geometry was determined as well (Manna and Vicsek 1991).

11.1.4. Pattern formation

Although the analogy of the equations describing Laplacian pattern formation (Sections 9.2 and 9.2) with DLA has been recognized from the very beginning of the investigations of diffusion-limited aggregation, the understanding of the relation of these two approaches has not been complete. The main difference between the above descriptions comes from two sources: i) the surface tension in Eq. (9.4) does not have an obvious counterpart in DLA, ii) perhaps more importantly, the equations of Laplacian pattern formation (9.1–9.4) are deterministic, while the diffusion-limited aggregation model is intrinsically stochastic.

One way to establish connection between the two kinds of phenomena is to compare the *average* behaviour of DLA clusters (Arnéodo et al 1989) with the exactly known (and experimentally supported) solutions for viscous fingering patterns in the Hele-Shaw cell with channel and sector-shaped geometries. In such systems the DLA clusters grown with reflecting boundary conditions are disordered tree-like structures, while for slow injection rates the corresponding Hele-Shaw pattern has a simple shape of a single finger. On the other hand, if the injection rate is large (this corresponds to small d_0 in (10.6)), the above finger becomes unstable, develops many branches and the viscous fingering pattern becomes reminiscent of a DLA cluster.

The main idea is to generate many clusters and test their average behaviour by calculating the occupancy profile which is defined as follows: n clusters of N particles are grown and simultaneously it is determined at each point of the grid that how many times it has been occupied by a particle of an aggregate. This number is divided by n to give $r(x, y)$, the mean occupancy of a point. The occupancy profile corresponding to the average behaviour then can be obtained from some condition, e.g., a line connecting the points for which the mean occupancy is half of the maximum can represent the overall features adequately.

The results for the sector shaped geometry for the angle $\theta = 90°$ are shown in Fig. 11.7. For this angle the analytical solution is shown in Fig. 11.7c. The dashed line shows how the standard Saffman-Taylor finger determined for the channel case (parallel walls) would look if it was mapped using conformal transformation into this geometry. The true solution (continuous line) is rather different, and the occupancy profile seems to fit the curve surprisingly well.

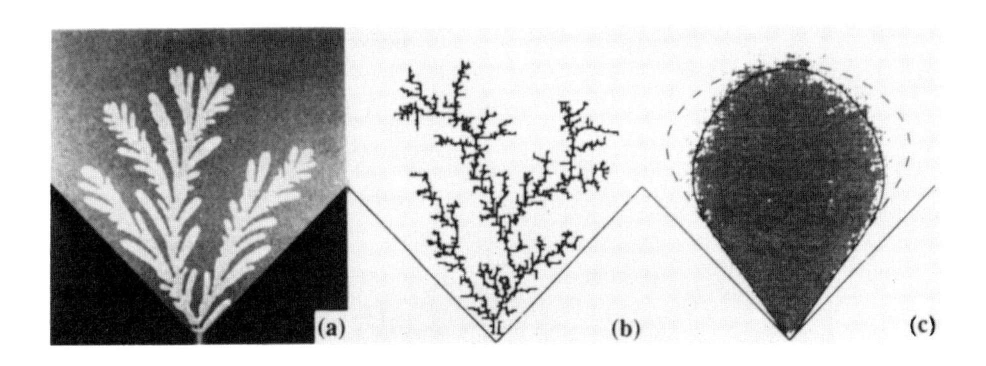

Figure 11.7. (a) Unstable viscous finger and (b) DLA cluster grown in a cell of angle $\theta = 90°$. (c) The grey region shows where the occupancy rate is larger than half of its maximum value for 420 clusters of mass 2000. The continuous line is the analytical solution, while the dashed line shows the conformal transform of the original Saffman-Taylor finger to this geometry which is a trivial (and incorrect) way to represent the solution for the sector shape cell (Arnéodo et al 1989).

Figure 11.7 suggests that the role of noise is relevant only with regard to the internal structure of the aggregates and the unstable viscous fingering patterns: the global shape is not affected by the randomness on a small scale. It is a very interesting fact that the average profile of the stochastic unstable structures is so well described by the *stable* solution of the problem. Another conclusion one can make is that the role of the viscous finger capillary length is played by the lattice constant (or particle size) in DLA simulations.

The analogy with Laplacian pattern formation can be pushed further by studying the effects of anisotropy. As discussed in Section 9.2.2, the method of noise reduction (NR) provides a convenient tool for taking into account in DLA the kind of anisotropy which in pattern formation is due to anisotropic surface tension. The investigation of the average occupancy profile of DLA clusters and experimental dendritic patterns in the channel (strip) geometry (Couder et al 1990, Arnéodo et al 1991) has led to a number of interesting results concerning the isotropic-anisotropic global shape crossover in DLA.

We recall that in noise-reduced DLA (NRDLA) a particle is added to the cluster at a given point only if previously $m-1$ trial trajectories have ended at that site. Ordinary DLA corresponds to $m=1$, and in that case the profile could be well fitted by the Saffman-Taylor solution with the relative finger width $\lambda = (y_m^+ - y_m^-)/W = 0.5$, where W is the width of the cell and $y_m^\pm$ are determined from the condition

$$y_m^\pm(x) = \frac{1}{r_{max}} \int_0^{\pm W/2} r(x,y)dy \tag{11.10}$$

(the pattern grows in the x direction). On the basis of the simulations Arnéodo et al (1991) find the finite-size scaling form

$$\lambda = \Lambda(1/m^{3/2}W)\,, \tag{11.11}$$

where the universal crossover scaling function $\Lambda(u)$ behaves as $\Lambda(u) \simeq 1/2$ for $u \gg u_c$ and $\Lambda(u) \sim u^{1/2}$ for $u \ll u_c$ (u_c is some critical value of u). This expression shows that the effective anisotropy (or the relative width λ of the needle-shaped occupancy profile) can be tuned by either changing m or the width of the strip W. This is true for the solutions of the Laplacian pattern formation problem in a strip with anisotropic surface tension as well (Kessler et al 1988). In this way a more explicit analogy between the noise reduction parameter m and the surface tension anisotropy has been established.

Finite size scaling analysis shows that the exponent 3/2 in expression (11.11) corresponds to a fractal dimension $D_{anis} = 3/2$ for DLA clusters with anisotropy and $D = 5/3$ for aggregates without anisotropy. According to the arguments, however, the former value can be observed only for extremely large systems because of the presence of an additive logarithmic term depending on the size. On the other hand, the question whether the exponent in (11.11) is exact or is a number close to 3/2 is still open.

In fact, a detailed study (Eckmann et al 1989, 1990) of NRDLA grown on square lattices in the circular geometry supports the idea that the asymptotic shape of such clusters is characterized by a finite relative width of the four arms. In addition, the angle of the tips is larger than zero which, in turn, corresponds to $D > 3/2$. These results follow from a first-order theory based on conformal mapping for the mass distribution in NRDLA clusters. In this approach the length of an arm of a NRDLA is denoted by R and the total mass of the cluster at a given moment is N. In addition, we denote by $(1/R)f(x/R)dx$ the probability that a particle hits an element of length dx at a distance $|x|$ from the origin. The zeroth-order approximation for the true shape is a cross. Correspondingly, the zeroth-order approximation to $f(y)$ is $f(y) = 4y/\pi(1-y^4)^{1/2}$, because this is the solution for the electric field for a cross-shaped conductor (and the gradient of the probability distribution determining the growth rate is analogous to the electric field). This probability diverges at the tip ($x \to R$ or $y \to 1$) as $(1-y)^{-1/2}$. From here it can be shown that $R(N) \sim N^{2/3}$ assuming that in our zeroth-order approximation $D = 3/2$. The first-order approximation is obtained by considering the total mass deposited at x, $M(x, R(N)$. This can be written (Eckmann et al 1989) as

$$M(x, R(N)) = \int_x^{R(N)} r^{-1} f(x/r)(dN/dr)dr = x^{1/2} \int_{x/R}^{1} \frac{f(y)}{y^{3/2}} dy \,. \quad (11.12)$$

Rewriting this we get

$$M(x, R(N)) = x^{1/2} F(x/N(R)) = R^{1/2} G(x/R(N)) \quad (11.13)$$

with f and G being uniquely determined by $f(y)$. Next, it can be argued that the envelope of the arms is given by the function

$$S(x; R(N)) = CM^2(x; R(N)) = CR(N)G^2(x/R(N)), \qquad (11.14)$$

where C is a constant. This is equivalent to a picture in which the width of the arms scales as their length and that the shape of the tip is some angle β depending in C.

This approach can be pursued further by iterating this procedure and showing that the shape $S(x; R(N))$ can be obtained as a fixed point of a functional equation. One more assumption then enables developing a renormalization group in the space of shapes (Eckmann et al 1990). The main results of such an approach are the following: i) The shape with zero tip angle ($\beta = 0$, $D = 3/2$) is not a fixed point, ii) lattice grown ordinary diffusion-limited aggregates in the asymptotic limit will be in the same universality class as noise-reduced DLA (confirming an earlier conjecture (Kertész and Vicsek (1986)). The intuitive picture behind this result is that in a renormalized DLA disk-shaped regions (introduced in the course of coarse graining) that were not filled with some minimal number of particles are removed which is equivalent to not growing them due to the noise-reduction rule.

Finally, the methods of taking into account the effect of surface tension in DLA type simulations have been refined in a few recent works. In particular, Fernandez and Albaran (1990) succeeded in introducing a modified algorithm to simulate viscous fingering with a special emphasis on the scaling of the fingers' width λ. They showed that the hydrodynamic scaling relation $\lambda = d_0^{-1/2} f(W d_0^{1/2})$ can be satisfied by appropriate choices for the attachment kinetics of the particles, where d_0 is the capillary number, W is the width of the strip on which the simulation is carried out, and f is some scaling function.

11.2. FRACTURE

The interest in the fractal aspects of fracture is very recent. Most of the studies have appeared after the manuscript of the first edition of this book has been completed, thus, this section is entirely new (has no counterpart in the previous parts of the book). Reviews on the statistical approaches to fractal crack formation can be found in the book edited by Herrmann and Roux (1990).

The breaking of solids as a consequence of an external stress is a phenomenon of great technological importance and has been studied intensively for many decades. Material failure processes due to crack propagation exhibit a rich phenomenology extending over a wide range of length scales from the atomic level to the overall size of the system. Here we shall be concerned with the growth of fractal cracks. Brittle fracture described by equations valid on the mesoscopic level will be considered and the standard approaches of statistical physics will be used to simulate the development of cracks having complicated geometry. These approaches include such methods and concepts as the use of lattice models, investigation of scaling and introduction of disorder. The motivation for including crack propagation into a book on fractal growth is demonstrated by Fig. 11.8 in which the complex morphology of a single crack obtained on the surface of a plexiglass plate is shown.

11.2.1. Equations

The best way to describe the motion of an elastic medium due to a stress field is to determine the displacement field $\vec{u}$ which gives for each volume element its position with respect to the equilibrium position it would have without any external force. In the linear elasticity theory the strain tensor is given by

$$\epsilon_{\alpha\beta} = (\partial_\alpha u_\beta + \partial_\beta u_\alpha)/2 \tag{11.15}$$

which is assumed to depend on the derivatives of $\vec{u}$ because of the condition of translational invariance. (11.15) is symmetrical to express the fact that in a homogeneous medium a single volume element cannot be rotated. The conjugate variable to $\epsilon_{\alpha\beta}$ is the stress tensor

$$\sigma_{\alpha\beta} = 2\mu\epsilon_{\alpha\beta} + \lambda\delta_{\alpha\beta}\sum_\gamma \epsilon_{\gamma\gamma}\,, \tag{11.16}$$

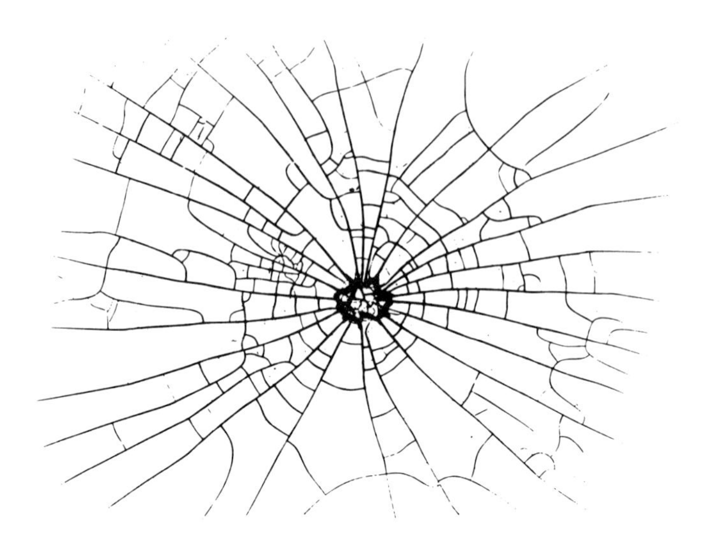

Figure 11.8. The crack displayed in this figure is an example for a fracture pattern slowly developing on a surface. The dimension of the structure was found to be in the region 1.7–1.8 (Ananthakrishna 1991).

where μ and λ are material constants called Lamé coefficients. The above linear relation is the well known Hooke's law. $\sigma_{\alpha\beta}$ is the β component of the force which is applied on the surface in the α direction of a volume element (see Fig. 11.9).

From the condition that in the quasi-stationary limit the sum of the forces exerted on a volume element has to be zero one obtains

$$div\ \bar{\bar{\sigma}} = \sum_{\alpha} \sigma_{\alpha\beta} = 0\,. \tag{11.17}$$

Inserting (11.15) and (11.16) into (11.17) we get

$$\nabla(\nabla \cdot \vec{u}) + (1 - 2\nu)\Delta\vec{u} = 0\,, \tag{11.18}$$

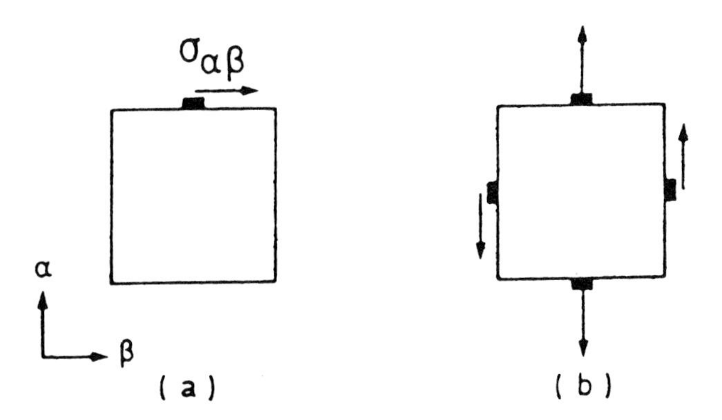

Figure 11.9. Schematic representation of the forces acting on a volume element. (a) definition of $\sigma_{\alpha\beta}$, (b) balance of forces in the α direction.

where ν is a constant (Poisson ratio). (11.18) is called the Lamé or Navier equation. It is a consequence of the conditions of invariance under translations and coordinate transformations.

The boundary conditions can be formulated on the basis of the following considerations. Imagine a piece of elastic material and a void (crack) in the bulk of it. We are interested in the development of the void's shape if a displacement $\vec{u}_0$ is imposed at the outer boundary of this piece of material. The stress perpendicular to the free surface of the void vanishes: $\sigma_\perp = 0$. The dynamics of the crack growth is determined by the velocity v_n normal to the surface. v_n depends on the stress field in a complicated way including several material dependent factors. In order to carry out actual calculations one unusually has to make a concrete assumption for this dependence. A possible common choice is $v_n = C(\sigma_\| - \sigma_c)^\eta$, where C is a constant, $\sigma_\|$ is the stress parallel to the surface of the crack, and σ_c is the cohesion force. With these comments, for the two dimensional case we arrive at the following boundary conditions determining a unique solution of the crack propagation problem

$$\vec{u} = \vec{u}_o \tag{11.19}$$

on the external boundary and

$$\begin{aligned} \partial_\perp u_\parallel + \partial_\parallel u_\perp &= 0 \\ (1-\nu)\partial_\perp u_\perp + \nu \partial_\parallel u_\parallel &= 0 \end{aligned} \tag{11.20}$$

on the internal boundary, where equations (11.15) and (11.16) were used to express the stress through the derivatives of the displacement field. The crack's surface moves with a velocity

$$v_n = C[\nu \partial_\perp u_\perp + (1-\nu)\partial_\parallel u_\parallel - \sigma_c] \,. \tag{11.21}$$

The Lamé equation together with the above boundary conditions form a moving boundary problem which is reminiscent of the equations (9.1–9.4) describing Laplacian pattern formation. This formal analogy suggests that the the methods developed for that case are likely to be useful in the description of fractal crack growth as well.

11.2.2. Lattice models of single cracks

In this section we shall consider two-dimensional lattice models of the formation of a single fractal crack (see, e.g., Kertész 1990). The simplest way of deriving a possible discretized form of equation (11.18) is to assume that the bonds of the lattice are Hookean springs which can freely rotate around the nodes. In this rather simplified *central force* model (Feng and Sen 1984) the shear mode does not exist on the square lattice and because of this the triangular lattice is used in the simulations of this model. A further, more realistic approach, the *beam model* (Roux and Guyon 1985) takes into account additional forces by assuming that the bonds are elastic beams rigidly connected at the nodes. These beams are also allowed to bend making the local twist at the nodes possible in accordance with the Cosserat elasticity in which the *asymmetric* terms in the expression for the strain are also considered.

After having made a choice for the elastic model one has to specify a rule which can be used as a criterion of breaking a bond. As mentioned before, the problem of the growth velocity of a crack as a function of the stress field is less well understood than, for example, the interface velocity in the case of Laplacian growth. A possible choice is (Louis and Guinea 1987) to assume that the probability p_i of breaking the ith bond is proportional to the force acting on the given bond (just as in the dielectric breakdown model the probability of adding a particle depends on the local value of the electric field). In the case of the central model this is equivalent to

$$p_i = \frac{|\delta_i|}{\sum_j |\delta_j|}, \qquad (11.22)$$

where δ_i is the strain of the ith spring. A simple generalization of the above expression can be obtained by using the η parameter $p_i \sim |\delta_i|^\eta$. In the beam model the local moments m_1 and m_2 occurring at the ends of the beams have to be also considered in the breaking rule. The expression

$$p_i \sim (f^2 + q \max(|m_1|, |m_2|))^\eta \qquad (11.23)$$

has been suggested (Herrmann et al 1989) in analogy with the von Mises yielding criterion (for $\eta = 1$). Here f is the compression force and q parametrizes the affinity of the breaking process to the bending mode.

A typical *realization* of these models includes the following procedures. i) A lattice of size on the order of 100 $\times$ 100 is defined. ii) One of the possible forms of an external stress or strain is introduced at the boundary (shear, uniaxial stress, uniform dilation or compression). Periodic boundary conditions can be used in the allowed directions. iii) The bond in the middle of the lattice is broken and the system is relaxed to its equilibrium. The conjugate gradient method (Batrouni and Hansen 1988) seems to be the most efficient, but simple overrelaxation also works. iv) One of the bonds from the surface of the crack is chosen randomly according to its p_i (bonds with larger p_i are selected with larger probability) and this bond is broken. With a new crack configuration the system is relaxed and the procedure is repeated until the desired size of the crack is reached. The process of generating cracks is quite time consuming and the largest configurations which can be obtained do not go beyond several thousand bonds. It may require several CPU hours on a supercomputer to generate a single crack consisting of 1000 bonds.

The most important remark one can make about the simulation results is that apparently all of the existing models lead to crack patterns which can be described by fractal geometry. The actual value of the fractal dimension D *depends* on a number of factors. First, the clusters of broken bonds are relatively small and, in the view of the slow crossovers typical for far-from-equilibrium growth, the determination of the asymptotic value of D can not be particularly accurate. Furthermore, the fractal dimension depends on the parameter η entering the breaking condition. Finally, for different definitions of the neighbouring bonds the numerical value obtained for D has also been found to be different. In general, D is smaller if a larger η is used, while D is increased if more bonds at the surface are considered as neighbouring ones. An example for the type of single cracks which can be obtained by simulations of the central force model on the triangular lattice is shown n Fig. 11.10.

If the randomness of the media is less relevant, the method of noise-reduction (Section 9.2.2) is expected to provide a better description of the growth mechanism. The application of noise-reduction to fracture models is straightforward (Fernandez et al 1988). The counters are put on the neighbouring bonds. After calculating the p_i-s trials are made according to (11.22) or (11.23) and a bond is broken as soon as it is selected m times. The new equilibrium is determined, but the counters are not changed. A transition from random to dendritic patterns has been seen, however, no needle type of morphology appeared up to $m \leq 1000$.

The infinite noise-reduction limit is deterministic crack propagation on a lattice. This process has been considered for the beam model on the square lattice (Herrmann et al 1989). The effects of the following breaking criteria have been investigated:
(I) The beam with the largest p is broken;
(II) The beam for which $q_0 = p + f_0 p_{-1}$ is the largest is broken, where p_{-1} is the value of p when the previous beam was broken and f_0 is a memory factor;
(III) On each beam a counter c is placed which is set to zero at the very beginning. Each time the probabilities p_i have been obtained, the quantity $\omega = (1 - c)/p$ is calculated and the beam with the smallest $\omega = \omega_{min}$ is

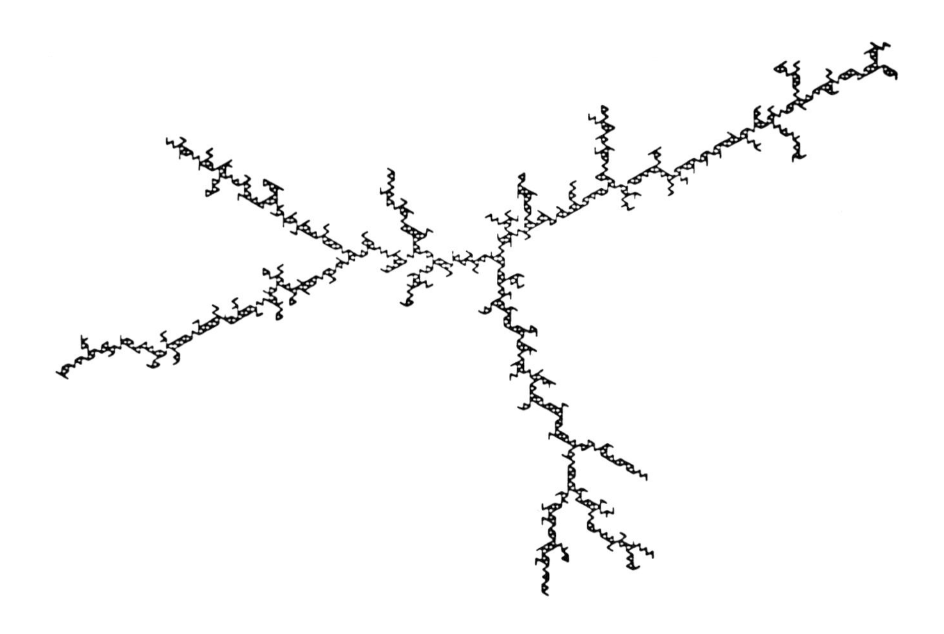

Figure 11.10. A typical crack generated using the central force model on a triangular lattice. The number of broken bonds is 1210, $\eta = 1$ and the linear size of the system is 125 lattice unit. The fractal dimension has been estimated to be about 1.35 (Meakin et al 1989).

broken. After this beam has been broken, every counter is set to $c = \omega_{min}p + f_1 c_{-1}$, where c_{-1} is the value the counter had before the breaking and f_1 is another memory factor.

The three different physical situations described by the above criteria can be summarized as follows: (I) corresponds to ideally brittle and the fastest rupture. (II) contains short time memory which is expected to play a role in cracks that propagate with a finite velocity and produce strong local deformations at the tip of the crack. (III) This model is designed to describe slow stress corrosion. The memory factors measure the time correlations and criterion (III) for $f_1 \to 0$ gives (II) with f_0, while in (II) the limit f_0 leads to (I).

Rule (I) results in branching, but nonfractal patterns. The clusters of broken bonds obtained for criterion (II) with $q = 0$ and $\eta = 0.7$ have been found to be fractals of dimension $D \simeq 1.25$.

Criterion (III) also leads to fractal cracks as is demonstrated in Fig. 11.11. This figure shows that unlike in DLA, the deterministic limit produces fractal structures in a natural way. For comparison an experimental picture is displayed as well. The origin of the fractal behaviour is the competition between the global stress perpendicular to the diagonal and the local stress which tends to follow a given straight crack due to the tip instability. This kind of the interplay of two different directions is only possible in a truly vectorial model and is obviously absent in the scalar model of diffusion-limited aggregation.

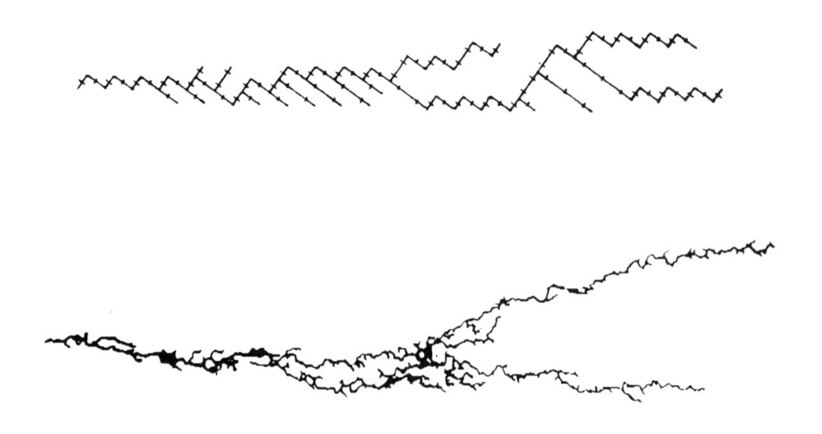

Figure 11.11. Numerical (top) and experimental (bottom) cracks. The simulated crack was obtained for criterion (II) with $f_0 = 1$ and $\mu = 0.2$ (Herrmann et al 1989).

11.2.3. Systems of cracks

In many cases of practical importance it is more appropriate to assume the presence of an ensemble of interacting cracks instead of a single crack discussed above. Let us consider a lattice of linear size L in two dimensions with periodic boundary conditions in the horizontal direction and fixed bus bars on the top and the bottom edges through which an external strain (dilation or shear) can be applied. Each bond is supposed to be ideally fragile, i.e., to have a linear elastic response with unit elastic constant up to a threshold force f_c at which it breaks. In addition, the thresholds are assumed to be distributed randomly according to some probability distribution $P(f_c)$ (Kahng et al 1988). A bond is irreversibly removed from the system once a force larger than its f_c is applied to it. In the case of the central force model the breaking rule is to choose the bond (anywhere in the system) with the maximum value of $|f|/f_c$. For the beam model two random thresholds are introduced and the bond for which p_i of equation (11.23) is the largest is broken. As the external force is increased a system of cracks is developed as bonds are breaking one after another until the system falls apart completely.

During the process of breaking the external force F and the external displacement λ can be monitored as a function of the number of broken bonds n. The relation between F and λ describes the breaking characteristics of the whole sample. In contrast to the behaviour of a single bond, the macroscopic response is nonlinear and after the maximum force has been attained the system can still be considerably stretched before it becomes disconnected. This behaviour can be observed experimentally if an external displacement (not a force) is applied.

Before the maximum is reached there are less statistical fluctuations and Fig. 11.12 supports the scaling law (Arcangelis et al 1989)

$$F = L^{\theta} \phi \left(\frac{\lambda}{L^{\rho}} \right) \tag{11.24}$$

with $\theta \simeq \rho \simeq 0.75$. Similarly, the number of broken bonds has been found to scale as

$$n = L^{\vartheta} \psi \left(\frac{\lambda}{L^{\rho}} \right) \tag{11.25}$$

with $\vartheta \simeq 1.7$. Both scaling forms have a very general applicability, and the

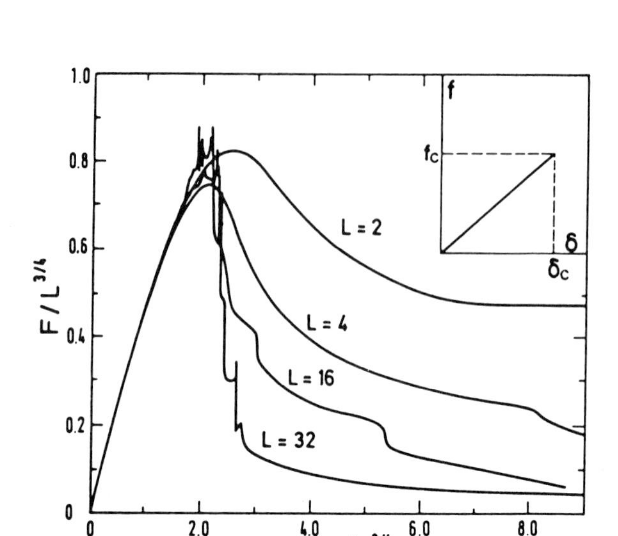

Figure 11.12. The dependence of the external force F on the external displacement λ in the beam model for various system sizes L (Arcangelis et al 1989). Note that both axes are rescaled by a factor $L^{-3/4}$.

corresponding exponents have been found to be the same for a wide range of parameter values and models.

The *multifractal* properties of the local forces f_i can be studied by calculating the various moments of their distribution $M_q = \sum_{bonds} f_i^q$ and investigating the L dependence of the corresponding normalized quantities $m_q = (M_q/M_0)^{1/q}$. The related numerical results demonstrate that for a variety of models $m_q \sim L^{y_q}$ with non-trivial q dependent exponents y_q for systems which are *just before* breaking apart.

Finally, an interesting model to simulate the fractal system of cracks developing in drying thin films has been proposed recently by Meakin (1988). This approach will be briefly discussed in Section 13.2.4 in the context of the related experiments.

11.3. OTHER MODELS

The number of possible and published models of self-similar fractal growth is enormous and, obviously, it would not be feasible to review here all of them. Instead, only a fraction of these models will be briefly described below which were selected on the basis of their potential applicability and relevance to some other important aspects of fractal growth.

i) **Models related to growing percolation**. One of the most typical experimental realizations of growing percolation is the advancement of a fluid front in a porous medium. The geometry of the front then depends on the capillary number and the wetting properties of the fluids. In Sections 5.1 and 5.2 the main features of the simplest growing percolation models have been discussed in terms of regular lattice models. In order to study the effects of the level of wetting in more detail Cieplak and Robbins (1988) have introduced a model in which the *wetting angle* can be tuned continuously. In their model a two-dimensional array of disks with random radii is used to simulate the porous medium. The disks are placed on an $L \times L$ triangular or a square lattice with unit spacing with L varying between 300 and 800. In this model the wetting angle θ can be fixed and the interface is assumed to be formed by a set of arcs connecting the consecutive disks with the proper wetting angle. Each arc must have a radius $r = \gamma/p$, where γ is the surface tension and p is the pressure difference. As the pressure difference is increased, some of the arcs become unstable and the front locally advances.

Simulations of this model show that as a function of the wetting angle a relatively sharp transition takes place in the mean width of the finger-like regions filled by the invading fluid. As θ is decreased from the nonwetting limit ($\theta = 180°$) and approaches a critical value θ_c, the widths of the invaded segments in slices made through the patterns have been found to diverge. These results are in qualitative agreement with the earlier experimental and simulational results discussed in Chapter 5, but represent further important information about the role of wetting in fluid flows which will also be relevant in the experimental studies of self-affine interfaces to be discussed in the last chapter.

Directed percolation (DP) is perhaps the simplest spreading percolation model. It is pertaining to a number of applications when one of the growth directions is preferred. Although the directed percolation clusters are known to be self-affine, their global fractal dimension has been conjectured to be non-trivial (Nadal et al 1984) in contrast to the behaviour of self-affine functions. This question has been recently investigated numerically using large scale simulations (Hede et al 1991).

In directed bond percolation bonds of a lattice are occupied with a fixed probability p. The bonds can be represented as arrows oriented according to an external direction and percolation against the arrows is forbidden. There is a critical point p_c in this system in dimensions $d \geq 2$ such that for $p > p_c$ the probability of an infinite cluster is non-zero. A cluster in DP is defined as the set of sites which can be reached from the origin via occupied directed paths. Typical clusters (an example is shown in Fig. 11.13) for $p \neq p_c$ are anisotropic and they are characterized by two different correlation lengths; $\xi_\parallel$ (parallel to the growth direction) and $\xi_\perp$ (perpendicular to it). As p approaches the critical point the two correlation lengths diverge with different exponents

$$\xi_\parallel \sim |p - p_c|^{-\nu_\parallel}, \qquad \xi_\perp \sim |p - p_c|^{-\nu_\perp}. \tag{11.26}$$

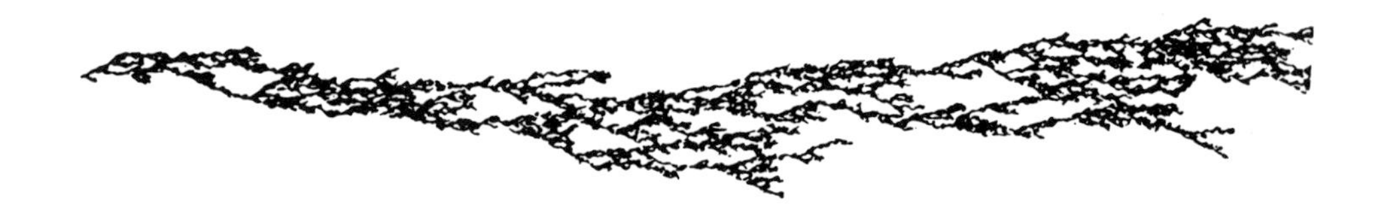

Figure 11.13. A typical directed bond percolation cluster at p_c after 3072 time steps (Hede et al 1991). The direction of growth is from left to right.

The fractal dimension D of DP clusters can be determined by using the box counting method. A correction to scaling analysis has resulted in the estimate $D = 1.765(10)$. This value favours the relation (Kinzel 1983)

$$D = 1 + (1 - \sigma\nu_{\|})/\sigma\nu_{\perp}\,, \tag{11.27}$$

where σ is a Fisher exponent in the scaling of the typical cluster size $s \sim (p_c - p)^{-1/\sigma}$ as p_c is approached from below. The above statement can be checked by inserting the precisely known values of the three exponents entering the rhs of (11.27).

Investigations of the structures built up by particles diffusing from a source (Sapoval et al 1985) have led to the interesting conclusion that the so-called *diffusion front* has fractal properties which can be related to the hull of percolation clusters. Let us consider the diffusion of particles from a line source on a lattice. After some time a smooth distribution of the concentration of particles (number of particles in a layer) is built up as a function of the distance x from the line source. For small x the concentration $p(x)$ is close to 1 and it drops to zero for large x. In between there is a region where $p(x)$ is about p_c, where p_c is the percolation threshold for the given lattice. Some of the particles are disconnected, however, most of them belongs to a single large cluster connected to the source line. The border between the unoccupied sites of the lattice and the above large cluster is the diffusion front (see Fig. 11.14).

The following important observations have been made concerning this model: a) If x_h denotes the average distance from the source of the points belonging to the front then $p(x_h) = p_c$. Using this result very accurate estimates of the threshold probabilities can be obtained. b) The diffusion front is a fractal object which in two dimensions has a fractal dimension numerically very close to 7/4. We recall that this value was shown to be equal to the fractal dimension of the hull of percolation clusters in $2d$.

ii) **Cluster growth through evaporation and aggregation** (Hayakawa and Matsushita 1989) represents an interesting model which is relevant to situations where the growing patterns in a system may interact

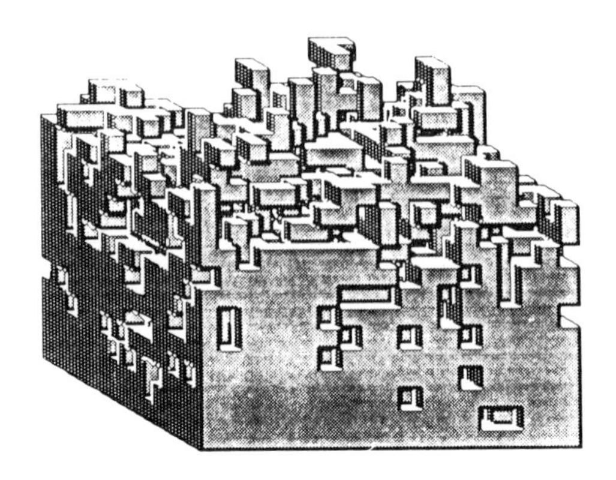

Figure 11.14. The diffusion front in a three-dimensional sample of size 19^3 (Rosso et al 1986).

without moving around. In this model the particles are at the beginning distributed randomly on a lattice. A particle is selected randomly and allowed to diffuse until it hits another one. The two particles form a rigid bond. Other particles are selected and at some point no more single particles are left. Then one of the randomly chosen "tip particles" (particles having only one bond) is allowed to evaporate (leave its cluster). This particle diffuses on the lattice until it sticks to the surface of one of the clusters (it may stick to the original one as well). The process of evaporation is repeated many times.

Several interesting results have been found for this model. The number of clusters consisting s particles at time t scales as $n_s(t) \sim s^{-2} f(s/t^z)$, where $f(x)$ is a scaling function and $z \simeq 1/2$ in $2d$ and $z \simeq 1/3$ in $3d$. These results are in overall agreement with the dynamic scaling theory described in Chapter 8, however, the values of z are specific to this model. A study of the total number of tip particles as a function of time shows large fluctuations which result in a power spectrum similar to that of the $1/f$ noise.

iii) **Julia sets** have been shown to be useful when investigating the structure and the multifractal nature of the growth probabilites of fractal clusters (Procaccia and Zeitak 1988, Bohr et al 1988). In this approach deterministically generated structures are used to mimic DLA type aggregates. Consider the mapping

$$z' = z^m + c\,, \tag{11.28}$$

where z and c are complex and m is an integer. For almost all points on the complex plane repeated iterations of (11.28) lead to either 0 or to ∞. The set of point that remains invariant under (11.28) is called the Julia set (see example 2.4 of Chapter 2). One way of obtaining a specific Julia set is solving for one of the fixed points $z^\star = (z^\star)^m + c$ and finding the preimages $(z^\star - c)^{1/m}$ (there are m of them). Each preimage has m further preimages, etc. There are infinitely many Julia sets of many kinds. Here we are concerned with those which are connected and have a dendritic structure. These can be easily found by choosing an appropriate value for c. Figure 11.15 shows examples for $m = 4$ and 5.

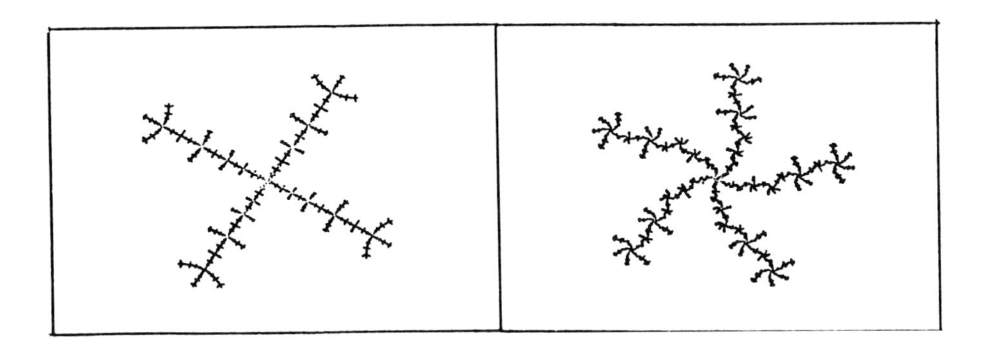

Figure 11.15. Julia sets of the mapping $z' = z^m + c$ for $m = 4$ (left) and 5 (right) with c satisfying $(c^m + c)^m = c^m$ (Procaccia and Zeitak 1988).

The most important fact about Julia sets regarding aggregation is that by iterating backwards from a fixed point and choosing randomly one of the

m preimages at each step, the density of the distribution of points obtained is a harmonic measure. This means that the rate of visitation of a little ball containing a part of the set by the iterates is the same as the probability that a random walker launched at infinity would hit that ball.

In the case of Julia sets the multifractal properties of the harmonic measure can be systematically investigated using numerical methods and analytical arguments. In particular, Bohr et al (1988) find that for $m = 2$ and some values of c the corresponding $f(\alpha)$ spectrum becomes one-sided and the largest value of the singularity exponent α_{max} diverges as it also has been suggested by recent simulations concerning the multifractal behaviour of the growth probabilities in diffusion-limited aggregation (Section 11.1.2).

11.4. THEORETICAL APPROACHES

A complete theory of fractal growth is expected to answer such questions as why a given physical process leads to fractals and what are the main geometrical features of the resulting structures. This program is supposed to be carried out using assumptions based on "first principles" only. During the past couple of years a few interesting theoretical approaches have been developed which in part satisfy the above criteria. These methods include i) the so-called fixed scale transformation and ii) the use of renormalization-group ideas.

i) The main idea of the *fixed scale transformation* approach (Pietronero et al 1988a,b) is to look for an iterative equation and its fixed point which is considered as corresponding to the fractal structure. In this sense the spontaneously selected geometry is looked at as an attractor in the abstract space of possible patterns. The basic features of the theory will be demonstrated using the strip geometry in two dimensions. The growth starts at the bottom of the cell and takes place in a direction parallel to the walls.

The intersection of a growing fractal structure of dimension D with a line perpendicular to the walls has a dimension $D' = D - 1$. Covering this cross-section with a set of boxes we assign a black dot to a box if it contains part of the cross-section and a white dot otherwise (see Fig. 11.16). When such a box is subdivided into two further boxes, two possible outcomes may occur: type 1 (one black and one white sub-box) and type 2 (both sub-boxes are black). Let us denote the corresponding probabilities by C_1 and C_2. The average number of black sub-boxes that appear at the next level of fine graining from one black box is

$$\langle n \rangle = \sum_i n_i C_i = C_1 + 2C_2 \,. \tag{11.29}$$

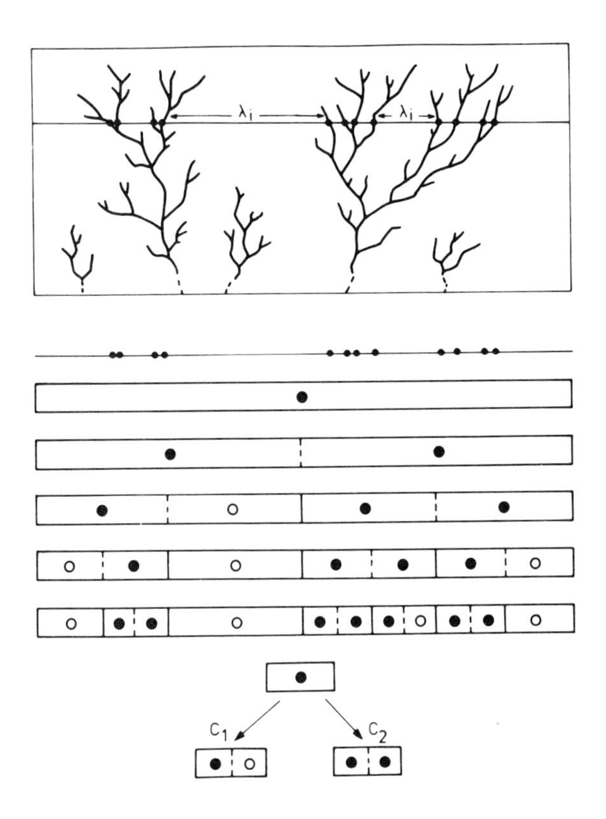

Figure 11.16. The possible elementary configurations during the process of fine graining of a DLA type pattern in the fixed scale transformation approach (Pietronero et al 1988).

Then it can be shown (Pietronero et al 1988a,b) that the fractal dimension of the whole structure is

$$D = 1 + \frac{\ln\langle n\rangle}{\ln 2} . \tag{30}$$

Thus, the central quantity to be studied is the distribution of elementary configurations $\{C_i\}$ which appears as one goes from one scale to the next in the process of fine graining.

Next we have to determine the appropriate iterative equations for the distribution $\{C_i\}$. Since our clusters are *intrinsically critical* one has a high degree of symmetry which has to be built into the equations. a) The scale invariance means that the values of C_1 and C_2 which are obtained when going from the scale l to $\frac{1}{2}l$ have to be the same as those from scale l' to $\frac{1}{2}l'$. b) For two different sections the same distribution of C_1 and C_2 as above has to correspond.

One can define the iteration based on property b) and then use the fixed point values for C_i at all other scales via equations (11.29) and (11.30). The fixed scale transformations are equations of the type

$$\begin{pmatrix} C_1^{(k+1)} \\ C_2^{(k+1)} \end{pmatrix} = \begin{pmatrix} M_{1,1} & M_{2,1} \\ M_{1,2} & M_{2,2} \end{pmatrix} \begin{pmatrix} C_1^{(k)} \\ C_2^{(k)} \end{pmatrix} , \tag{11.31}$$

where the matrix element $M_{i,j}$ defines the conditional probability to have a configuration of type i followed by one of type j in the growth direction within the "frozen" part of the structure. Since $M_{1,1} + M_{1,2} = 1$ and $M_{1,2} = M_{2,2} = 1$, the fixed point condition gives

$$C_1 = \left(1 + \frac{M_{1,2}}{M_{2,1}}\right)^{-1} \tag{11.32}$$

and from (11.29) and (11.30) D can be calculated. The details about the explicit calculation of $M_{i,j}$ can be found, for example, in Pietronero et al (1988a,b). The fixed scale transformation method has been successfully applied to a number of growth models giving reasonable estimates for the fractal dimension.

ii) The *position-space renormalization-group* method (PSRG) represents a standard theoretical tool in the description of critical phenomena (some of the possible approaches are discussed in Section 4.3). The PSRG method has led to a number of interesting results on fractal growth as well (Nagatani 1988, Nagatani and Stanley 1990, Nagatani et al 1991). However, its application to far-from-equilibrium processes typically involves heuristic assumptions.

When deriving the renormalization transformation small cells having the geometry of the square or the diamond hierarchical lattice are used. As a prototype of fractal growth the dielectric breakdown model is considered (Section 6.3). At the beginning of the growth process each bond is occupied by a resistor of unit conductance. A constant voltage is applied between the top and the bottom of the cell. The main idea behind the PRSG is that three types of bonds are assumed to be present before and after an iterative step corresponding to the renormalization is made: a) breakdown bonds which constitute the fractal pattern, b) growth bonds at the perimeter of the pattern, and c) unbroken bonds consisting of the original resistor. Then, the renormalization equation is written down for the effective conductance of the growth bonds by taking into account the probabilities of the possible configurations after a growth event.

The most important advantage of the PSRG approach is that it can incorporate the various extra mechanisms and factors which can be introduced into a particular growth model. By studying the flow diagram of the resulting transformations one can make statements about the relevance of a given aspect of the growth process. To mention an example, crossovers from DLA to Eden cluster geometry or from Eden to DLA can be detected due to finite viscosity and effects of chemical dissolution, respectively (Nagatani et al 1991).

In a PSRG study by Barker and Ball (1990) the *role of the anisotropy* of the underlying cell versus the *noise* represented by the random walkers has been investigated. Renormalizing in real space always involves some shape of lattice, and since this kind of anisotropy is known to have a strong effect on the geometry of DLA clusters one is forced to study the effect of this bias.

To do this two parameters are introduced: p controls the anisotropy of the sticking rules and ϵ quantifies the noise. With appropriate choices for p large scale growth along the axes or the diagonals of the square lattice can be made preferred. The parameter ϵ is essentially inversely proportional to the noise-reduction parameter m.

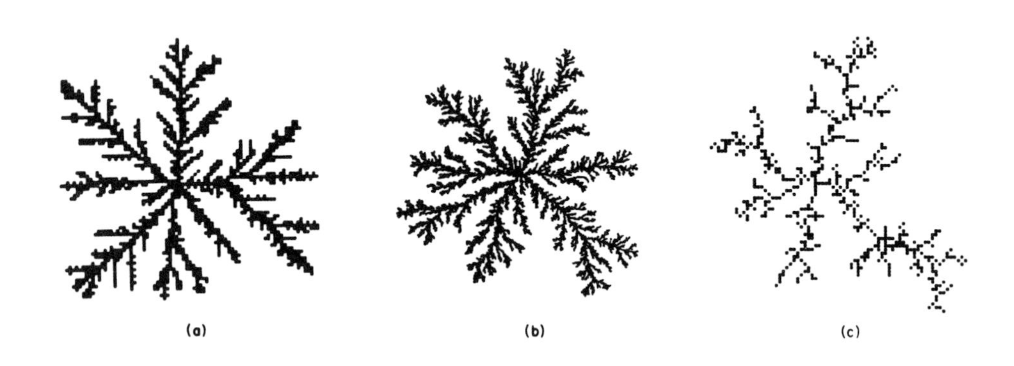

Figure 11.17. (a) A DLA cluster of radius 50 grown at the unstable fixed point. It should correspond to a coarse-grained version of a large isotropically grown cluster which is shown in (b). For reference a standard DLA cluster of radius 50 is shown in part (c) (Barker and Ball 1990).

The transformations for these parameters are derived on the basis of changing the length scale on which the cluster is viewed. Two stable and one unstable fixed points can be identified on the $p - \epsilon$ plane; each fixed point corresponds to a kind of self-similar cluster. Naturally, the two stable fixed points correspond to cross-like shapes, one with arms along the axes, the other one along the diagonals of the square lattice on which the study has been carried out. The unstable fixed point is the most interesting, since the clusters grown with parameters of this point $p^\star$, $\epsilon^\star$ are statistically isotropic for all sizes (see Fig. 11.17). One of the interesting features of this fixed point is that $\epsilon^\star$ is two orders of magnitude smaller than $\epsilon = 1$ corresponding to ordinary DLA. This fact explains why very large clusters are needed to show sensitivity to lattice anisotropy in diffusion-limited aggregation.

Chapter 12

DYNAMICS OF SELF-AFFINE SURFACES

While in the period between 1982–87 it was the growth of self-similar fractals which attracted most of the interest, during the last few years the investigation of self-affine growing surfaces has gone through a spectacular explosion of activity. A significant progress has been made in our understanding of the dynamics of kinetic roughening due to the convergence of important new results from computer simulations, analytical theories and experiments. The field of self-affine growth represents a rare example where these three major approaches of physics can be applied successfully to a class of far-from-equilibrium phenomena. Because of the exciting new developments, kinetic roughening will be reviewed here in more detail, including the various simulational and theoretical results. Readers who need more specific information are advised to consult the review of Krug and Spohn (1990) and a very recent edited book on *Dynamics of Fractal Surfaces* (Family and Vicsek 1991).

If the conditions of the growth process are such that the development of the interface is *marginally stable* and the fluctuations are relevant, the resulting structure is a rough surface and can be well described in terms of self-affine fractals (Mandelbrot 1982). Marginal stability means that the interface is neither stable nor unstable. In other words, the fluctuations (which die out quickly for a stable surface and grow exponentially for an unstable one) survive without drastic changes for a long time in the case of marginally stable growth.

In many cases rough surfaces are generated by a growing interface which advances as new parts are added according to some dynamical process. Examples include, crystal growth, vapor deposition, electroplating, spray painting and coating, and biological growth. Fractally rough surfaces may also be formed during the removal of material, as in chemical dissolution, corrosion, grinding, erosion, blasting, wear and all types of polishing. The growth of interfaces is intimately related to a variety of other processes, including the propagation of flame fronts, the long time behavior in randomly stirred fluids, impurity roughening and pinning of interfaces, and the problem of directed polymers in random media. These relations provide powerful connections among seemingly different phenomena, which can be exploited in the development of various approaches for understanding the evolution of rough interfaces.

12.1. DYNAMIC SCALING

Dynamic roughening of interfaces is an example of a far-from-equilibrium phenomenon without a complete theory based entirely on first principles. However, the analysis of the scaling behaviour of the time and spatial dependence of the surface properties has led to the development of a general dynamic scaling approach for describing growing interfaces (Family and Vicsek 1985). This formalism, which is based on the general concepts of scale invariance and fractals, has become a standard tool in the study of growing surfaces and has been applied to the study of a variety of theoretical models of growing interfaces, and some recent experiments.

Consider the time evolution of a rough interface in a d-dimensional space starting from an initially flat surface at time $t = 0$. Let us concentrate on a part of the surface having an extent L in $d-1$ dimensions perpendicular to the growth direction. The surface typically can be described by a single-valued function $h(\vec{r}, t)$ which gives the height (distance) of the interface at position $\vec{r}$ at time t measured from the original $d-1$ dimensional flat surface. During growth the interface heights fluctuate about their average value $\bar{h}(t) \simeq t$ and the extent of these fluctuations characterizes the width or the thickness of the surface. The root mean-square of the height fluctuations $w(L, t)$ is a quantitative measure of the surface width and is defined by

$$w(L,t) = [\langle h^2(\mathbf{r},t)\rangle_r - \langle h(\mathbf{r},t)\rangle_r^2]^{1/2}\,. \tag{12.1}$$

Note that from now on we adopt the notation w for the width, because this has been widely accepted in the literature recently (in Chapter 7 we used σ for the same quantity). Since we assume that the surface is a continuous function, its width has to saturate to some value after a relaxation time τ. The only scale in the problem is the linear size of the substrate L (there is no intrinsic time scale), thus, w depends only on some power of L and a ratio of the form t/L^z, where z is an exponent describing how the relaxation time depends on the system size. Correspondingly, the dynamic scaling of the surface width has the form (Family and Vicsek 1985, see also Chapter 7) $w(L,t) = L^\alpha f(t/L^{\alpha/\beta})$, where $\alpha/\beta = z$ is the dynamic scaling exponent. In the limit where the argument of the scaling function $f(x)$ is small, $x << 1$, the width depends only on t and this implies that for small x the scaling function $f(x)$ has to be of the form $f(x) \sim x^\beta$. Saturation at large times means that $f(x)$ goes to a constant in that limit. Thus,

$$f(x) \sim \begin{cases} x^\beta, & \text{if } x \ll 1 \\ \text{Const}, & \text{if } x \gg 1\,. \end{cases} \tag{12.2}$$

An alternative approach to the characterization of self-affine surfaces changing in time is the determination of various *correlation functions.* The most convenient quantity is the so-called height-height correlation (or height difference) function $c(r,t)$ which is defined as

$$c(r,t) = \langle |h(\vec{r}',t') - h(\vec{r}' + \vec{r}, t' + t)|\rangle_{\vec{r}',t'}\,, \tag{12.3}$$

which is the average height difference measured for a time difference t at two points whose coordinates on the substrate are separated by $\vec{r}$. For surfaces isotropic in the directions perpendicular to the growth the correlation function depends only on the absolute value of this vector, r. On the basis of the scaling behaviour of self-affine functions, $c(r,t)$ scales the same way as the width,

$$c(r,t) \sim r^{\alpha} f\left(\frac{t}{r^{z}}\right) \tag{12.4}$$

with an $f(x)$ analogous to (12.2), however, in this case L is replaced by r. Correspondingly, for long times and for $r << L$ the correlation function $c(r,0)$ behaves as

$$c(r,0) \sim r^{\alpha}\,, \tag{12.5}$$

and for fixed r and short times

$$c(0,t) \sim t^{\beta}\,. \tag{12.6}$$

The above discussed expressions can be used to determine the exponents α and β for experimentally observed surfaces. The schematic behaviour of $c(r,0)$ and $c(0,t)$ is shown in Fig. 12.1. The average height difference can be considered as the correlation length $\xi_{\perp}$ in the growth (perpendicular to the substrate) direction. According to (12.2) and (12.4)

$$\xi_{\perp} \sim t^{\beta}, \quad \text{for} \quad t \ll \tau \qquad \text{and} \qquad \xi_{\perp} = Const, \quad \text{for} \quad t \gg \tau\,, \tag{12.7}$$

where the relaxation time τ is proportional to L^{z}. The spread of fluctuations in the direction parallel to the substrate is characterized by the correlation length $\xi_{\|}$. This shows how far the effect of a perturbation can get after some time t along the surface. On the basis of the dynamic scaling form (12.4)

$$\xi_{\|} \sim t^{\alpha/\beta}, \quad \text{for} \quad t \ll \tau \qquad \text{and} \qquad \xi_{\|} = L, \quad \text{for} \quad t \gg \tau\,. \tag{12.8}$$

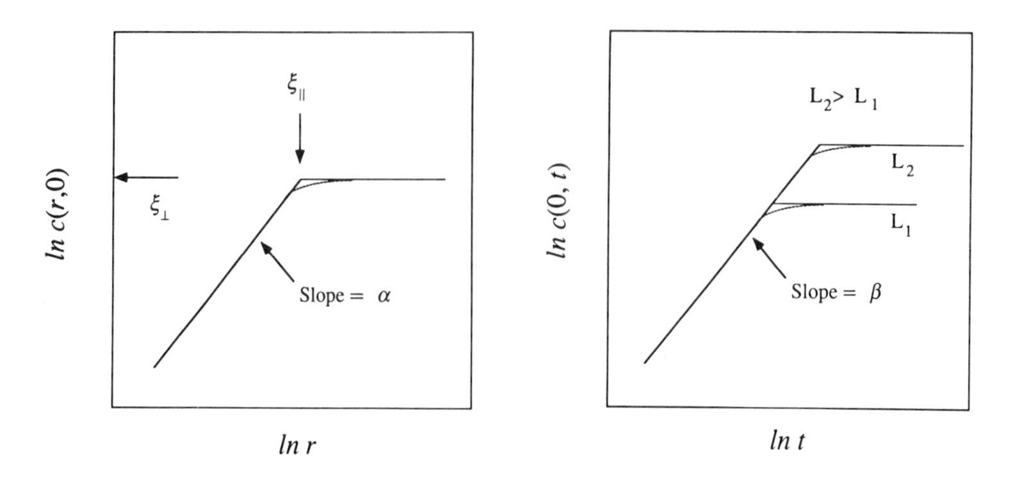

Figure 12.1. Schematic plots of the height correlation functions $c(r, 0)$ and $c(0, t)$.

As we shall see most of the studies in the field of self-affine growth concentrate on the determination of the exponents α and β entering the above definitions characterizing the time development of surfaces.

12.2. AGGREGATION MODELS

As demonstrated in Chapter 7 one of the most successful approaches to understanding the dynamics of growing interfaces has been the study of simplified numerical models. A characteristic feature of cluster models for simulating marginally stable interfaces is that stochasticity is naturally built into the rules which are used to generate the aggregates.

Ballistic deposition models (Section 7.3) represent a popular computer method of investigating the scaling behaviour of self-affine surfaces. An important recent development in this area has been the introduction and extensive study of solid-on-solid type of models. Many generalizations of the single-step model by Meakin et al (1986) have been considered with remarkable success as concerning the improvements of the estimates for the critical exponents.

i) **Solid-on-solid and related models.** The main feature of these models is that no overhangs are allowed to exist in the deposit's structure. The particles are assumed to rain down along randomly positioned vertical trajectories as in ordinary ballistic aggregation, however, they are incorporated into the deposit only if a) the height difference between any two columns is smaller or equal to a given number of k lattice units (mostly $k = 1$), b) no overhangs occur, i.e., a particle always has to sit on top of another one. This model leads to perfectly compact clusters which has a surface with slopes smaller or equal to k.

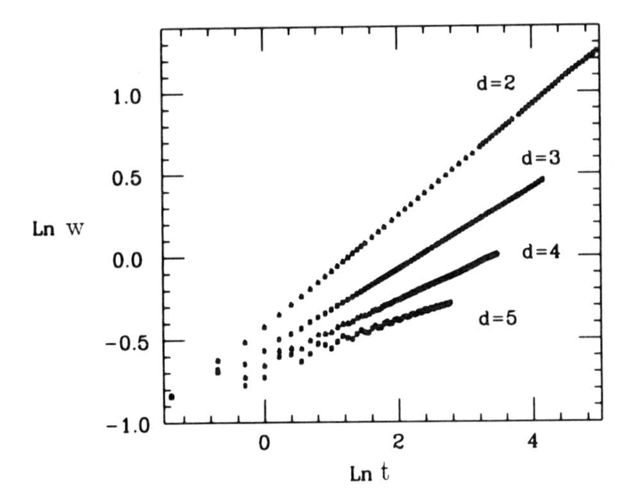

Figure 12.2. The time (average height) dependence of the width in the restricted solid-on-solid model for surfaces grown in $d = 2$ to 5 dimensions (Kim and Kosterlitz 1989). Estimates for the exponent β can be obtained from the slopes of the straight lines fitted to these data.

The extensive simulations of the above restricted solid-on-solid model has resulted in accurate estimates for β in dimensions $d - 1 = 2 - 4$ (Kim and Kosterlitz 1989). The quality of the data is demonstrated by Fig. 12.2, where the growth of the surface width is displayed as a function of time for the above range of dimensionalities. From the slope of the lines fitted to the data the following values were obtained for the exponent $\beta(d)$

$$\beta(3) = 0.25 \pm 0.005, \qquad \beta(4) = 0.20 \pm 0.01, \qquad \beta(5) \simeq 0.17\,. \qquad (12.9)$$

These results have led the authors to conjecture that the values of the exponents in any dimension can be given by the formulae

$$\beta(d) = 1/(d+1) \qquad \text{and} \qquad \alpha(d) = 2/(2+d) \qquad (12.10)$$

which for the dynamic exponent gives $z = 2(d+1)/d+2)$. Conjectures made for exponents generally represent a challenge for those who are involved in simulations and in theoretical approaches. Now we turn to the description of the results obtained by the hypercube stacking model which is closely related to the restricted solid-on-solid models.

In the hypercube stacking model the surface of a stack of hypercubes on a $(11\cdots)$ substrate is considered. This means that the hypercubes are piled up in such a way that one of their corners points downward; in $d = 2$ this corresponds to square shaped particles which are put on a square lattice rotated by 45° (the diagonals point vertically). The model is defined as a sequence of stacking events; each time a hypercube is added to the surface at a deposition rate p^+ or removed with a rate p^- (Liu and Plischke 1988, Forrest and Tang 1990). The solid-on-solid condition with $k = 1$ is assumed at all stages. One of the important aspects of this model is that there exists a mapping of $h(\vec{r})$ to a Potts-spin configuration which is given by $\sigma(\vec{r}) \equiv h(\vec{r})$ mod d. The Potts-spin configuration $\{\sigma(\vec{x})\}$ obtained in this way can be shown to be equivalent to the ground state of a chiral Potts model with a well defined Hamiltonian. This fact turns out to be very useful, since in this way the extremely efficient methods which have been worked out for simulating spin systems can be utilized. In particular Forrest and Tang (1990) used a multispin coding, parallel processing algorithm to obtain accurate estimates.

The careful evaluation of the simulations carried out in systems of sizes $L = 11520^2$ in $d = 2+1$ and $L = 2 \times 192^3$ in $d = 3+1$ dimensions have led to the values

$$\beta(3) = 0.24 \pm 0.001 \qquad \text{and} \qquad \beta(4) = 0.180 \pm 0.005 \tag{12.11}$$

which together with the scaling law $\alpha + \alpha/\beta = 2$ uniquely determine the other two exponents as well. Two comments are in order concerning (12.11): a) The studied system sizes are considerably larger than in any of the previous investigations, b) the estimates (12.11) are in *disagreement* with the conjectures (12.10).

ii) Models with surface diffusion. A common process that is known to occur in vapor deposition and sedimentation processes is the local diffusion of the newly arriving particles along the deposited surface. Surface diffusion leads to surface relaxation and its effects are similar to surface tension in liquid surfaces. A deposited particle is allowed to diffuse on the surface within a finite distance from the column in which it was dropped, until it finds a position which is considered to be optimal within the particular model. At this point the particle sticks to the top of that column and becomes part of the aggregate. Simulations indicate that surface fluctuations are independent of the length of the distance over which the particles are allowed to diffuse, as long as it is finite. One possible rule is (gravitation driven restructuring) that the diffusing particle stops at the nearest column with a local height minimum. This case was briefly discussed in Section 7.3. It can be described by a linear theory leading to $\beta = 1/4$ in $d = 2$ and $\beta = 0$ for $d > 2$.

During vapour deposition of molecules the gravitational force is negligible compared to the binding forces acting on the particles at the surface. This situation is especially relevant from the point of molecular beam epitaxy (MBE) experiments which have very important applications in semiconductor industry. The right rule in any simulational model under such conditions is that the particle should terminate its motion at a column with the highest binding energy (largest number of occupied nearest neighbours, Fig. 12.3). If only the nearest neighbours to the original landing position are checked (Wolf and Villain 1990) the model corresponds to deposition at not too high temperatures where the particles move only a short distance to find a favourable site before another particle is deposited on the top of them. An alternative

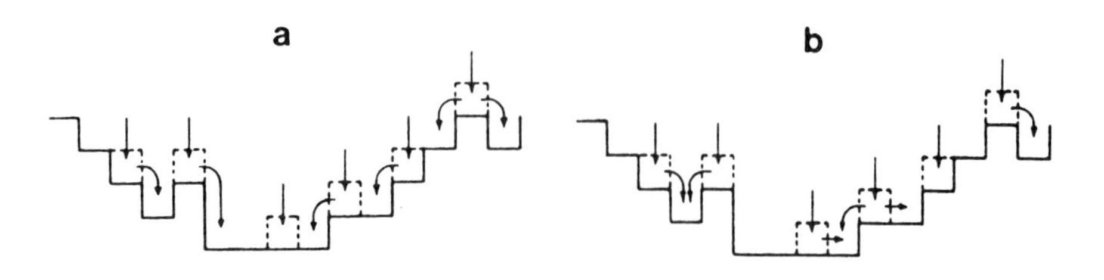

Figure 12.3. 6 possible moves of a newly arrived particle on the surface. (a) Gravitation force dominated model. (b) Binding force dominated model (the particle ends up in a column where it touches the largest number of occupied sites, Wolf and Villain 1990).

choice is to let the particle hop with a probability to the nearest position which has a binding energy corresponding to a kink at the surface within some distance typically larger than one lattice unit (Das Sarma and Tamborenea 1991). Obviously the two approaches are very closely related. In spite of the seemingly small difference in the rules compared to the previous (gravitation dominated) relaxation models, the obtained surfaces have rather different scaling behaviour and visual appearance as is demonstrated by Fig. 12.4.

The simulation of the above models have resulted in the estimates

$$\beta(2) = 0.365 \pm 0.015, \qquad \alpha(2) = 1.4 \pm 0.1 \quad \text{(Wolf and Villain 1990)}$$
$$\beta(2) = 0.375 \pm 0.005, \qquad \alpha(2) \simeq 1.47 \quad \text{(Das Sarma and Tamborenea 1991)}$$

Clearly these results are consistent with each other. Furthermore, they are in agreement with the expressions

$$\beta(d) = (5 - d)/8 \qquad \text{and} \qquad z = 4 \tag{12.12}$$

which can be easily obtained by Fourier transformation of the linear equation

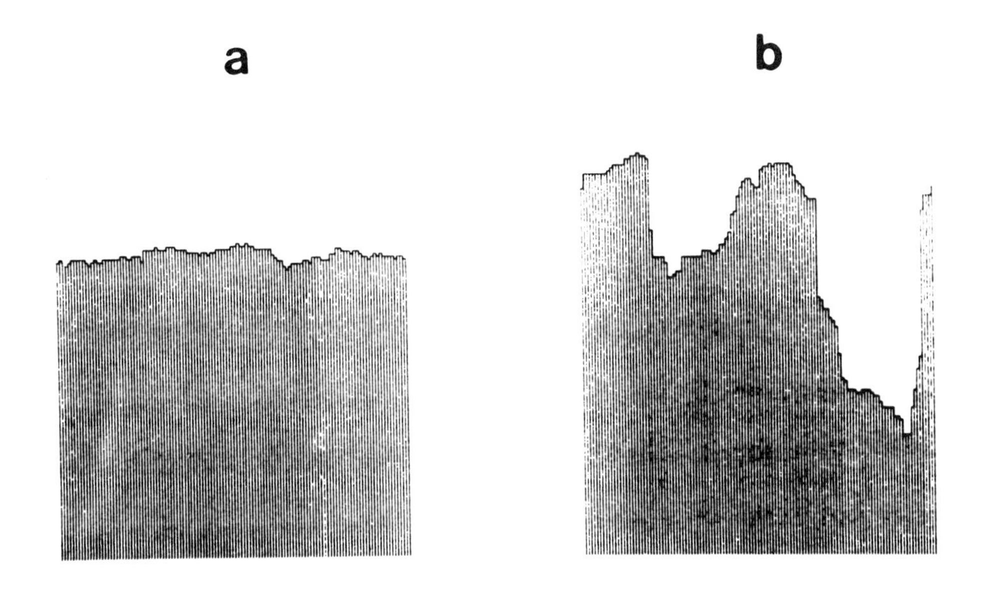

Figure 12.4. Surface configuration in the steady-state in the (a) gravitation and (b) binding force models for $L = 120$ (Wolf and Villain 1990).

for the surface velocity $\partial h/\partial t = \nu \nabla^4 h + \eta$, where η is the noise and ν is related to surface diffusion. This agreement is quite surprising: since $\alpha > 1$, the relative slopes along the surface in an infinite system diverge as $t \to \infty$ and the applicability of a linear approximation cannot be justified. Further investigations are needed to resolve this question including simulations and theoretical studies along the lines of the recent works by Lai and Das Sarma (1991) and Golubovic and Bruinsma (1991).

iii) Other models and results. Finally, a few additional recently published models will be mentioned. A question that is often of relevance in aggregation models is what happens when there is a nonzero flux of particles arriving at the surface of a growing aggregate. In the corresponding model (Baiod et al 1988) the authors study such a model of ballistic deposition and find that the dynamic scaling of the interface is not affected by the flux density. This remains true in the off-lattice case as well (Jullien and Meakin 1989). The above results are particularly interesting for cellular automaton simulations of interface growth, where all or a large fraction of the surface sites are updated simultaneously.

If the direction of the incoming particles may change randomly within some interval of angles, the so called shadowing instability results in a complicated structure with an array of deep grooves and peaks depending on the particular model used. A rich variety of behaviors has been obtained in the related papers (see, e.g., Bales and Zangwill (1989), Tang, Alexander Bruinsma (1990) and Roland and Guo (1991)).

12.3. CONTINUUM EQUATION APPROACH

The dynamics of growing rough surfaces had almost exclusively been investigated by simulations until a continuum equation for the velocity of the interface was proposed by Kardar, Parisi and Zhang (KPZ) (1986). During the last couple of years the continuum equation approach has attracted great interest and has been developing through a) numerous studies of the original equation and b) investigations of closely related other equations containing further terms (in addition to those proposed originally by KPZ). The KPZ equation is for the time dependence of the height on the spatial derivatives of the surface (Section 7.4). It is written down in a coordinate system moving with a velocity v equal to the normal velocity of the growing surface. Its usual form is

$$\frac{\partial h}{\partial t} = \nu \nabla^2 h + \lambda/2 (\nabla h)^2 + \eta(\vec{x}, t) . \qquad (12.13)$$

The noise term $\eta(\vec{x}, t)$ in the original equation was assumed to be uncorrelated, white and have a Gaussian distribution, so that $\langle \eta(\vec{x}, t) \rangle = 0$, and

$$\langle \eta(\vec{x}, t) \eta(\vec{x}', t') \rangle = 2D \delta^d (\vec{x} - \vec{x}') \delta(t - t') , \qquad (12.14)$$

where D is a constant. In the above equations the notation is slightly changed from that used in Section 7.4 in order to follow the presently common way of denoting the various quantities. The behaviour of the KPZ equation for the type of noise (12.14) has been investigated by both renormalization group techniques and numerical simulations. According to the theory, in $d = 1+1$, $\beta = 1/3$ and $\alpha = 1/2$. From the point of the renormalization group the dimension $d = 2+1$ is critical. Above this dimension there are two phases in the phase diagram of the system; one is dominated by the weak, the other one by the strong coupling fixed point (more about this will be given in the next section).

In a complicated physical system the assumption for the noise of the form (12.14), corresponding to the presence of short-range correlations only, may be an oversimplification. Instead of (12.14), Medina et al (1989) considered the cases when the noise term is allowed to have *long-range correlations* in space and/or time. The spectrum $D(k,\omega)$ of this kind of noise can be defined through

$$\langle \eta(\vec{k},\omega)\eta(\vec{k}',\omega')\rangle = 2D(k,\omega)\delta^{d-1}(\vec{k}+\vec{x}')\delta(\omega+\omega')\,, \tag{12.15}$$

having power law singularities of the form

$$D(k,\omega) \sim |\vec{k}|^{-2\rho}\omega^{-2\theta}\,. \tag{12.16}$$

There are several possible cases depending on the actual values of λ and d. The associated scaling behaviors can be studied by dimensional analysis and renormalization group calculations. The results can be summarized as follows: In dimensions higher than a critical d_c weak and strong noise lead to different scaling exponents, while for $d < d_c$ any amount of noise is relevant resulting in the strong coupling behaviour. For weak noise ($\lambda \simeq 0$) the exponent α has the modified value $\alpha(\lambda \simeq 0) = (3-d)/2 + \rho + 2\theta$, while z is unchanged $z(\lambda \simeq 0) = 2$. For $d < d_c = 3 + 2\rho + 4\theta$ these results are not valid as λ becomes a relevant parameter. If $\theta = 0$ but $\rho > 0$, two regimes are found. For small ρ the standard exponents α_w and z_w obtained by white noise remain valid. When ρ becomes larger than $\rho_c(d) = \alpha_w + (d-1-z_w)/2$ the long-range part of $D(k)$ takes over and the new exponents can be obtained by a simple Flory type scaling

$$\alpha(d) = (3 - d + 2\rho)/3 \qquad \text{and} \qquad z(d) = (3 + d - 2\rho)/3\,. \tag{12.17}$$

Furthermore, the scaling law $\alpha + z = 2$ remains valid since it can be shown to follow from Galilean invariance. When long-time temporal correlations are also present the scaling behaviour becomes rather complicated and several regimes appear. Moreover, in this case the Galilean invariance is lost and $\alpha + z = 2$ breaks down.

In addition to considering different types of noises one can also assume special forms for the *externally imposed perturbations.* This method may also serve a test of the phenomenological KPZ theory. Wolf and Tang (1990) considered the case when the deposition rate κ is uniform everywhere except on regularly spaced lines along the substrate

$$\kappa = \kappa_0 + \kappa_1 \sum_i \delta(x - L/2 - iL)\,. \tag{12.18}$$

It is expected that after averaging out the shape fluctuations the mean profile $H(x,t) = \langle h(\vec{r},t)\rangle$ depends only on the coordinate x. Wolf and Tang (1990) obtained the macroscopic shape $H(x,t)$ resulting from the above perturbations analytically for the KPZ equation. Simulations of the single-step model of Meakin et al (1986) confirmed the predictions of the KPZ equation and, as an important product of the approach, gave explicit values for the non-linearity parameter λ corresponding to the particular growth model when described by the continuum equation.

An interesting idea is (Krug and Spohn 1988) to study the dynamics of a generalized KPZ equation *without noise.* This is a relevant problem because this type of deterministic equation describes the relaxation of an initially rough surface to a flat one. The generalization is made through the introduction of the non-linearity term in the form $|\nabla h|^\zeta$. For ζ=2, the equation reduces to a deterministic version of the KPZ equation. In fact, using symmetry arguments it can be shown that ζ=2 is the only possible value in the case of stochastically growing surfaces. However, for a deterministic model other values of ζ are possible and can be considered.

In this study it is the initial surface which is assumed to be random, in contract to the flat starting geometry used in stochastic growth simulations. On the basis of scaling arguments, it is shown by Krug and Spohn (1988) that $z = \alpha(1 - \zeta) + \zeta$. Note that for ζ=2 one recovers the above mentioned scaling relation $\alpha + z = 2$ valid for stochastic growth. For $\zeta \geq (2-\alpha)/(1-\alpha)$ the nonlinearity is irrelevant, because the diffusive relaxation is faster than the effect of the nonlinear term. It is remarkable that the scaling properties of the noise-driven and the deterministic cases are so analogous in the sense that the spatio-temporal behaviour of the interface in both type of systems can be described in terms of dynamic scaling with the same exponents.

An interesting theoretical approach to the KPZ equation has been made by generalizing the *Kolmogorov type arguments* originally developed for the description of the inertial range of turbulence (Hentschel and Family 1991). The basic idea is that an equation of the kind (12.13) exhibits scaling only if each separate term (including the noise), when coarse grained over length scales, must be of the same order of magnitude or negligible. The validity of a scaling regime can then be found in a self-consistent manner from the region of length scales over which the intrinsic assumptions apply. The main task is to estimate the magnitude of the individual terms just as in Flory theories. During the application of this approach to the KPZ equation an assumption has to be made for the scaling behaviour of the noise. With an appropriate (however, rather arbitrary) choice for the noise, exponents coinciding with the conjecture of Kim and Kosterlitz (1989) can be derived.

Naturally, the KPZ equation can be solved *numerically*. The corresponding large scale computations resulted in estimates for the exponents α and β which were in good agreement with those obtained for the discrete aggregation models discussed in the previous section (see, e.g., Amar and Family 1990a).

Finally, a natural way to extend the applicability of the continuum equation treatment is the introduction of *additional terms* into the equation or to use *different type* of terms. The latter approach was undertaken by Sun et al (1989) who studied the dynamics of driven interfaces subject to a conservation law. One of the most profound questions regarding the dynamic scaling of growing surfaces is how various universality classes (if they exist) are defined and what features determine a particular class of interface growth processes. As can be seen in Chapter 7 very different models lead to the same exponents belonging to the "KPZ universality class".

Sun et al (1989) have considered the equation

$$\frac{\partial h}{\partial t} = -\nabla^2[\nu\nabla^2 h + \lambda/2(\nabla h)^2] + \eta(\vec{x}, t)\,, \tag{12.19}$$

where the noise is such that

$$\langle\eta(\vec{x}, t)\eta(\vec{x}', t')\rangle = -2D\nabla^2\delta^d(\vec{x} - \vec{x}')\delta(t - t')\,. \tag{12.20}$$

To motivate the conservation law, they consider a surface with a fixed number of atoms, where due to a driven flux the atoms migrate from site to site. Using a dynamic renormalization group approach, they show that the exponents α and z can be calculated exactly in d dimensions. In particular, they find that there is an upper critical dimension d_c=3, where the surface roughens logarithmically. They also find that the exponent identity $\alpha + z = 2$ is changed to $\alpha + z = 4$, giving the classical value z=4 in $d \geq 3$. Simulations of the conserved version of the restricted solid-on-solid model of Kim and Kosterlitz (1989) in two dimensions led to the values of α, β and z that are consistent with the analytical results, $\alpha = (3-d)/3$, $\beta = (3-d)/(9+d)$, and $z = (9+d)/3$, in d=2.

Lai and Das Sarma (1991) has carried out a *systematic* analysis of the possible linear and nonlinear terms which can be incorporated into a an equation written down in the spirit of the KPZ approach. These terms correspond to the effects of various kinds of relaxation processes and forces driving the interface. Dynamic renormalization group analysis is used to calculate a new set of exponents which are supported by numerical simulations. A detailed study of a number of 1+1 dimensional models of surface dynamics has been carried out by Rácz et al (1991), involving analytical and numerical calculations. Among other effects, a phase transition between a flat phase and a grooved phase has been observed. They argue that the models satisfying detailed balance (like growth described by the equation $\partial h/\partial t = -\lambda/2(\partial^4 h/\partial^4 x) + \eta(\vec{x}, t))$ all have scaling exponents $z = 4$ and $\alpha = 1/2$ in $2d$.

12.4. PHASE TRANSITION

Since the KPZ equation (12.13) has two terms depending on the derivatives of the surface, the possibility of a crossover between two morphologies is given. This transition is supposed to take place between the phase governed by the linear term and the phase described entirely by the nonlinear term. The terminology here is adopted from the studies of equilibrium phase transitions although we consider far-from-equilibrium processes. However, as we shall see the analogy is rather deep and the nature of the equilibrium and non-equilibrium transitions has many similar features.

The renormalization group analysis predicts (Kardar et al 1986, Medina et al 1989, Tang et al 1991) the following picture (illustrated in Fig. 12.5). Up to a critical dimension $d_c = 3$ (two-dimensional substrate) the nonlinear term dominates except if it is not present at all ($\lambda = 0$). This means that for $\lambda = 0$ the growth is described by the Edwards-Wilkinson linear theory (thus, in $d = 2$ $\beta = 1/4$, $z = 2$), while for any nonzero λ the strong coupling fixed point dominates the dynamics leading to $\beta = 1/3$ and $z = 3/2$ in $d = 2$.

The situation above d_c is somewhat different. In this region there is a finite, dimension dependent critical value of λ at which the smooth linear phase changes over to the rough phase corresponding to the nonlinear term. Less is known about what happens exactly at d_c. In fact, most of the computer simulations devoted to detecting a transition at d_c has been subject to criticisms of various origin. The renormalization group in this case suggests an *exponentially slow* logarithmic to power law crossover (Tang et al 1991), and some of the simulations seem to support this result.

Before reviewing the numerical results on phase transition in surface growth models let us discuss what are some of the major expected features of a non-equilibrium morphological crossover. In the general case the crossover takes place between two phases characterized by different scaling exponents.

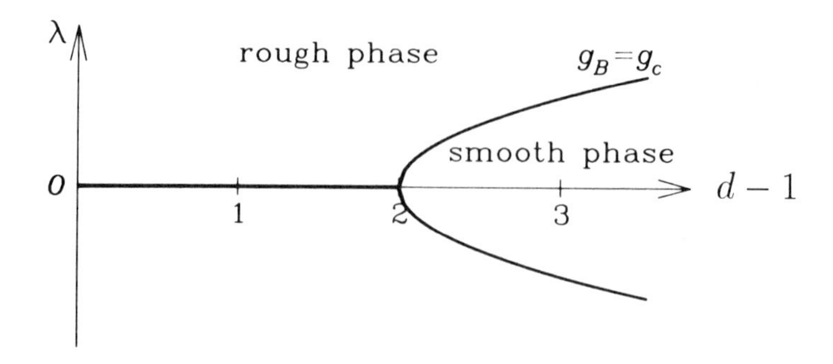

Figure 12.5. Schematic phase diagram of the KPZ equation from a one-loop renormalization group analysis, where the transition points are denoted by thick lines (Tang et al 1991).

A well understood specific situation is when the transition is triggered by a diverging length scale (Kertész and Wolf 1989) $\xi_r \sim |p - p_c|^{-\nu_p}$, where ν_p is some exponent and the transition takes place at p_c as the parameter p is approaching its critical value. Under such circumstances the dynamic scaling (7.4) for the width has to be modified according to the scaling Ansatz

$$w \sim \xi_r^{\alpha'} F(L/\xi_r, t/\xi_r^{z'}), \tag{12.21}$$

where the anomalous values of the roughening and the dynamic exponents are denoted by α' and z', respectively. The usual dynamic scaling is recovered only in the limit $L \gg \xi_r$ and $t \gg \xi_r^{z'}$, when (12.21) becomes equivalent to (7.4). At $p = p_c$ the length ξ_r diverges and the width scales as

$$w \sim L^{\alpha'} \tilde{f}(t/L^{z'}). \tag{12.22}$$

The above phenomenological scaling theory has been tested using a polynucleation growth model. In short, this model corresponds to a simultaneous deposition of particles followed by a relaxation step when a particle is added at each kink position along the surface. This model can be treated in terms of directed percolation (Kertész and Wolf 1989) and the exact values of the anomalous exponents can be obtained. The result $\alpha' = 0$ has been verified by computer simulations which indicate that

$$w(L,t) \sim (\log t)^{1/2} \qquad \text{and} \qquad w(L, t=\infty) \sim (\log L)^{1/2}. \tag{12.23}$$

Another interesting aspect related to non-equilibrium phase transitions is the effect caused by the underlying lattice structure present in most of the simulations and experiments. In order to take into account the periodicity in the vertical direction Hwa et al (1991) investigated the equation

$$\mu^{-1}\frac{\partial h}{\partial t} = F + \nu\nabla^2 h - y\sin(2\pi h) + \lambda/2(\nabla h)^2 + \eta(\vec{x}, t), \tag{12.24}$$

where F is the driving force, μ is the microscopic mobility of the interface and y is a parameter. Renormalization group calculations predict the existence of a roughening transition for every finite F and y for some λ even in $d = 3$.

Next we turn to the discussion of *computer simulation* results concerning a phase transition in models described by the KPZ equation. The existence of a transition from the smooth to the rough phase in dimensions $d > 3$ seems to be well established (e.g., Forrest and Tang 1990, Yan et al 1990, Pellegrini and Jullien 1990).

The situation is less clear in $d = 3$, although this is the physically relevant case. There are several papers in which the existence of a phase transition is indicated. The problem of interest here is the question whether the observed transition is a trivial one or it takes place for a parameter value of the models which corresponds to a nonzero value of λ in the language of the KPZ equation. By a trivial transition here one means that with tuning the parameters of a model it is in principle possible to achieve a situation in which the behaviour of the model becomes linear ($\lambda = 0$). (For example, $\lambda > 0$ for ballistic deposition, but $\lambda < 0$ for the solid-on-solid models.) Then the smooth phase is reduced to a single point as is predicted by the renormalization group for $d < 3$.

Phase transitions were reported to occur in the following models. Amar and Family (1990b) considered a modified solid-on-solid model in which the probability to grow at a given site depended on the change of energy (depending on a temperature-like variable and the number of neighbours) upon adding a site to the deposit. By varying the temperature a sharp transition was observed between two phases described by different exponents accompanied by a flat phase at the point of the transition. Arguments based on the KPZ equation suggested (e.g., Huse et al 1990) that this transition is trivial in the above sense of making λ equal to zero by tuning the temperature. The difference in the values of the exponents is not explained by these arguments; it may be due to lattice effects or to an extremely slow crossover.

Yan et al (1990) studied a ballistic deposition model in which the particles were allowed to diffuse. The effect of diffusion was taken into account in a manner which destroyed the discrete nature of the model in the growth direction. As the parameter p of the model (controlling the rate of aggregation versus diffusion) was changed, a marked decrease in the effective exponents was observed at a critical value of p signalling a non-trivial phase transition since the exponents remained small beyond p_c. Similar results were reported by Pellegrini and Jullien (1990) who investigated ballistic deposition of sticky and nonsticky particles.

The complexity of the situation is demonstrated by Fig. 12.6, where results obtained for the hypercube stacking model (see Section 12.2) are shown (Forrest and Tang 1990). The quantity $w^2(2t) - w^2(t)$ is plotted versus the time t in order to exclude the effect of the intrinsic width. The data are for various $p^- = m/64$ values for fixed p^+. It can be clearly seen that approaching the equilibrium case ($m = 32$, $\lambda = 0$) the crossover time at which the asymptotic behaviour sets in diverges rapidly. As a result, effective exponents different from the asymptotic value can be associated with large fractions of the data. The situation in $3d$ is different (bottom); $m = 8$ seems to be a critical value, for larger m the fluctuations of the interface decay in time, the surface remains flat.

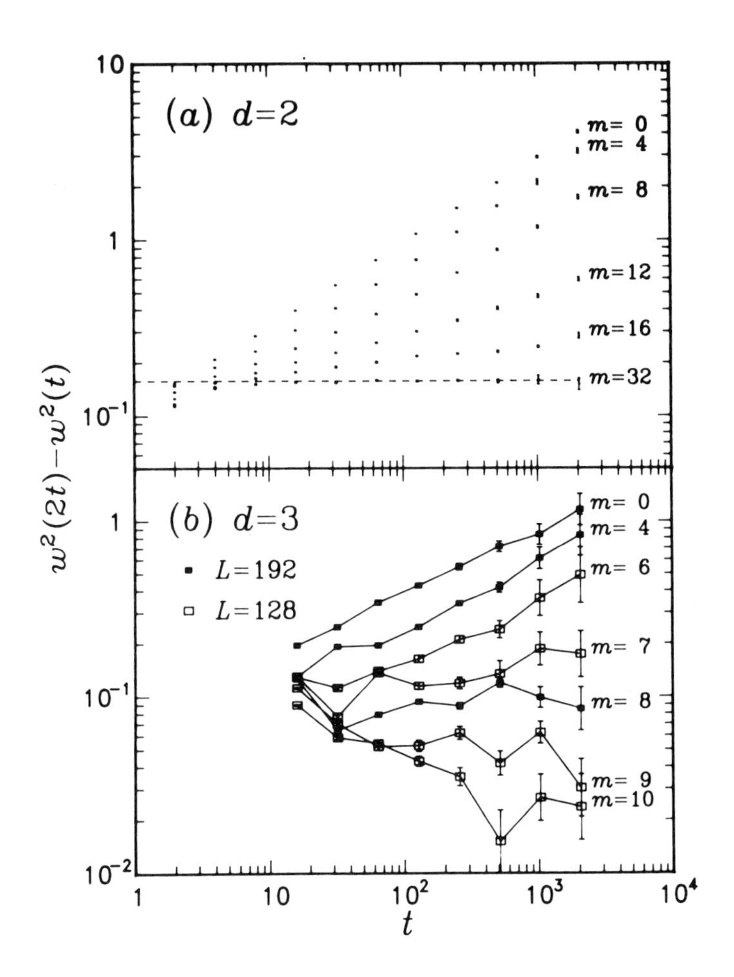

Figure 12.6. Results obtained for the hypercube stacking model for (a) $d = 2$ and (b) $d = 3$. The quantity $w^2(2t) - w^2(t)$ is plotted versus the time t in order to exclude the effect of the intrinsic width. The data are for various $p^- = m/64$ values for fixed $p^+ = 1/2$ (Forrest and Tang 1990).

12.5. RARE EVENTS DOMINATED KINETIC ROUGHENING

The computer simulations of simple aggregation models and the dynamic renormalization group treatment of the KPZ equation resulted in the predictions $\alpha = 1/2$ and $\beta = 1/3$ for the exponents describing the scaling of marginally stable interfaces growing in two dimensions. However, the available *experimental* results are in clear disagreement with the above values since the measurements of the surface roughness in quasi two-dimensional two-phase viscous flows (Rubio et al 1990, Horváth et al 1991) and in bacterial colony formation (Vicsek et al 1990b) led to considerably larger values for α (see the next chapter).

A plausible resolution of this problem is based on the observation that the above mentioned universal values of the exponents have been obtained under the assumption (or in the simulations in the presence of) a Gaussian white noise. It is natural to think that changing the qualitative behaviour of the noise in the KPZ equation may lead to a new type of scaling behaviour. As was discussed (Section 12.2), long range spatial or temporal correlations can change exponents. On the other hand, one faces difficulties when trying to find the origin of such kind of noises.

Recently it has been proposed (Zhang 1990a) that, in contrast to most problems in statistical physics, in the case of growth phenomena the microscopic details can influence the large scale behaviour in a substantial way, thus violating the naive universality concept. To show this a non-Gaussian distribution of the noise amplitudes is assumed and shown to lead to non-universal exponents in a surface growth model. Following Zhang (1990a) let us consider a noise which is independently distributed on each site (lattice regularization is assumed) according to the distribution density

$$P(\eta) = \frac{\mu}{\eta^{1+\mu}} \quad \text{for} \quad \eta > 1 \qquad \text{and} \quad P(\eta) = 0 \qquad \text{otherwise.} \qquad (12.25)$$

This distribution has finite first and second moments (the second moment diverges as $\mu \to 2$, while the expectation value of η remains always small), but it does not have an absolute cutoff. Correspondingly, very large values of η may appear with a finite probability. These are the *rare events* which are supposed to change the behaviour qualitatively. There exists, however, a qualitative estimate for the largest expected noise amplitude η_{max} after N random numbers are generated according to the distribution (12.25). This can be shown by noting that in order to generate numbers satisfying (12.25) one can use random numbers r uniformly distributed on the interval [0,1] and calculate η from the expression

$$\eta = \frac{1}{r^{1/\mu}}. \tag{12.26}$$

If N numbers are generated uniformly on the unit interval, the probability for a number to fall into the interval $(0, 1/N)$ (or the probability that r is on the order of $1/N$) is some finite value. This leads to the estimate

$$\eta_{max} \sim N^{1/\mu}. \tag{12.27}$$

It follows from the above that in an actual simulation one needs a random number generator which has more significant digits than $\log_{10} N$. Since on a supercomputer it is common to go up with N to 10^{11} this condition is not trivially satisfied by a 32 bits random number generator!

There is a *theoretical* result for the exponents α and β (Zhang 1990b, Krug 1991). Let us consider (Krug 1991) a part of the interface having a linear extension x and assume that the roughness is stationary on this scale, i.e., $x \ll \xi_{\|}(t)$, where $\xi_{\|}(t) \sim t^z$ is the correlation length along the substrate (see Section 12.1). The number of individual growth events taking place within the given part of the surface over a time interval τ is $N = x^d \tau$. The time needed to create a spontaneous fluctuation of amplitude $\xi_{\perp}$ in the direction of growth is

$$\tau_c \sim \xi^{\mu}/r^d. \tag{12.28}$$

The nonlinear term in the KPZ equation drives the fluctuation $\xi_{\perp}$ over the distance x during some time τ_c. This spreading time scale can be obtained by balancing the nonlinear term against the time derivative (Krug and Spohn 1990)

$$\tau_s \sim x^2/\xi_{\perp}. \tag{12.29}$$

During stationary growth fluctuations with $\tau_c \sim \tau_s$ give the main contribution to the roughness, since larger fluctuations will spread to rapidly to have a significant statistical weight. Thus, comparing (12.28) and (12.29) we get $\xi_\perp \sim x^\alpha$ with

$$\alpha = \frac{d+2}{\mu+1}. \tag{12.30}$$

Since the scaling relation (7.28) $\alpha + \alpha/\beta = 2$ is not affected by the noise distribution for the other exponents we have

$$\beta = \frac{d+2}{2\mu - d}. \tag{12.31}$$

In obtaining the above relations the cumulative effect of small fluctuation has been neglected, thus, the theory is expected to be less accurate in the large μ limit. Since α becomes equal to its universal value at a d-dependent μ in each dimension the existence of a critical μ_c can be conjectured, beyond which the exponent is universal. For example, in two dimensions $\mu_c = 5$.

To study growth subject to rare events Zhang (1990a) proposed a model which is discrete along the substrate, but is continuous in the growth direction. Perhaps the simplest variant of the Zhang model in two dimensions is represented by the growth rule

$$h(x,t+1) = \max[h(x,t)+\eta(x,t);\ h(x+1,t)+\eta(x+1,t);\ h(x-1,t)+\eta(x-1,t)]. \tag{12.32}$$

Simply stated this means that at every time step a) the local noise $\eta(x,t)$ is added to every site and b) each site then takes a new value equal to the maximum of itself and its two neighbours. In the original version two sublattices were considered and only the maximum of the neighbours were used to update the height at site x and η was assumed to be distributed according to (12.25). However, the global scaling behaviors of these versions is expected to be the same. The geometry of the surfaces produced by the sublattice version of the algorithm (12.32) is demonstrated in Fig. 12.7. Naturally, any other distribution can be used for η. In higher dimensions the rule can be written as

$$h(\vec{x}, t+1) = \max_{\vec{e}_i}[h(\vec{x}+\vec{e}_i, t) + \eta(\vec{x}+\vec{e}_i, t)], \tag{12.33}$$

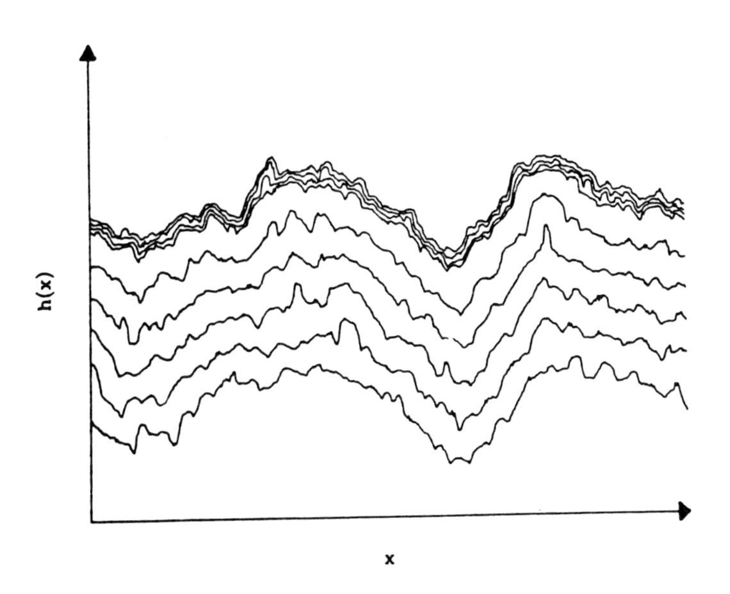

Figure 12.7. Successive snapshots of an interface evolved according to the sublattice version of rule (12.32). The longer time intervals are 30 time steps, the shorter ones are 5 time steps (Zhang 1990).

where one of the vectors $\vec{e}_i$ is equal to zero, while the rest run over the nearest neighbours of $\vec{x}$.

In spite of the fact that the Zhang model is very recent there are already several papers published on the numerical simulation of models related to (12.32). The most precise estimates for the scaling exponents (Amar and Family 1991, Bourbonnais et al 1991a) have been obtained on parallel supercomputers (Connection Machine) capable of handling many thousand operations simultaneously. Figure 12.8 shows the dependence of the exponent α on the parameter μ as determined from large scale simulations. For comparison the theoretical prediction is also shown. As expected, the agreement is less good for large μ. The actual value of the critical μ_c cannot be precisely determined from these results indicating that (if it exists at all) μ_c is likely to be somewhere in the region between 5 and 8. Simulations of the three-dimensional version (Bourbonnais et al 1991b) also gave a reasonable agreement with the theory. In all cases Gaussian noise results in universal exponents.

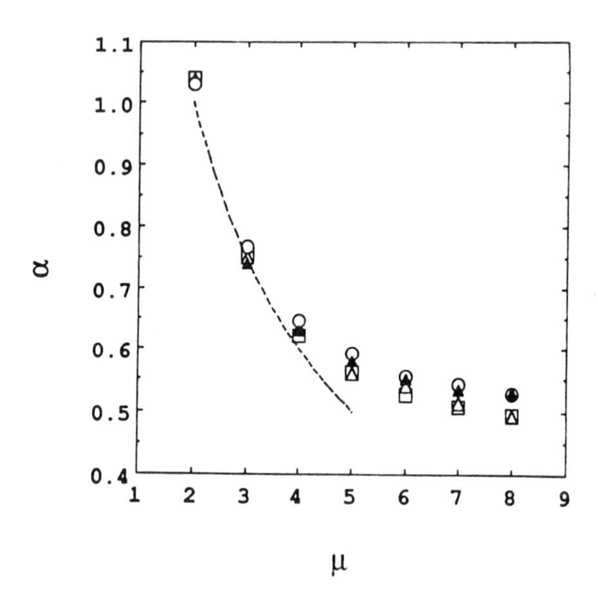

Figure 12.8. Surface roughness exponent α as a function of μ in the Zhang model (open squares) and three additional, closely related models (Amar and Family 1991). The dashed line indicates the theoretical prediction of Zhang (1990) and Krug (1991).

The idea of a power law distribution for the noise amplitudes can be applied to systems which are related to, but are different from the model (12.32) discussed above. Marconi and Zhang (1991) studied the directed polymer problem (see, e.g., Kardar and Zhang 1987) which can be shown to have scaling properties equivalent to those exhibited by surface roughening processes. In their numerical study the distribution of the energies associated with the lattice sites was assumed to have the form (12.25) and the exponent corresponding to α was found to be in reasonable agreement with the theory and related simulations.

Another variant of the Zhang model can be realized by ballistic deposition of rods with a power law distribution of lengths (Buldyrev et al 1991). In the two-dimensional simulations of this model the growth is assumed to proceed by the following rules: a) Choose randomly a position x along the substrate of length L at which the rod will be dropped, b) choose a random length l for the rod according to (12.25) and, finally, c) attach the rod to the surface using the ballistic aggregation criterion so that the new height at x becomes

$$h(x, t+1) = \max[h(x,t),\ h(x+1,t)-1,\ h(x-1,t)-1] + l. \qquad (12.34)$$

The structure of the deposit generated by this model is shown in Fig. 12.9, where the effect of the rare events is visualized. Calculation of the width w

Figure 12.9. Ballistic deposition of rods with a power-law length distribution for $\mu = 3$ and $L = 512$ at time $t = 281$ (Buldyrev et al 1991).

and the correlation function $c(x,t)$ were used to obtain estimates for β and α. A finite size scaling type extrapolation was used to get as precise estimates as possible. The main results are summarized in Fig. 12.10. In contrast to the Zhang model based on simultaneous updating, the present sequential model resulted in exponents which are somewhat below the theoretical values (12.30) and (12.31). These simulations suggested that the critical value of μ should be close to 5.

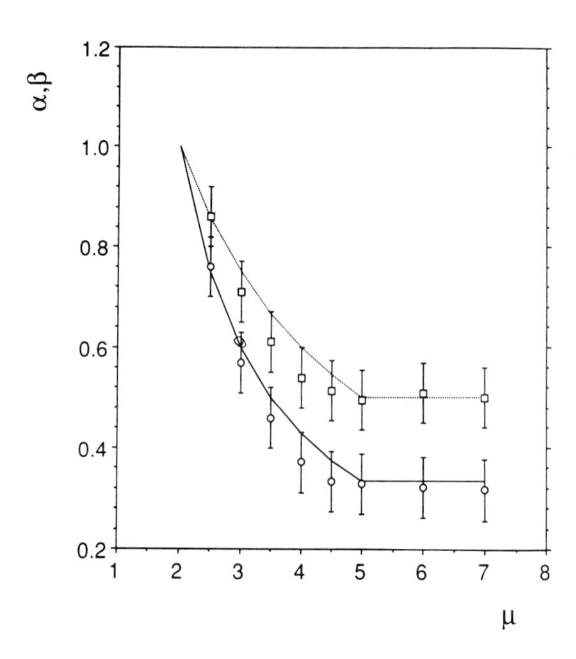

Figure 12.10. Estimates for the exponents α (upper) and β in the ballistic deposition variant of the Zhang model (Buldyrev et al 1991). The theoretical values are shown as dotted and continuous lines.

In summary, the numerical results in all cases are close to the theoretical ones, although the actual estimates are a bit scattered, sometimes out of the error bars. In addition, the scaling law $\alpha + \alpha/\beta = 2$ seems to be always satisfied within a few percent.

Of course, the importance of the Zhang model can only be justified by its relevance to real experimental systems. This aspect will be discussed in the next chapter, where it will be shown that there exists experimental evidence for the power law scaling of the surface fluctuations.

12.6. MULTIAFFINITY

The new concept of multiaffine geometry has been introduced very recently (Barabási and Vicsek 1991) in order to provide a more complete description of the geometry of fractal surfaces. This has been done in the same spirit as fractal measures can be used to characterize the detailed features of self-similar fractals.

The multiscaling properties of the self-affine function $h(x)$ of a single variable can be investigated by calculating the *qth order* height-height correlation function which we define as

$$c_q(x) = \frac{1}{N}\sum_{i=1}^{N} |h(x_i) - h(x_i + x)|^q, \tag{12.35}$$

where $N \gg 1$ is the number of points over which the average is taken, and only terms with $|h(x_i) - h(x_i + x)| > 0$ are considered. This correlation function exhibits a non-trivial multiscaling behavior if

$$c_q(x) \sim x^{qH_q} \tag{12.36}$$

where H_q changes continuously with q at least for some region of the q values. Note that in our case qH_q is the natural choice for the exponent in the right-hand-side of Eq. (12.36) (instead of $(q-1)H_q$ as for fractal measures). A scaling assumption of a form analogous to (12.36) has been used in the studies of fully turbulent flows to describe the nature of velocity correlations (see, e.g., Nelkin 1990). For self-affine functions defined on a finite interval the above scaling is expected to hold in the $x \ll 1$ limit, while in the discrete case (where there exists a lower cutoff length Δx as, e.g., in the lattice growth models) Eq. (12.36) is satisfied in the $x \to \infty$ limit. It can be shown that a continuous spectrum of H_q values is not consistent with the expression $h(x) \simeq \lambda^{-H} h(\lambda x)$ which is valid for standard self-affine functions with a single exponent H (see Section 2.3.2). For self-affine functions with multifractal properties this expression is modified since the local scaling of the height differences depends on x in analogy with the local scaling of the measure in a box for fractal measures.

For a continuous surface, which without loss of generality can be considered as defined on $[0, 1]$, $c_q(x)$ can be computed as follows. One makes a partition of the $[0, 1]$ interval into N equal parts and sums the qth powers of the height differences $\Delta h = |h(x_i) - h(x_i + x)|$ at points separated by $x = 1/N$. However, a more general approach requires that the limits $x \to 0$ and $N \to \infty$ are taken independently. (Earlier studies of the velocity structure functions (Frisch and Parisi 1985) implicitly assumed the special case $x = 1/N$.)

As mentioned above we shall assume that, when evaluating (12.35), x and N may be related in a way different from $x \sim 1/N$ (Barabási et al 1991a). Let us consider the dependence $N \sim x^{-\phi}$. For the commonly used partition we have $\phi = 1$. For $\phi < 0$ in the $x \to 0$ limit $N \to 0$ which violates the condition that N has to be large. The case when the number of points N is fixed is equivalent to $\phi = 0$. The choice of a particular partition has no effect on the H_q spectrum. However, as will be shown, ϕ enters the relations between the multifractal spectra.

In many cases non-trivial multiplicative processes can generate non-uniform surfaces having infinitely many singularities. These singularities appear in the scaling of Δh in the vicinity of different points on the surface, and this scaling can be different from point to point. This property can be described by an exponent γ_i trough the following relation

$$|h(x_i) - h(x_i + x)| \sim x^{\gamma_i} \tag{12.37}$$

in the limit $x \to 0$. The non-integer exponent γ corresponds to the strength of the local singularity of the signal. Although γ depends on the actual position, there are many intervals of size x with the same index γ, and their number is expected to scale with x as

$$N_{\gamma_i}(x) \sim x^{-h(\gamma_i,\phi)}, \tag{12.38}$$

where $h(\gamma, \phi)$ for $\phi = 1$ is viewed as the fractal dimension of the subset of the points with the same exponent γ. For other ϕ values $h(\gamma, \phi)$ simply describes the scaling of the number of places with a given local singularity.

In analogy with the standard multifractal formalism (Section 3.2) in the limit $N \to \infty$ the correlation function (12.35) can be written as

$$c_q(x) \simeq \frac{1}{N} \int_{(\gamma')} x^{\gamma' q} x^{-h(\gamma',\phi)} \rho(\gamma') d\gamma'. \tag{12.39}$$

Following the steps of the derivation given in Section 3.2 it is straightforward to show that the relationships among the various exponents introduced above have the forms

$$h'(\gamma) = q, \tag{12.40}$$

$$h''(\gamma) < 0, \tag{12.41}$$

and from the comparison with (12.36)

$$qH_q = \phi + q\gamma(q) - h(\gamma(q), \phi), \tag{12.42}$$

$$\gamma(q) = \frac{d}{dq}(qH_q). \tag{12.43}$$

Before proceeding to examples we mention the differences and similarities of these relations with those used in the standard multifractal formalism to link the generalized dimensions D_q and the $f(\alpha)$ spectrum.

First, a similarity between the correlation function $c_q(x)$ and the generating function $\chi(q) = \sum_i p_i^q$ for a measure p_i is obvious, however the scaling of the latter is described by an exponent $(q-1)D_q$, where $(q-1)$ appears due to the normalized condition for p_i, but in the former such a normalization does not exist, since the scaling exponent is qH_q. Another difference is the term ϕ in relation (12.42) which appears as a result of the non-trivial scaling of N as a function of x in the limit $x \to 0$. However, the general technique is very similar: like in the case of the multifractal formalism, H_q is related to $h(\gamma, \phi)$ by a Legendre transformation. It is possible to extend the formalism to the case when the distribution of height differences is normalized and the corresponding D_q and $f(\alpha)$ spectra can be defined (Barabási et al 1991a).

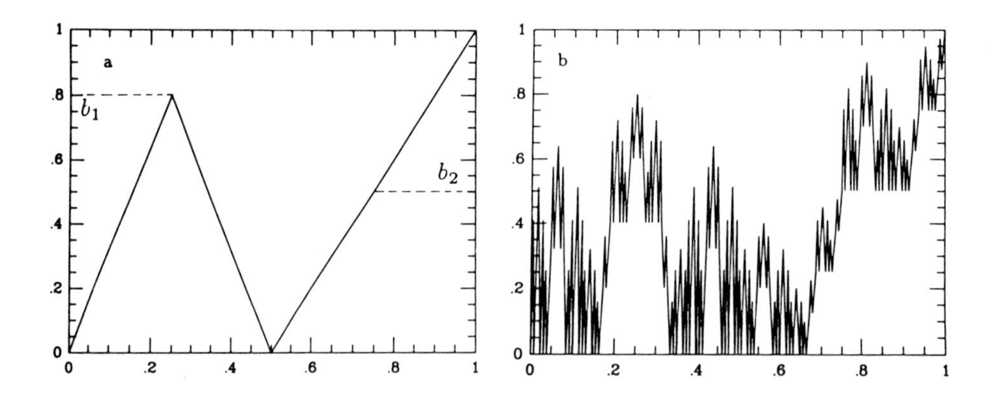

Figure 12.11. (a) The generator and (b) the fourth stage of the deterministic multi-affine model for surfaces with multifractal properties (Barabási and Vicsek 1991).

Let us examine the above derived relations on an exactly solvable deterministic model (Barabási and Vicsek 1991). The iteration procedure, which is a generalization of the construction proposed by Mandelbrot (see Fig. 2.12), is demonstrated in Fig. 12.11. In each step of the recursion the intervals obtained in the previous step are replaced with the properly rescaled version of the generator which has the form of an asymmetrical z made of four intervals. During this procedure every interval is regarded as a diagonal of a rectangle becoming more and more elongated as the number of iterations

k increases. The basis of the rectangle is divided into four parts and the generator replaces the intervals in such a way that its turnovers are always at analogous positions (at the first generator and the middle of the basis). The function becomes multi-affine in the $k \to \infty$ limit. Depending on the parameter b_1 very different structures can be generated, the b_2 is fixed to be 0.5.

The H_q spectrum can be calculated for this construction exactly assuming that the scaling properties are entirely determined by the behavior of the function over intervals of length 4^{-k}. Denoting with $N(\Delta h)$ the number of boxes in which $|h(x) - h(x + \Delta x)| = \Delta h$, we have $N(b_1^n b_2^{k-n}) = 2^k C_k^n$, where $n = 0, ..., k$. Thus $c_q(\Delta x) = \sum_{n=0}^{k} 2^{-k} C_k^n b_1^{nq} b_2^{(k-n)q}$ with $\Delta x = 4^{-k}$. Since this expression for c_q can be written as $c_q(\Delta x) = [(b_1^q + b_2^q)/2)^k$ we have

$$H_q = \frac{\ln[(b_1^q + b_2^q)/2]}{q \ln(1/4)}. \qquad (12.44)$$

In the present approach the roughness exponent introduced earlier is $H = H_1$. The above H_q spectrum is shown in Fig. 12.12, where the results of the corresponding numerical test are also shown. The construction of Fig. 12.11 can be generalized easily to include more than two parameters and randomness (Vicsek and Barabási 1991) resulting in surfaces strikingly similar to the simulated ones.

The significance of the above described concept of multiaffinity depends on its applicability to actual growth processes. In the previous section we argued that it is of crucial importance to understand the structure and dynamics of surfaces in the presence of power law distributed noise. Below the multiaffine aspects of such surfaces will be discussed for the example of the Zhang model in 1+1 dimensions (Barabási et al 1991b).

Figure 12.13 presents data for $L = 2^{16}$ at time $t = 602890$, i.e., deep in the saturation regime; an average over five runs was taken. The important conclusions one can draw from these log-log plots showing the qth order correlation function for various q as a function of q are the following: i) The initial part of the data sets for each q exhibits scaling behaviour with a

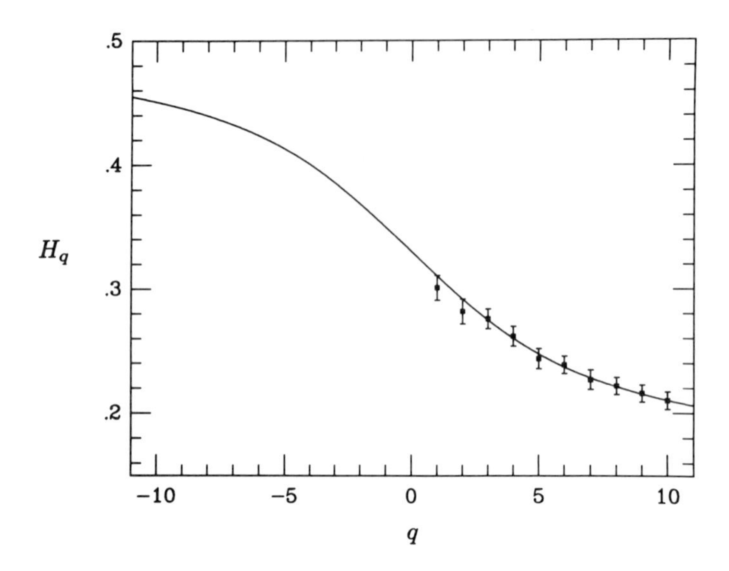

Figure 12.12. The H_q spectrum calculated from Eq. (12.44) for the self-affine fractal of Fig. 12.11 with $b_1 = 0.8$ and $b_2 = 0.5$ (continuous line). The theoretical result is compared with the numerically determined data obtained using Eqs. (12.35) and (12.36) for a prefractal generated in the 9th step of the construction. The numerical calculations can be carried out only for $q > 0$ because of the large uncertainties (Barabási and Vicsek 1991).

unique slope depending on q, i.e., multifractal scaling is present, ii) this kind of scaling crosses over into the uniform scaling behaviour for x exceeding some characteristic crossover length $x_\times$.

In Fig. 12.14 the function H_q is presented as measured in the multifractal scaling region of Fig. 12.13. The deviations from the simple scaling behavior are clear; changes in H_q as a function of q are rather dramatic.

Feature ii) means that in addition to the characteristic length $\sim t^{\beta/\alpha}$ always present in kinetic roughening before the system size L is reached at saturation, a new characteristic length $x_\times$ occurs in surface growth dominated by rare events. For $x < x_\times$ the qth order correlations show multifractal scaling behavior. For $x_\times < x < t^{\beta/\alpha}$ conventional scaling sets in while no

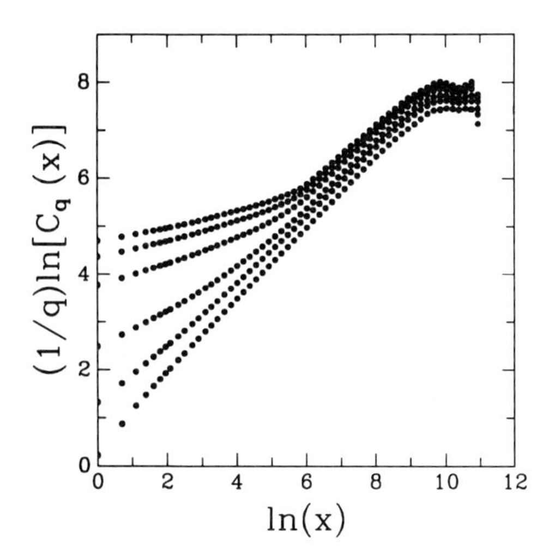

Figure 12.13. The qth order correlation functions after the surface width has saturated ($L = 2^{16}$ and at $t = 602890$ sweeps) for q =1, 3, 5, 7 and 9 with q increasing from the bottom to top (Bourbonnais et al 1991).

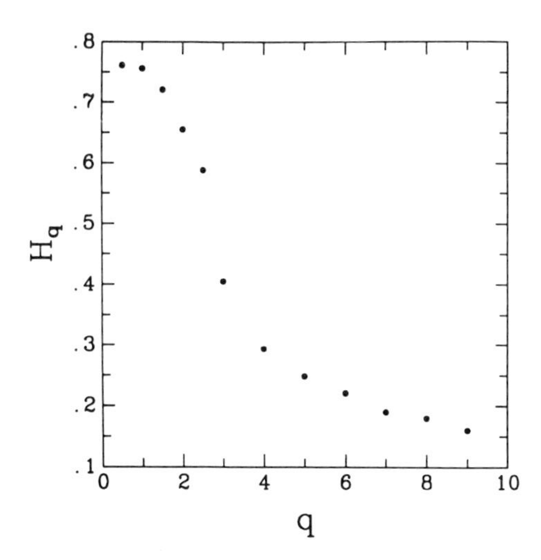

Figure 12.14. The exponent H_q vs. q as taken from runs described in Fig. 12.13. Simple scaling would imply no variation with q. The relatively sharp change for $q \sim 3(= \mu)$ is an indication of a "phase transition"(Bourbonnais et al 1991).

correlations are present for $x > t^{\beta/\alpha}$. The crossover length $x_\times$ depends on L. Estimation of the size dependence of the crossover length gives $x_\times \approx 5$, 20, 60, 160 for $L = 2^{12}$, 2^{14}, 2^{15}, and 2^{16}, respectively. Although $x_\times$ is quite small, this way it is demonstrated that as L is increased, the relative size of the region over which the multifractal scaling behaviour is observed becomes larger.

Chapter 13

EXPERIMENTS

There are less papers available on the experimental aspects of fractal growth phenomena than works using other approaches to the question of clarifying the main features and mechanisms corresponding to a given type of growth process. This, however, does not mean that the experiments play a less important role in the progress of our ideas on the reasons behind the development of complicated patterns in nature. The point is that in most of the cases the realization of an appropriate experiment is a very time consuming, costly adventure which typically requires much more effort than carrying out the simulation of a relatively simple computer model.

The interpretation of the experimental results is also sometimes a highly non-trivial task. Recent theoretical concepts focusing on the interplay between the macroscopic driving force and the microscopic interfacial dynamics have been published in the excellent reviews by Kessler et al (1988) and Ben-Jacob and Garik (1990).

The latest developments in the area of experimental studies are related mainly to studies of new systems concerning both the growth mechanism and the material of the growing phase. In particular, real systems exhibiting dynamic scaling of the surface roughness have been first considered during the last years only. The fractal analysis of fracture patterns represents also a relatively new approach. As an example for the use of novel type of materials in the experiments on fractal growth here we mention the investigations of biological growth of various origin.

13.1. SELF-SIMILAR GROWTH

Many examples of self-similar fractal growth were discussed in Chapter 10, where the experimental studies of the most essential types of such phenomena (viscous fingering, electrodeposition, crystallization dielectric breakdown, etc.) were reviewed. In the present section mostly new kinds of processes will be considered (fracture and biological growth). However, in the first subsection the latest results on diffusion-limited growth will be given as well. In fact, the development of self-similar patterns in biology has also been argued to be related to diffusion-limited growth, but the complexity and the very different nature of the involved phenomena justifies the treatment of biological pattern formation in a separate subsection.

13.1.1. Diffusion-limited growth

Viscous fingering. The development of finger-like structures in the Hele-Shaw cell (Section 10.1.1) has remained a popular subject of investigations devoted to fractal growth. One of the basic questions in this area is whether the simplest arrangement, the case of radial fingering, leads to Laplacian patterns with a well defined fractal dimension or results in a random dense branching morphology with a trivial $D = d$. The most recent, very careful studies of this problem (Couder 1988, May and Maher 1989) have concluded that these patterns are fractals and the corresponding dimension is in the region $D = 1.75$–1.85. The behaviour of the effective fractal dimension as a function of the pattern size is shown in Fig. 13.1. These values are somewhat puzzling because it is definitely above the accurate estimates for the fractal dimension of diffusion-limited aggregates considered to be the prototypes of growth phenomena dominated by the Laplace equation in the presence of noise. It is still not completely clear whether (or why) the deterministic version of Laplacian growth leads to a universality class different from DLA.

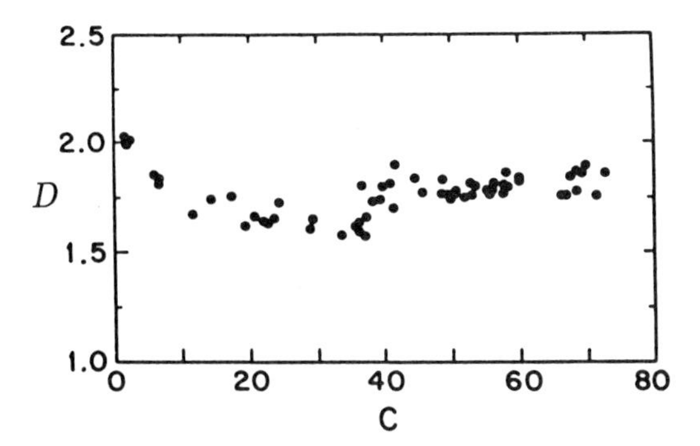

Figure 13.1. The effective fractal dimension D of viscous fingering patterns in the radial Hele-Shaw cell as a function of the applied driving force C (May and Maher 1989). Larger pressures correspond to larger patterns if the linear size of a pattern is measured using the average finger width as unit.

The various aspects of deterministic versus stochastic growth in viscous fingering have been considered in a couple of recent works. As discussed in Section 11.1.4 in the limit when the fingering pattern has a random geometry in the longitudinal cell, the average behaviour (obtained by calculating the mean occupancy of a given region, where averaging over many experiments is made) corresponds to that of a single deterministic Saffman-Taylor finger (Arneodo et al 1989, Couder et al 1990, Arneodo et al 1991).

The degree of the randomness of fingering patterns in the radial cell has been analyzed (Horváth and Vicsek 1989) using techniques common in the investigations of dynamical systems. This approach is expected to be useful since during the past decade powerful methods have been elaborated for the treatment of chaotic signals observed as a function of time in chaotic systems. To characterize the patterns one determines $\kappa(s)$ which is the local curvature of the interface as a function of the arc length s (distance measured along the surface). Figure 13.2 shows a series of curvature data corresponding to the kind of fingering patterns shown in Fig. 10.2b.

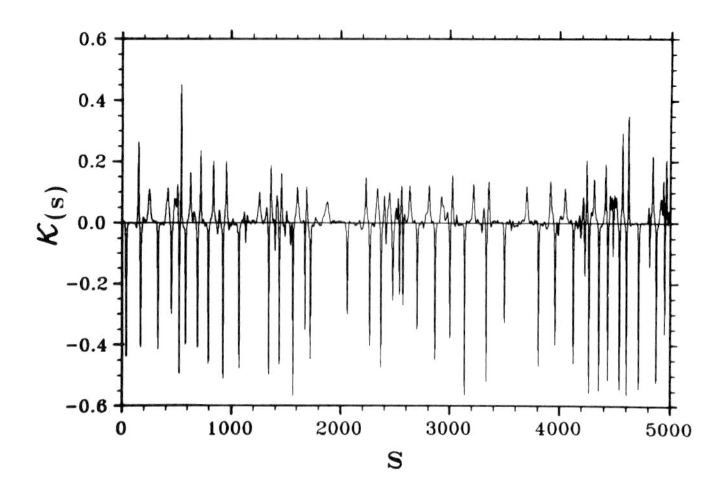

Figure 13.2. Part of a long sequence of surface curvature $\kappa(s)$ data versus the arc length s measured along the interface (Horváth and Vicsek 1989).

In the related analysis $\kappa(s)$ is considered as an *analogue of the time series* measured in the experiments on dynamical systems. The result of the associated calculations indicate that the interfaces obtained in the unstable regime of viscous fingering are highly stochastic; there is no sign of a low-dimensional chaotic behaviour, although the analysis shows that the fingers are less disordered than a completely random pattern. The reason for the stochasticity is the instability of the interface: perturbations beyond a surface tension dependent wave length grow indefinitely. These perturbations are present in the *initial condition*, i.e., the starting shape of the interface is never perfectly symmetrical. Apparently, the non-linear mixing of the many Fourier modes of the initial interface produces high-dimensional behavior. In the absence of stabilizing effects (e.g., anisotropy), there is no mechanism which would drive the system into any of the regular shapes. Note that in systems with low-dimensional temporal chaos the onset of chaotic behaviour usually takes place as one of the parameters of the system is changed. In our case the role of this parameter is possibly played by the anisotropy.

Solidification. The most important factor in determining the shape of a dendritic crystal is the distribution of the field-like quantity controlling the growth process around the growing phase. The knowledge of the actual behaviour of this distribution is crucial for understanding fractal growth in real systems. In the isothermal solidification experiments taking place in supersaturated solutions it is the concentration of the solidifying molecules which governs the development of the dendrites. In an interesting experiment (Raz et al 1989) on the crystallization of NH_4Cl the spatial dependence of this concentration has been determined by measuring the refractive index of the solution. To obtain the results described below the following method has been used.

The ammonium chloride crystals are grown from a supersaturated solution in a thin cell. The solute concentration field in the liquid around the growing crystal is measured from the shift of the fringes which cross the field when the cell is observed in an interference microscope. This technique measures the integrated quantity of solute in the cell along a line normal to the parallel plates. A typical result for the concentration field around a dendritic pattern is shown in Fig. 13.3. The most interesting feature of this figure is exhibited by the lines of equal concentration ending at various points of the surface. If in the boundary condition (9.4) the capillary number and the kinetic coefficient were both very small $\Delta u \simeq 0$, lines close to the surface would be parallel to it since the value of the concentration along the surface would be everywhere approximately equal to the saturation value corresponding to the given temperature of the sample. As under the conditions of the experiment the capillary number can be neglected (Raz et al 1989), it is the kinetic term which dominates the growth. The detailed investigation of the concentration patterns results in the conclusion that the kinetic term is very nonlinear. This is consistent with a model in which the growth along the surface is also dominated by a nonlinear nucleation mechanism.

In order to study the fractal nature of dendritic crystals Couder et al (1990) grew ammonium bromide crystals at small Peclet numbers in a Hele-Shaw type cell. Such crystals look very much like clusters generated in large scale simulations of noise-reduced DLA. The fractal dimension was obtained by image processing parts of the large dendrites far from their tip. Using the box counting method the estimate $D \simeq 1.58$ was obtained. This value is in a

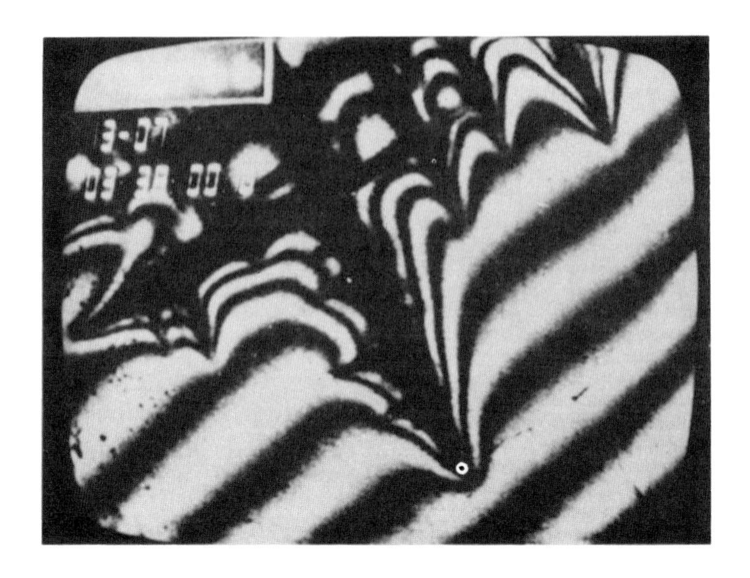

Figure 13.3. Interferogram of the concentration field around a growing NH_4Cl dendrite (Raz et al 1989). The stripes, just like equipotential lines, correspond to places with roughly the same concentration.

remarkable agreement with the theoretical and simulational predictions for the fractal dimension of noise-reduced DLA clusters.

Deposition. Finally, diffusion-limited patterns which can be found easily in nature will be discussed. In quarries it is no rarity to find black or reddish brown designs of tree-like structures on the surface of limestones. These patterns are known as *mineral dendrites.* In any glossary of geology they are defined, for instance, as: "Plant-shaped surface deposits of hydrons iron or manganese oxides found along joint, bedding and cleavage planes in rocks". An example is shown in Fig. 13.4. What we know about their origin is that they are chemical deposits (oxides) that formed when at some point in the geological past the limestone was penetrated by a supersaturated solution of manganese or iron ions. The question of the actual mechanism of fractal pattern formation in such phenomena is still open. For example, one would like to know how the shape and density of the deposites depends on the external parameters like concentration gradient or reaction rate.

Figure 13.4. Manganese oxide deposit formed as a result of a reaction-diffusion process on the surface of a limestone found in Bavaria (and common in many other regions). This mineral dendrite has a fractal dimension close to 1.78 (Chopard et al 1991).

Very recently an analysis of the shape of samples of different origins has been carried out and the results have been compared with numerical work obtained from a simple reaction-diffusion lattice-gas model (Chopard et al 1991). The dimension can be determined by digitizing the pictures and using the sand box method (Chapter 4). The result for the type of dendrites (found in Bavaria) shown in Fig. 13.4 is that the fractal dimension is about 1.75. A set of pictures taken from a dendritic pattern which grew within the plane of a crack inside a quartz crystal originated from Brasil led to the estimate 1.5.

These results can be explained by the following reaction-diffusion model (Chopard et al 1991): Consider two species of particles, A and B, which diffuse on a square lattice. When an A and a B particle are simultaneously on the same site they react forming C $(A + B \to C)$. The C particles also diffuse until they precipitate into D $(kC \searrow D)$. The D particles are motionless and the sites where they are created will be marked black for all successive time-steps. Two conditions will lead C to transform into D: (i) when at least k particles of type C simultaneously meet on the same site, a saturation occurs and the C precipitate into D; (ii) when a C particle becomes nearest neighbour to a black site D, it aggregates to the cluster. Such a reaction-diffusion process will cause a reaction front to build up if the A and B species are initially separated in space. If the concentrations of A and B are different, this reaction front moves across the system, leaving clusters of D particles in its wake. Simulations of the model have shown that the resulting clusters are reminiscent of mineral dendrites and their fractal dimension depends on the concentration of A and B in a range similar to the one described above for the observed pictures.

13.1.2. Fracture

In this section a collection of experimental results on self-similar crack propagation will be given. In some cases the development of fractal fracture patterns is understood (e.g., in terms of a related computer model leading to very similar geometry), while there are crack networks whose fractal origin is less clarified.

Surface cracks. Let us start with a short description of the beautiful radial fracture pattern shown in Fig 11.8 (Ananthakrishna 1991). The branching crack displayed in this figure is a good example for a fracture pattern slowly developing on a surface subject to stress. It was obtained by using a chemical to control the stress concentration on the surface of a plexiglass plate. Under the action of a drop of the chemical in a localized region, a high stress concentration is induced due to the adsorption of the atoms on the surface of the plexiglass. Due to the symmetry of the affected region cracks nucleate and propagate radially.

Since the plexiglass is isotropic, no preferred direction of growth exists. Note, however, that the angle between the side branches and the main branches is always almost exactly 90° (this is the only angle which can be selected because of the isotropy). The growth of side branches is either spontaneous or induced by the chemical moving along the already existing cracks. From digitizing Fig. 11.8 the dimension for the structure was found to be in the region 1.7–1.8.

Stress corrosion cracks. It is known that different cracking processes produce patterns with different fractal dimensions. Thus, it seems useful to concentrate on one experimentally well explored mechanism and to try to reconciliate experimental and numerical data specifically for this mechanism. Stress corrosion cracking in metals is a good candidate since it is known to be relatively well defined and also a lattice model has been proposed (Herrmann et al 1989) for its description (see Section 12.2). This model generates fractal cracks also without noise and in this case rather precise estimates can be obtained for the fractal dimension D in two dimensions. In fact, it is found that D depends continuously on a parameter η which describes the mesoscopic breaking rule: the susceptibility for a bond to break increases like the local stress to the power η. No *a priori* derivation is known for the mesoscopic rules from the microscopic breaking mechanisms. Therefore, η can only be determined as a parameter by comparing to experimental data.

In stress corrosion cracking a chemical agent impregnates the microfractures and corrodes the crack tips where the stress intensity is highest. So the crack grows at a lower applied stress with a velocity controlled by the chemical reaction. Usually ductile materials become brittle. Two-dimensional cuts through stress corrosion cracks in alloys (an example is shown in Fig. 13.5a) are typically tree-like and quite branched.

A careful analysis of various experimental stress corrosion cracks in alloys and leads to the conclusion (Horváth and Herrmann 1991) that they are all fractals and have roughly the same fractal dimension of $D = 1.38 \pm 0.05$. Comparing this value to the theoretical predictions for D as a function of η (Herrmann et al 1989) one sees that η should be between 1 and 1.2 if the damage caused by the corroding agent has a finite life-time. If the damage is permanent η would be about 20% lower. It is an interesting fact that the local breaking rule has an η of the order of unity. This is in fact the exponent one derives from first principles (Gibbs–Thompson relation) for the problem of solidification from an undercooled melt which is a scalar analog to the moving boundary problem that can be formulated for the growth of cracks in an elastic medium.

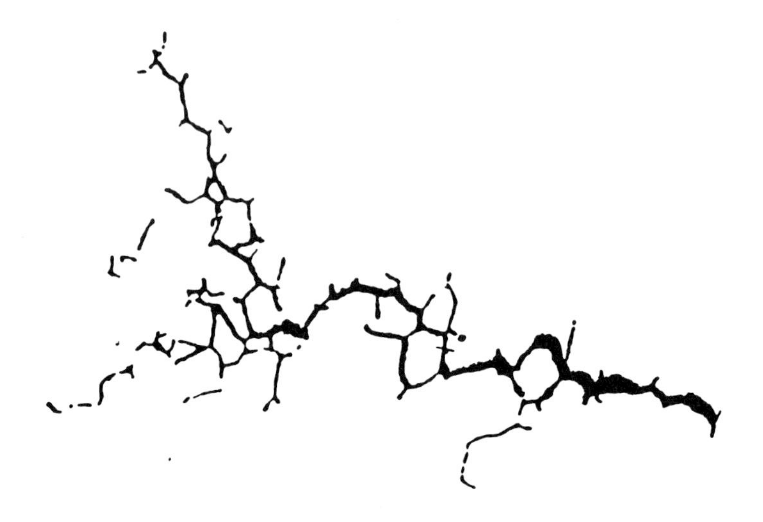

Figure 13.5. Example for an intergranular stress corrosion crack which are common in metal alloys (Brown 1972).

Fracture patterns in thin films. Perhaps the most common example for fracture patterns is the rich network of cracks which are induced when the surface of a material is shrinking. For instance, this phenomenon takes place when an area covered originally with wet mud dries out (a sight upsetting farmers).

To investigate this type of crack propagation and geometry Skjeltorp (1988) used uniformly sized sulfonated polystyrene spheres of effective diameter 3.4 μm dispersed in water. By confining the spheres between planar glass plates it is possible to form a polycrystal with grains containing typically 10^5–10^6 spheres. The fracturing process takes place as the result of particle shrinking during the slow drying of the spheres reducing their diameter to about 2.7 μm. The whole process can be characterized by a strong bonding between the spheres and a relatively weak bonding to the glass surfaces. The results are shown in Fig. 13.6a-d, where a typical picture taken at the final stage of the growth is shown for increasing magnification. The corresponding fractal dimension is determined from the box counting method. In Fig. 13.6e the number of boxes needed to cover the cracks is plotted as a function of the box size. The data fall onto a straight line with a slope $D \simeq 1.68$.

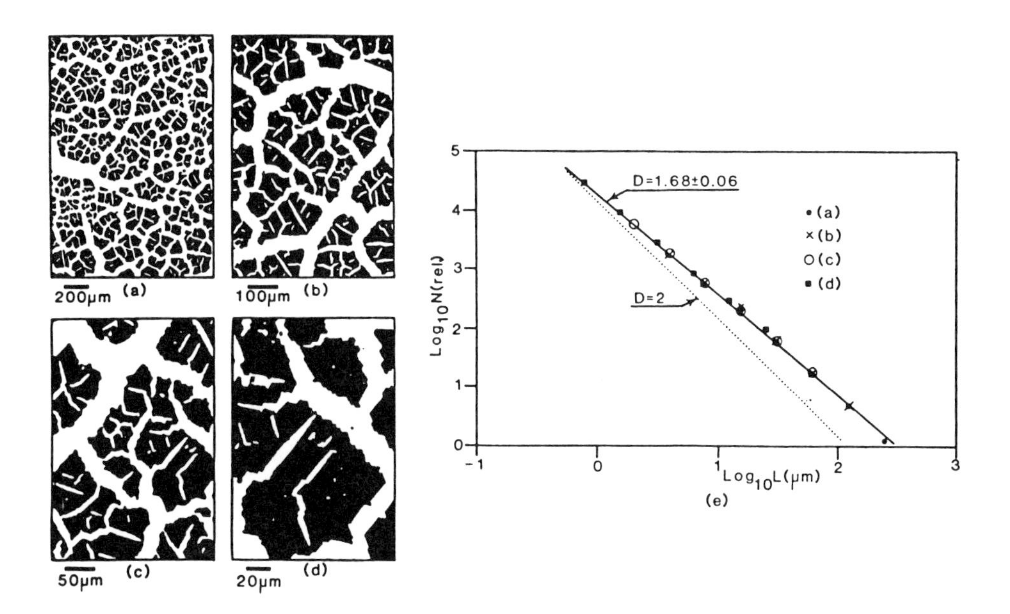

Figure 13.6. A typical picture taken at the final stage of the crack propagation in the system of uniformly sized sulfonated polystyrene spheres (Skjeltorp 1988). (c)-(d) shows parts of (a) for increasing magnifications. In (e) the number of boxes needed to cover the cracks is plotted as a function of the box size.

A computer model by Meakin (1988) is very useful in finding clues when trying to interpret the above behaviour. The elastic film is represented by a network of nodes and springs as described in Section 12.2. However, in the present case the elastic network is attached to a rigid underlying substrate with bonds which have a small force constant. To account for the fact that the polystyrene may move during drying, the attachment position of the weak bonds to the surface should be allowed to shift if the tension in these bonds becomes larger than a critical value. This model results in cracks which resemble those observed in the above experiments. The main observation is that without a coupling to the substrate no fractal network of cracks can develop. In the absence of the weak bonds the cracking stops after a few cracks relieve the stress.

Cracks in viscoelastic media. Van Damme (1989) originally studied viscous fingering in clay suspensions. This is a complex system where the more viscous wet clay can be considered as a non-Newtonian fluid. According to the experiments there is a morphological transition (Lemaire and Van Damme 1989) as a function of the increasing density of the substance (Fig. 13.7). At the transition the speed of crack propagation is increased dramatically. Simultaneously, the fractal dimension changes from 1.7 to 1.4.

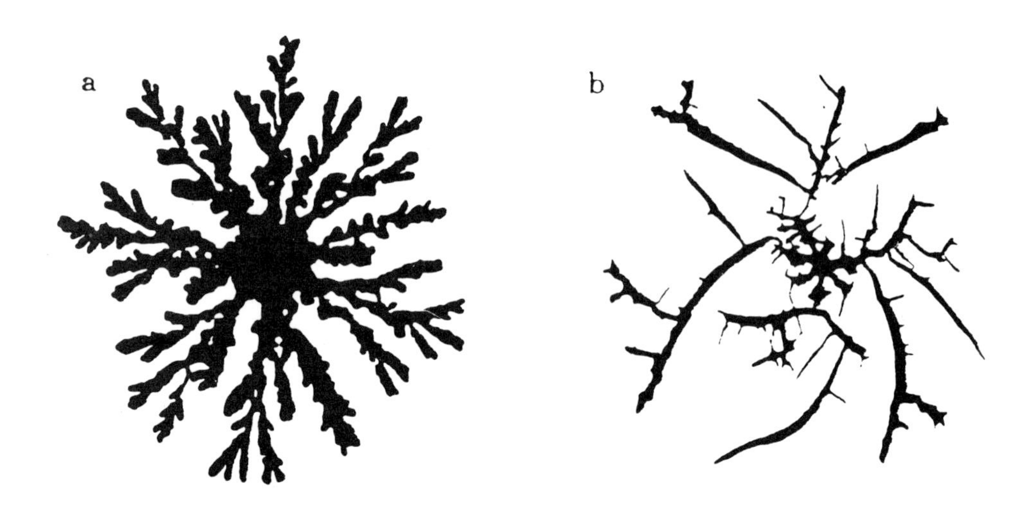

Figure 13.7. Crossover from (a) viscous fingering to (b) cracking in a viscoelastic medium (Lemaire and Van Damme 1989). Water was injected into a quasi-two-dimensional layer of bentonite/water paste of concentrations (a) 0.08 and (b) 0.20.

In addition, the branching angle which is in the fingering regime about 30° becomes approximately 90° (the same angle as in the first example in this section). These observations indicate that the growth mechanism changes sharply at a critical concentration of the more viscous material.

In the limit when the characteristic time corresponding to the propagation velocity of the pattern is much smaller than the elastic relaxation time, breaking phenomena are likely to occur even in a system which for slower motion displays viscous fingering. This is the reason behind the transition demonstrated in Fig. 13.7.

13.1.3. Biological growth

During the last two years the interest in the fractal aspects of biological patterns has considerably increased. In particular, it has been suggested by several works (to be discussed below) that a variety of biological processes can be described in terms of Laplacian growth. The suggestion that physical phenomena such as diffusion-limited growth may account for the structure of patterns of biological origin is highly nontrivial and raises a number of fundamental questions. Among these are: is there an intimate connection between simple physical processes and the morphologies observed in the living world? In which phenomena (if in any) are the physical aspects dominant over the genetic ones? The results described in the following do not answer these questions, however, they provide the first relevant data for a future conceptual framework.

Bacteria colony formation. A striking evidence for DLA type growth in a biological system can be observed during the growth of bacteria colonies on the surface of agar (gel containing nutrient) plates (Matsuyama et al 1989, Fujikawa and Matsushita 1989, 1991). The experiments can be realized with various kinds of bacteria, here we shall consider the growth of *Bacillus subtilis* colonies (following Fujikawa and Matsushita 1989, 1991). It is easy to isolate a strain of such bacteria from our everyday food (after having it unattended for a couple of days).

The agar has to be prepared according to standard techniques. The surface is inoculated at a point. The plates are stored in a humidifier and then incubated in a thermostatic room regulated at $35 \pm 1°$. After several days of incubation a DLA like pattern develops on the agar surface if the humidity and concentration of the nutrient is kept in a proper region. However, on the basis of the shape of the pattern it would be too early to assume that a diffusion-limited mechanism is responsible for the morphology.

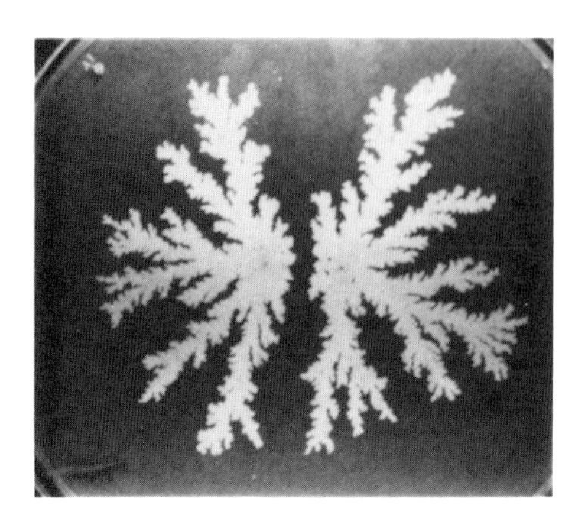

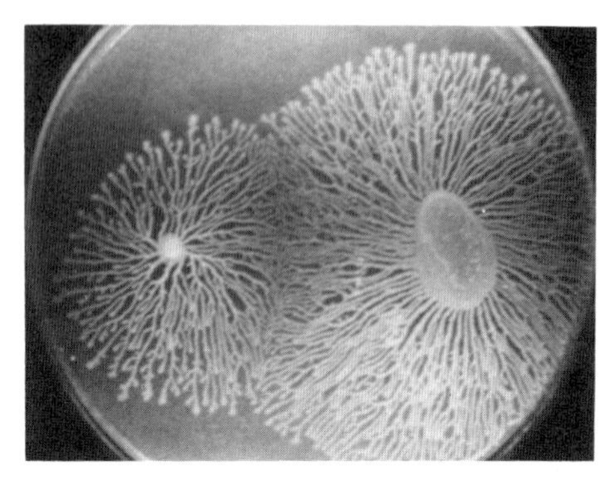

Figure 13.8. Bacterial colony morphologies after inoculating the surface at two points separated by 4 cm (Fujikawa and Matsushita 1991). In the case of low nutrient concentration nutrient (left) there exists an effective repulsion between the two colonies. For larger nutrient concentrations (right) there is no such repulsion.

Several tests can be carried out to test the idea. i) Increasing the amount of nutrient results in a crossover to a dense structure indicating that if there is plenty of food no fractal structure is grown since the effect of screening is diminished. Screening here works in the sense that the most advanced parts of the colony use up the nutrient in the DLA regime and the parts getting behind do not have access to enough food to grow. ii) An important evidence can be obtained by inoculating the surface at two points (Fig. 13.8). In case of low concentration of the nutrient there exists an effective repulsion between the two colonies (Fig. 13.8, left). A very similar picture would be obtained in a simulation of diffusion-limited aggregation with two seeds. The reason is that the growing patterns screen the region between them. Again, for larger nutrient concentrations there is no such screening (Fig. 13.8, right). A schematic representation of the possible morphologies as a function of the gel density and the amount of nutrient is shown in Fig. 13.9.

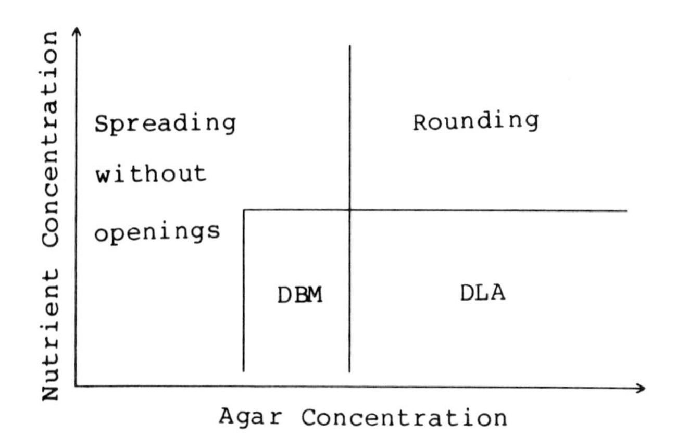

Figure 13.9. Schematic phase diagram of the colony morphologies developing on agar surfaces in terms of the peptone and agar concentrations (Fujikawa and Matsushita 1991).

The spontaneous multiplication of *Ashbia gossipi* (Obert et al 1990a) and *Sordaria macrospora* (Obert et al 1990b) also leads to complicated structures. The elongated bacteria in the *Sordaria macrospora* colonies prefer to stay parallel to each other so that the pattern develops long branches with occasional side branching (see Fig. 13.10). At early stages (Fig. 13.10a) the structure looks like a fractal aggregate and can be characterized by an effective dimension in the range from 1.3 to 1.6. At later stages the empty regions typical for a fractal pattern gradually become filled and the resulting morphology corresponds to a dense structure with a rough surface. Indeed, (Fig. 13.10b) is very reminiscent of the off-lattice ballistic aggregate shown in Fig. 7.4.

Retinal vessels. The formation of retinal vessels in the developing human eye has a complicated mechanism. On the other hand, the resulting branching pattern is very similar to diffusion-limited aggregates (Fig. 13.11). This question has been addressed (Masters et al 1989, Family et al 1989, Mainster 1990) by carrying out a fractal analysis of the corresponding pictures. The evaluation of the data led to the estimate for the dimension of the retinal vessels $D \simeq 1.7$ in a good agreement with the dimension of the DLA clusters grown in two dimensions.

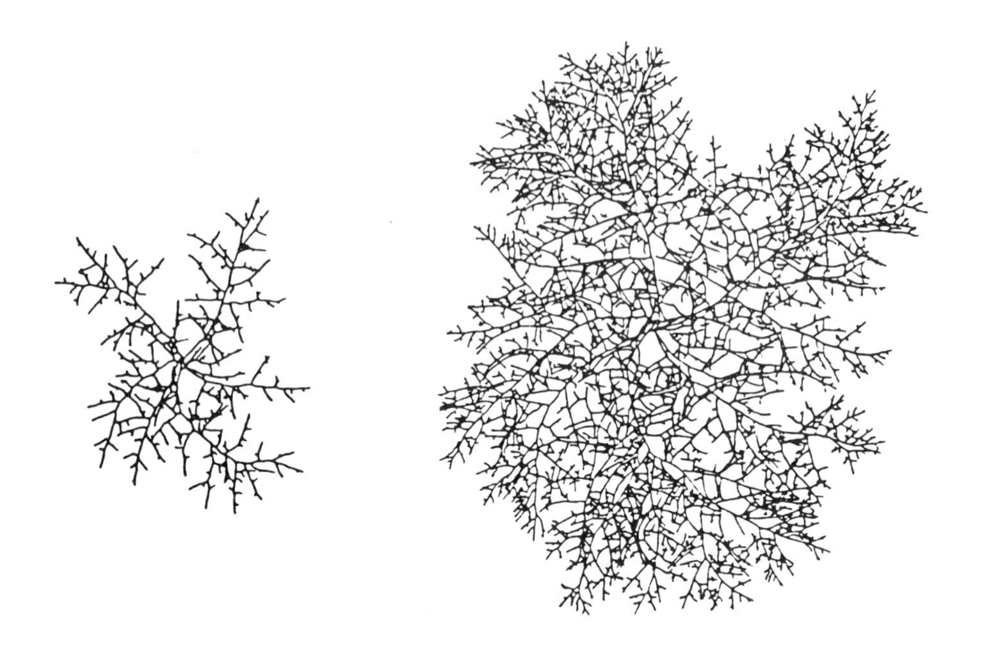

Figure 13.10. Patterns observed during the growth of the colony formation of the fungus *Sordaria macrospora* (Obert et al 1990b) after 19h 45′ (left) and 23h 45′ (right).

According to a current hypothesis for the development of the inner retinal vasculature there is a relationship between the growth of vessels and the maturation of the photoreceptors. During the developmental stage the maturing photoreceptors consume progressively more oxygen, decreasing the available amount at the inner retina. The migrating spindle cells (the use of a few less known technical terms is not avoidable here due to the nature of the topic) in the inner retina sense this diminished concentration of oxygen and release angiogenic factors. The angiogenic factors diffuse in the plane of the retina and stimulate the growth of new retinal blood vessels. Therefore, diffusion plays an essential role in the formation of the vasculature in the eye. However, it should be pointed out that for a typical DLA situation the

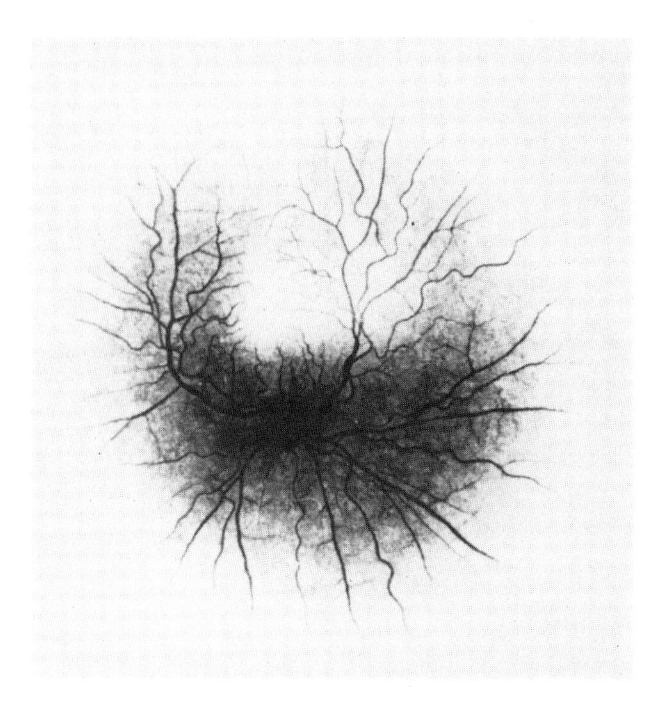

Figure 13.11. Picture of the network of retinal vessels in the eye of a patient (Family et al 1989).

source of the factors facilitating the growth has to be far from the pattern, while in the case of the retina this condition does not seem to be satisfied.

Neurite outgrow. The shape of neurons (single nerve cells) is usually very complicated and can be described in terms of fractal geometry (Smith et al 1989, Caserta et al 1990). In a study which concentrated on comparing the structure of retinal neurons developing *in vivo* and *in vitro* Caserta et al (1990) found that the fractal dimensions of the neurons growing in the two type of environments are different. The digitized image of an *in vivo* nerve cell (taken from a cat's retina) and the corresponding fractal analysis is shown in Fig. 13.12. From the slope of the correlation function in Fig. 13.12b it was concluded that $D \simeq 1.66$ which is very close to the dimension of diffusion-limited aggregates. The authors suggest that a mechanism analogous to DLA is likely to be responsible for the morphology.

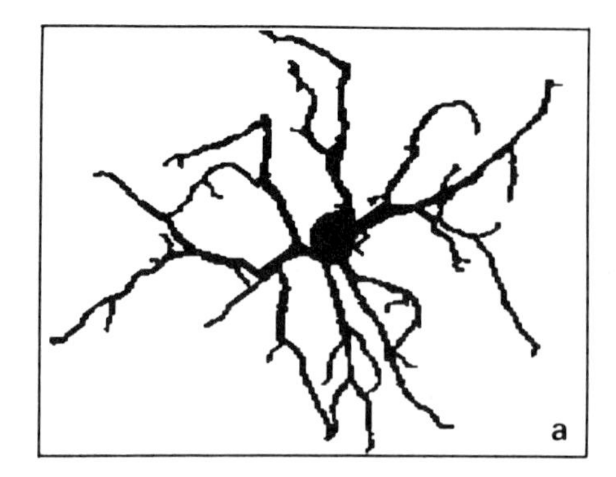

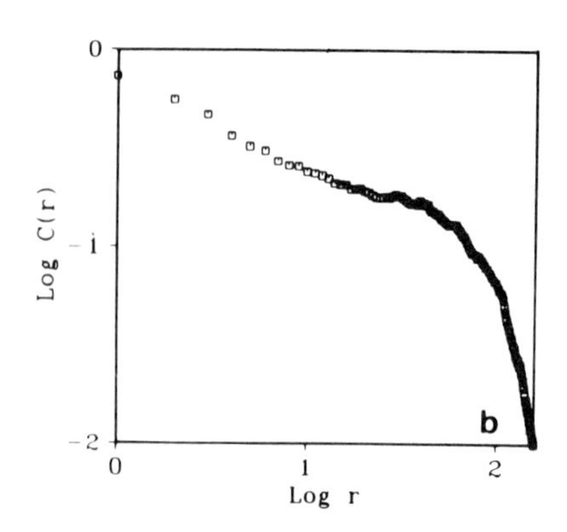

Figure 13.12. (a) The digitized image of an *in vivo* nerve cell (taken from a cat's retina) and (b) the corresponding fractal analysis (Stanley, 1989, Caserta et al 1990). From the slope of the correlation function in Fig. 13.12b the fractal dimension can be estimated to be about 1.66.

Interestingly, the *in vitro* neurons were found to have a significantly smaller dimension (about 1.4). Apparently the environmental factors can influence the fractal structure of retinal nerve cells considerably.

13.2. SELF-AFFINE GROWTH

Although there are many examples of self-affine surfaces in nature, until the last year, no experimental studies had been devoted to the dynamics of growing fractally rough interfaces. Very recently three different kinds of growth phenomena have been investigated experimentally in order to examine the scaling properties of real surfaces. The main conclusion one can draw on the basis of these studies is that the development of rough surfaces under experimental conditions is dominated by factors which are not fully taken into account in the simplest models of self-affine growth.

13.2.1. Two-phase viscous flows

The penetration of a wetting fluid into a porous medium is a very common phenomenon in our immediate environment. One example is the kind of spots produced on the ceiling by a leaking roof. Another one is shown in Fig. 13.13, where the regions in a paper towel penetrated by coffee are shown. These type of processes are called two-phase viscous flows in porous media (one of the phases is air in the above examples).

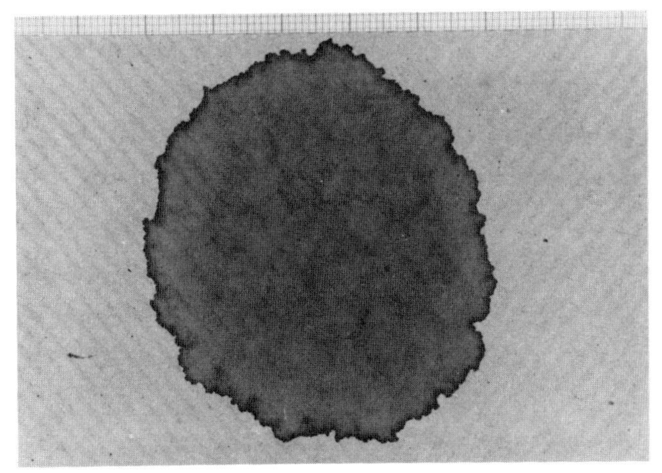

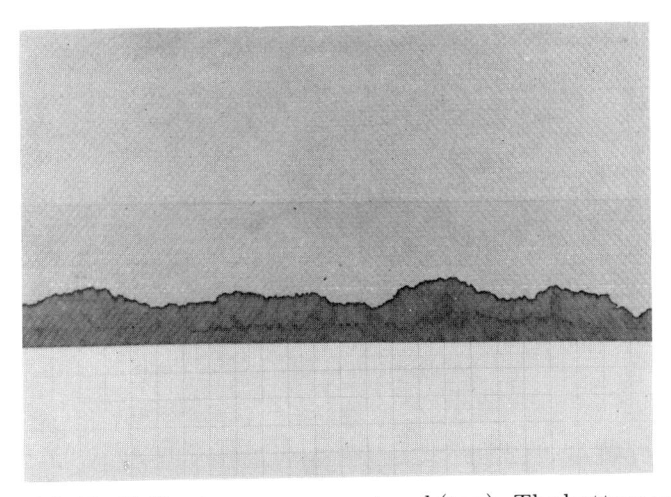

Figure 13.13. Coffee drop on a paper towel (top). The bottom pattern can be obtained by immersing a paper towel into coffee kept in a dish.

To study such phenomena under well controlled conditions the experimental setup usually consists of a linear Hele-Shaw cell (see Fig. 10.1) made of parallel plexiglass or glass plates (Rubio et al 1989, Horváth et al 1991a). In the following the apparatus used by Horváth et al (1991a) will be described. The linear sizes of a typical cell are 24 cm$\times$100 cm. Glass beads of diameter about 220 μm are packed between the plates. The beads are spread randomly and homogeneously in one layer and glued to the lower plate. The upper plate is placed directly on the beads and iron rods and clamps are used to prevent the lifting of the plates. Colored glycerol with 4 vol% of water is slowly introduced at a fixed flow rate into the system along a line at one of the shorter sidewalls.

In such a system the motion of the fluid-air interface is influenced by several factors. In fact, changing the fluids and the flow rates a number of different regimes can be achieved including DLA and invasion percolation regimes resulting in self-similar fractal structures (Section 10.1.2). The relevant control parameter is the modified capillary number Ca$= V\mu b^2/k\gamma$, where V is the average velocity of the interface, μ is the dynamic viscosity of the injected fluid, b is the width of the layer of glass beads, γ is the air-fluid interfacial tension and k is the permeability. The capillary number measures the ratio of the viscous pressure $V\mu b/k$ to the capillary pressure γ/b on the scale of the pore size. In a wide range of parameter values the pressure drop in the fluids can be kept much smaller than across the interface. Thus, in the experiments with *wetting* fluids it can be achieved that at slow flow rates the stabilizing effect of the pressure distribution in the glycerol and the instabilities caused by capillary effects can be neglected. This is why the experiment can be considered as a realization of marginally stable growth. The fluctuations are presented by the random distribution of voids between the glass beads.

The dynamics of the interface evolution is studied by recording the surface images on a videotape. In this way a sampling rate of 0.28 sec can be achieved. Figure 13.14 shows the development of the fluid-air interface by plotting the digitized image of the meniscus as it evolves. The shift between the patterns is proportional to the time lapse between the pictures. The roughening of the interface is manifested by the appearance of large mountains and valleys. The three-dimensional appearance of the patterns is

Figure 13.14. Roughening of the fluid-air interface in the experiment on the penetration of a wetting fluid into a layer of glass beads (Horváth et al 1991a). The flow takes place from left to right and for each recorded interface the most advanced parts are plotted darker. This method results in a virtual three-dimensional character of the picture.

obtained by making the brightness of the surface at a given point proportional to its height.

The dynamic scaling of the growing interface is demonstrated by the data displayed in Fig 13.15, where the logarithm of the square root of the height-height correlation function for $x = 0$ is plotted against $\ln t$. The slope of the straight line fitted to the data indicates that the roughening process is characterized by the exponent $\beta \simeq 0.65$. A similar analysis of the spatial correlations leads to the estimate $\alpha \simeq 0.81$. The latter number is higher,

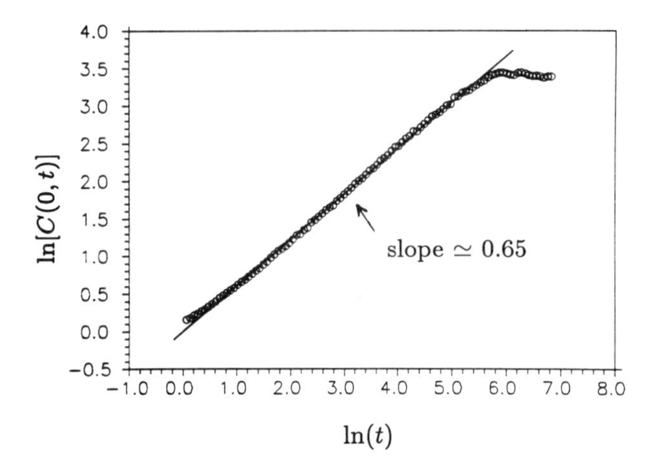

Figure 13.15. Determination of the exponent β from the time correlation function $c(0,t)$ calculated for surfaces similar to those shown in Fig. 13.14.

but in a qualitative agreement with $\alpha \simeq 0.73$ obtained in the independent measurements of Rubio et al (1989).

These results demonstrate that the evolution of the interface in two-phase fluid flows in porous media is governed by *dynamic scaling.* The values of the surface exponents of α and β appear to be different from the corresponding values obtained in the simplest models of interface growth. A remarkable feature of the results is that although the values $\beta \simeq 0.65$ and $\alpha \simeq 0.81$ are non-universal, they *satisfy the scaling law* $\alpha + \alpha/\beta = 2$ within the error bars.

The question of non-universality of the experimentally determined exponents has been addressed by Zhang (1990a) who suggested that a power law distribution of the perturbations may lead to a behaviour different from the Gaussian one. The relevance of this proposition to the experimental situation can be tested by using various approaches to extract the *noise spectrum* from an extensive set of digitized images of the fluctuating fluid-air interface in the above described experiments.

Let us define reference interfaces $h(x,t_2)$ for each of the surfaces in the data set $h(x,t_1)_{t_1\in[0..T]}$ (Horváth et al 1991b). These reference interfaces are defined using the condition that the shift $\Delta h = \langle h(t_2)\rangle_{x\in[1..L]} - \langle h(t_1)\rangle_{x\in[1..L]}$ between the average heights of the surface pairs has to be a fixed value. Due to the constant injection rate this corresponds to $t_2 = t_1 + \tau$ with τ proportional to Δh.

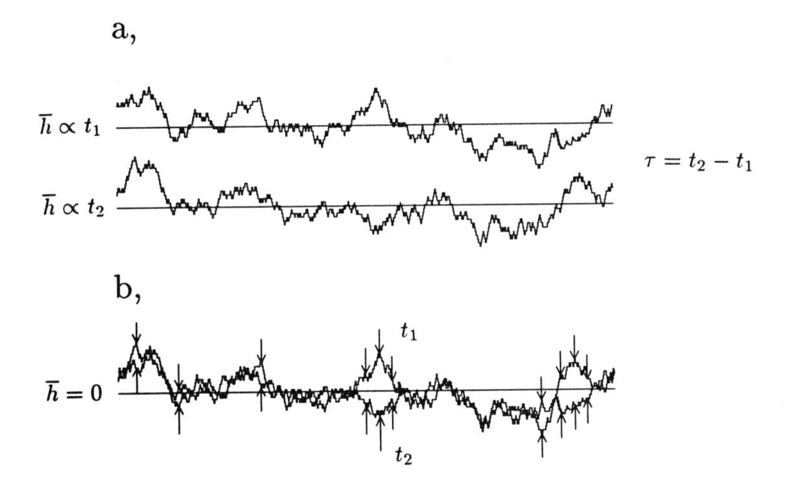

Figure 13.16. A plausible definition of the noise for experimental interfaces is the difference of the heights at a given position along the substrate measured at small time intervals, assuming that a correction for the small average velocity is taken (the surfaces are shifted so that their averages coincide, Horváth et al 1991b).

Using this definition one calculates the deviations from the average position of the interface $\tilde{h}(x,t) = h(x,t) - \langle h(t)\rangle_x$ for a set of t_1 and t_2 values, where averages are taken over $x \in [1..L]$. Then the noise is defined as the difference of these deviations normalized by the shift (see Fig. 13.16)

$$\eta(x,t_1) \equiv \frac{[\tilde{h}(x,t_2) - \tilde{h}(x,t_1)]}{\Delta h} \tag{13.1}$$

$\eta(x,t)$ is then determined for different experimental runs. The open circles in Fig. 13.17 show the noise distribution for $\eta(x,t) > 0$ averaged over the experiments. The experimental data points can be fitted by a straight line on a log-log plot indicating an algebraic decay of the noise amplitudes. The exponent of the corresponding power-law behavior $P(\eta) = c\eta^{-(\mu+1)}$ is $\mu = 2.67 \pm 0.2$.

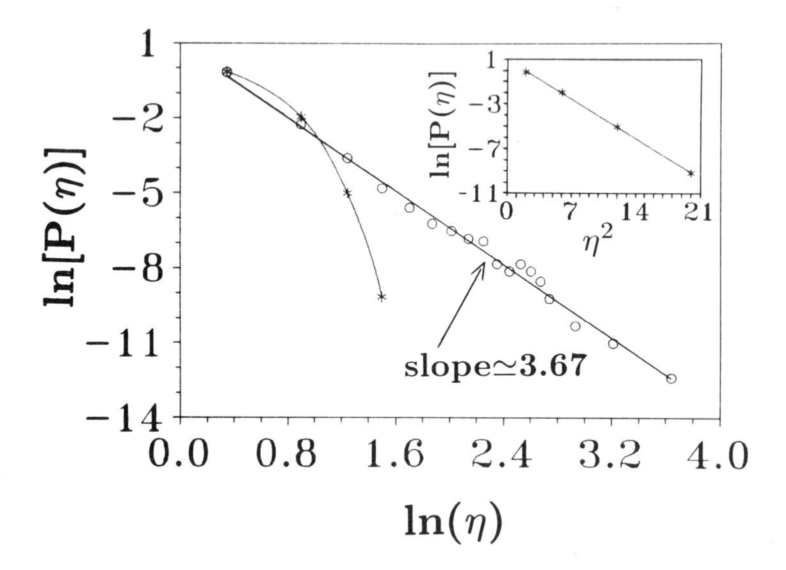

Figure 13.17. The noise distribution $P(\eta)$ for $\eta(x,t) > 0$ averaged over the experiments on two-phase flow in porous media (open circles). The experimental data points can be fitted by a straight line on a log-log plot indicating an algebraic decay of the noise amplitudes (Horváth et al 1991b). The stars in the main part and in the inset show the results of a similar analysis which was carried out for the restricted solid-on-solid deposition model. As expected the noise distribution is Gaussian in a model exhibiting universality.

For μ=2.67 the numerical simulations of the Zhang model and its variants (Section 12.5) give estimates for α close to 0.8, in remarkably good agreement with the experimental value for the present system $\alpha \simeq 0.81$. Although the agreement between the theory and the measured value of the exponent μ is excellent, it is not obvious how the input noise arising from the presence of the porous media is transformed by the surface dynamics into the output noise measured experimentally from the set of digitized interfaces.

A possible interpretation of the experimental results can be made through a model consisting of an array of disks (Cieplak and Robbins 1988) in which the microscopic geometry of the porous medium as well as the wetting or non-wetting characteristics of the fluids are taken into account. As the contact angle θ is reduced, the fluid becomes more wetting. The characteristics of the interface dramatically changes as θ passes through its critical value, θ_c. The simulations indicate (Martys et al 1991) that in the non-wetting limit ($\theta < \theta_c$) the interface is a self-affine fractal with a roughness exponent $\alpha = 0.81 \pm 0.04$ in close agreement with the experiments.

In the above model the surface advances in jumps due to single instabilities. Let us denote the invaded area after such a jump by A. It is possible to determine the distribution of A for a range of pressure values below θ_c. Close to a critical pressure the distribution of A has been found to follow a *power law*. This result is indirectly related to the measured power law distribution of the local jumps of the interface and is consistent with the experimental finding.

13.2.2. Deposition

The deposition of atoms or molecules onto substrates is an important technological technique which is used to grow thin solid films. Depending on the growth conditions vacuum deposition processes may lead to a variety of microstructures determining the physical properties of the films. Until now the fractal aspects of deposits have been studied almost exclusively by computer simulations and theoretical approaches. Here we describe the first available experimental results (Pfeifer et al 1989, Krim et al 1991, Chiarello et al 1991) on the fractal scaling of the surfaces of vacuum deposits.

The roughness of a deposit is manifested on the atomic scale which makes its direct study difficult because of the experimental limitations. Three techniques have been considered to estimate the roughness exponent α of the surfaces: i) adsorption measurements, ii) scanning tunneling microscopy and iii) x-ray reflectivity. Obviously, scanning tunneling microscopy gives the most direct information about the geometry, while the advantage of the other two methods is that they also provide data about a given aspect of the actual physical behaviour of the samples. A disadvantage of the adsorption method is that the relation of the measured quantities to the roughness exponent is rather indirect and the corresponding various theoretical considerations (Pfeifer et al 1990, Kardar and Indekeu 1990a, 1990b) lead to different estimates.

In the following x-ray reflectivity and nitrogen adsorption isotherm study of a variety of film surfaces will be discussed. The two kinds of measurements can be simultaneously carried out if the films are grown on a quartz crystal microbalance. The x-ray reflectivity data can be evaluated using the method of Sinha et al (1988). The conditions during the preparation of the surfaces were close to those which are assumed in the standard simulational models. A typical Ag deposit surface recorded by scanning electron microscopy is shown in Fig. 13.18.

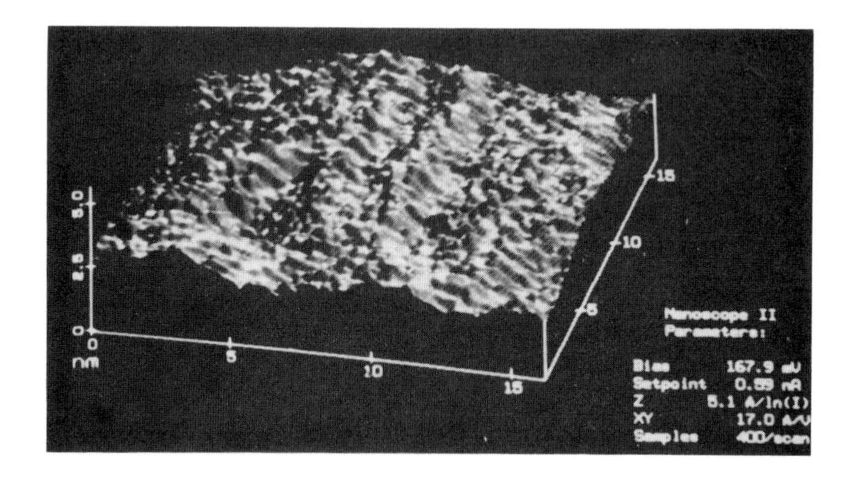

Figure 13.18. The microscopic structure of a vapour deposited Ag surface as determined by scanning tunneling microscopy (Krim et al 1991).

The Ag and Au films were deposited by collimated thermal evaporation at 10^{-9} torr. The surface of the quartz crystal had to be "overstone polished" since a test measurement showed that the roughness of the underlying crystal's surface may lead to an apparent self-affinity of the thin films. As a result of the experiments three types of surface morphologies were observed.

First silver atoms were deposited onto a substrate held at a low temperature (80K) and then the samples were warmed to the room temperature.
i) The deposition of a 1500Å thick Ag layer at a rate 0.5Å/sec at a normal angle (90° measured from the plane of the substrate) produced films with a non-fractal (smooth) surface.
ii) The same procedure was repeated for a close to normal angle of incidence (85°). In this case *both* the adsorption and the x-ray reflectivity measurements resulted in an estimate for the roughness exponent

$$\alpha \simeq 0.5. \tag{13.2}$$

iii) The deposition of 1500Å thick Au layers at a temperature 500K led to a surface structure which was not flat, but could be described in terms of a single new length scale instead of fractal scaling.

In conclusion, the investigated metal atom deposits were found to have a self-affine surface only when the process of deposition was realized at a low temperature and at an angle different from the normal to the substrate. The result 0.5±0.1 for the roughness exponent is close to, but somewhat larger than the corresponding values $\alpha = 0.36 - 0.40$ obtained by computer simulations of simple ballistic deposition models. Since the surface of the films becomes smoother (α becomes larger) when warmed to room temperature the measured roughness exponent is an upper limit to the low temperature value. Naturally, the determination of α for surfaces kept at low temperatures throughout the investigation is of considerable interest.

13.2.3. Biological growth

In Section 13.1.3 examples of self-similar growth in various biological systems were discussed. There is, however, only a single published work which is devoted to the self-affine scaling of living matter although this kind of geometrical behaviour is likely to be very general in biology. In fact, one of the most studied model leading to self-affine surfaces is the so-called Eden model (1961) which was originally proposed to characterize bacteria colony formation.

Consider a bacteria colony growing on an agar surface as it was described in Section 13.1.3. The morphology of the developing pattern depends on a number of factors. In particular, in the limit when the growth is dominated by the diffusion of the nutrient the colonies have a self-similar shape (Fujikawa and Matsushita 1989). The reason for self-similarity in these cases is an instability: the bacteria in the most advanced parts of the surface use up most of the nutrient. As a consequence, large empty regions appear within the structure where the nutrient concentration is too small for growth.

If the supply of nutrient is not limited, it is the stochastic sequence of the spontaneous multiplication of cells which is expected to play a major role in the development of the surface of a colony.

To see whether the growth of bacteria colonies results in self-affine fractal surfaces several series of experiments have been carried out (Vicsek et al 1990b) using two kinds of bacteria: *Escherichia coli* and *Bacillus subtilis.* In the following we shall describe the observations concerning the surface development of *E. coli* colonies in more detail. Again, the best choice is the strip geometry, where the growth starts from a line, because this is the situation which is described by dynamic scaling (Section 12.1). For this purpose the inoculation has to be made along a straight line segment on the surface of the nutrient rich agar prepared according to standard methods (LB medium moniatis).

The growth of bacteria colonies is known to be a rather complicated process. During the growth various macroscopic features of the morphology can be observed (e.g., Shapiro 1987). In addition, a number of microscopic changes take place including the alignment of the cells at the surface followed by the appearance of filaments and multicellular arrays. The colonies in the present experiments (Vicsek et al 1990) also displayed a non-trivial structure. Although the bulk part could be considered as being homogeneous on a length scale comparable to the size of the whole colony, inside the patterns dense tree-like structures could be observed corresponding to regions richer in bacteria and resembling the structures obtained in experiments on dense branching morphology and computer simulations of ballistic deposition. The analysis of the data is made as follows.

After 3-6 days pictures of the developing colonies are taken and subsequently digitized using a video image processor. The surface of the patterns is defined as the set of points which are at the highest position (measured from the line of inoculation) in each column of the array of occupied pixels representing the digitized image of the colonies. A part of a representative colony is shown in Fig. 13.19. Figure 7.2 displaying Eden clusters can be used for comparison.

Figure 13.19. Picture of a representative colony of *Escherichia coli* bacteria taken on the fourth day after inoculation (Vicsek et al 1990). This colony developed on a thick (5mm) layer of nutrient rich agar which was prepared according to standard methods.

The roughness exponent H can be calculated by determining the standard deviation $\sigma(l)$ (or the width w) of parts of the interfaces with linear extension l. Figure 13.20 shows how the average standard deviation depends on the length l of the intervals it was calculated for. The plot has three regions and it is the middle part which corresponds to self-affine scaling. The absence of scaling at the very beginning of the curve is due to the lower cutoff length (pixel size) inherent to any digitizing technique. The plato for large l values is a result of the fact that the self-affine scaling of our samples can be observed on length scales smaller than at which $\sigma(l)$ saturates. This characteristic length is closely related to the limited height of our samples. From the slopes of the straight lines fitted to the data similar to those displayed in Fig. 13.20 an average value $\alpha = 0.78 \pm 0.07$ is obtained for the roughness exponent describing the self-affine scaling of the colony surfaces.

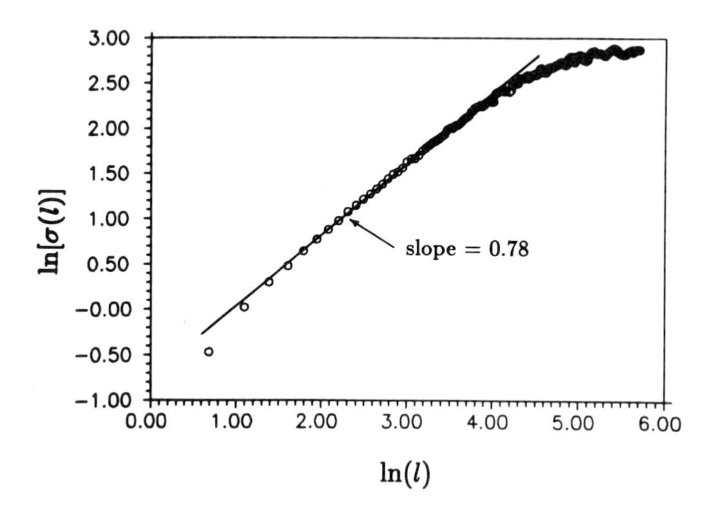

Figure 13.20. The dependence of the surface width $\sigma(l)$ calculated for the surface displayed in Fig. 13.19 on the length l of the intervals on which it was determined. The plot has three regions and it is the middle part which corresponds to self-affine scaling. The roughness exponent $H \simeq 0.78$ describing the self-affine scaling of the colony surfaces was determined from the slope of the straight line fitted to the middle part of the data.

The above results demonstrate that the investigated bacteria colony surfaces are self-affine structures on intermediate length scales in analogy with the behaviour of the surface of Eden clusters. The value we obtained for α is larger than $H = 0.5$ obtained for the simplest computer models. This discrepancy is likely to be due to the fact that bacteria colony formation is a much more complex phenomenon than the processes realized in most of the simulations. One important difference is the quasi-two-dimensional character of the colonies: the samples consist of many layers of bacteria on top of each other. This feature is expected to lead to a smoother overall surface which is also seen in simulations of the noise-reduced Eden model. The study of the self-affine nature of bacteria colony surfaces illustrates that refined theoretical approaches such as correlated or non-Gaussian noise are needed to describe realistic growth phenomena.

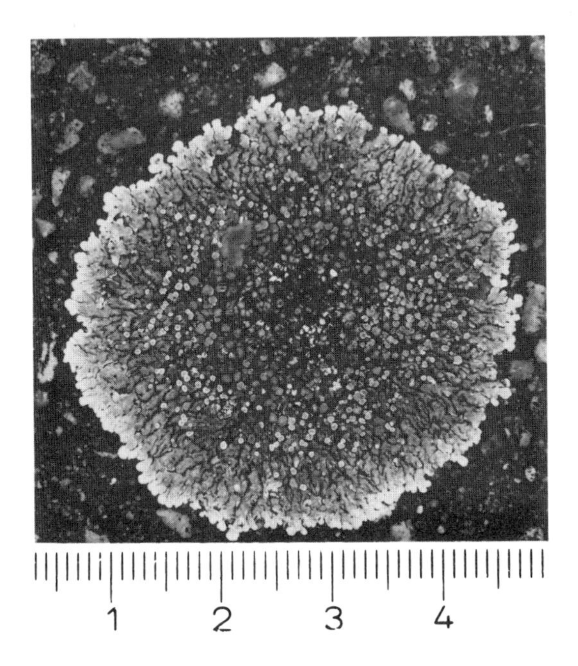

Figure 13.21. Picture of a plant called lichen which is composed of a fungus whose mycelium forms a matrix for algae living in a symbiotic relation with the fungus. Lichens can be found under a wide variety of climates growing on the surface of rocks and trunks. For this pattern the correlation function $c(\theta)$ scales as a function of θ with an exponent about 0.55 (Vicsek and Wolf 1991).

Finally we mention that other patterns of biological origin should also be tested in the future for self-affinity. Figure 13.21 shows the picture of a lichen which has a shape very much reminding Eden clusters (Fig. 7.1) or coffee drops (Fig. 13.13). The proper quantity to be calculated for such a pattern is the correlation function

$$c(\theta) = \langle |R(\theta') - R(\theta' + \theta)| \rangle_{\theta'}, \tag{13.3}$$

which is the average difference of the radii R corresponding to two surface points separated by an angle θ, where the radii and the angle is measured from the centre of mass of the structure. Preliminary results for $c(\theta)$ (Vicsek and Wolf 1991) indicate that it scales for lichens as θ^α with $\alpha \simeq 0.55$.

13.2.4. Fracture

Breaking of a three-dimensional object typically leads to a rough surface. This is a very general process which takes place (sometimes rather unfortunately) on an everyday basis in our environment. Naturally, any related result has important industrial applications. This is why we discuss this phenomenon, although it cannot be considered as a classical example for a growth process. The rough nature surfaces obtained by breaking a piece of metal is demonstrated in Fig. 13.22.

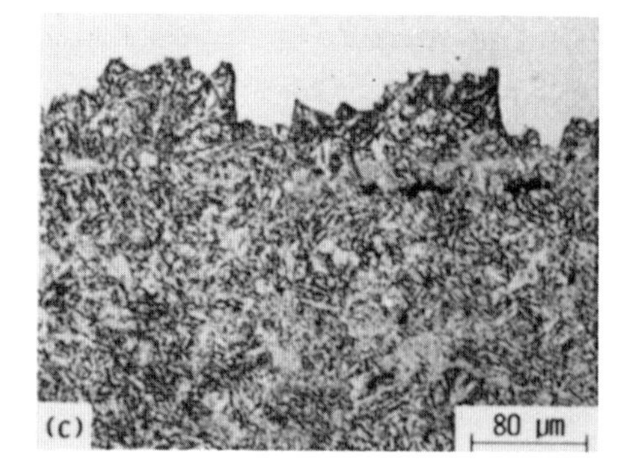

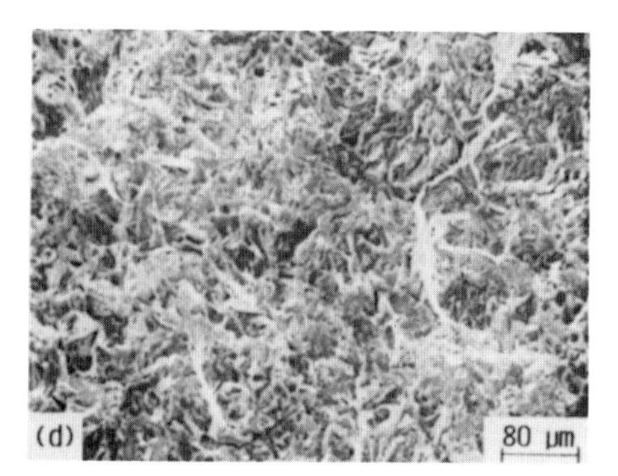

Figure 13.22. Optical micrograph of the profile (left) and the scanning electron micrograph (right) of the corresponding fractured surface. This surface was obtained by quasicleavage fracture of hydrogen charged A533B steel at $-196°$C (Dauskardt et al 1990).

The first study devoted to the question of the fractal aspects of fractured surfaces (Mandelbrot et al 1984) presented an elegant method of determining the fractal dimension D of the surface and indicated that there is a correlation between the fracture toughness and D. On the other hand, a recent work (Bouchaud et al 1990) concentrating on this correlation reported that for a variety of rupture modes and materials the observed fractal dimensions were the same within the error bars.

In commercial aluminium alloys fracture proceeds by two competing ductile processes: i) void formation and then linking up of these voids by transgranular rupture and ii) intergranular rupture. Changing the heat treatment results in a wide variation of the fracture characteristics and a transition between these modes. In the experiments of Bouchaud et al (1990) a variety of aluminium plates were investigated under different conditions. The fractal dimension of the surface was determined from image analysis using the following technique.

The fractured surfaces were coated with a thick layer of electrodeposited nickel. Then the samples were polished in a direction parallel to the overall plane of the surface. These cuts (see Fig. 13.23) had islands of aluminium atoms (where the surface bulged out and the polishing removed both the coating and some aluminium) embedded into regions corresponding to nickel. Such a polished surface can be well studied with an electron microscope using a backscattered electron contrast.

By calculating the correlation function corresponding to the perimeters of the aluminium islands the determination of the roughness exponent is straightforward since the expressions 2.18 and 7.21 can be used for this purpose. Let us denote by D_{cross} the fractal dimension of the perimeters which are obtained as a cross-section of the fractal surface with a plane. According to 7.21, $D_{cross} = 2 - \alpha$. On the other hand, the dimension D_{cross} of a fractal embedded into two dimensions is related to the exponent ξ describing the algebraic decay of its pair correlation function $c(r) \sim r^{-\xi}$ through $D = 2 - \xi$ (note that in 2.18 α plays the role of ξ and that $d = 2$ since the perimeters are embedded into a plane). From here

$$\alpha = \xi. \tag{13.4}$$

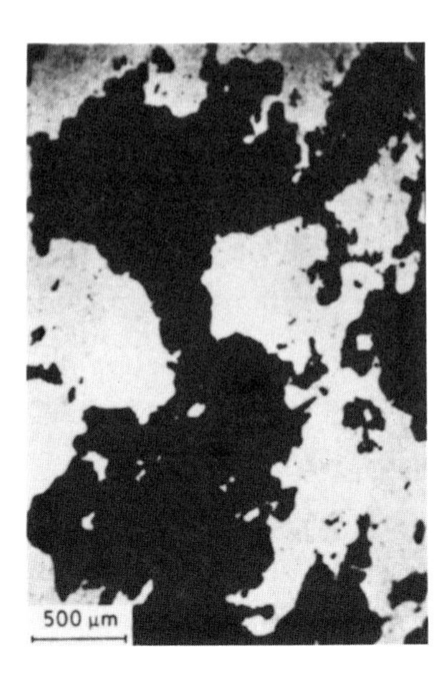

Figure 13.23. Micrograph showing the polished surface of a fractured surface previously coated by nickel. When observed with a backscattered electron contrast the aluminium alloy appears in black and the coating in white (Bouchaud et al 1990).

The numerical evaluation of the pictures similar to Fig. 13.23 resulted in estimates for α scattered in the narrow region between 0.75 and 0.82. Other investigations (Mandelbrot et al 1984, Dauskardt et al 1990) also reported results centered around $D = 3 - \alpha = 2.2$ which, according to the conclusion of Bouchaud et al (1990), may turn out to be a universal value.

REFERENCES (PART IV)

Aharony, A. and Feder, J., editors, 1989 *Fractals in Physics* (North-Holland, Amsterdam)

Amar, J. G. and Family, F., 1990a *Phys. Rev.* **A41**, 3399

Amar, J. G. and Family, F., 1990b *Phys. Rev. Lett.* **64**, 543

Amar, J.G. and Family, F., 1991 *J. Phys.* **A24**, L79

Ananthakrishna, G., 1991 private communication

de Arcangelis, L., Hansen, A. and Herrmann, H. J., 1989 *Phys. Rev.* **B40**, 877

Arnéodo, A., Couder, Y., Grasseau, G., Hakim, V. and Rabaud, M., 1989 *Phys. Rev. Lett.* **63**, 984

Arnéodo, A., Argoul, F., Couder, Y. and Rabaud, M., 1991 *Phys. Rev. Lett.* **66**, 2332

Argoul, F., Arnéodo, A., Grasseau, G. and Swinney, H. L., 1988 *Phys. Rev. Lett.* **61**, 2558

Argoul, F., Arnéodo, A., Elezgaray, J., Grasseau, G. and Murenzi, R., 1989 *Phys. Lett.* **135A**, 327

Argoul, F., Arnéodo, A., Elezgaray, J., Grasseau, G. and Murenzi, R., 1990 *Phys. Rev.* **A41**, 5537

Avnir, D., editor, 1989 *Fractal Approach to Heterogeneous Chemistry* (John Wiley, New York)

Baiod, R., Kessler, D., Ramanlal, P., Sander, L. M. and Savit, R., 1988 *Phys. Rev.* **A38**, 3672

Bales, G. S. and Zangwill, A., 1989 *Phys. Rev. Lett.* **63**, 692

Ball, R. C. and Brady, R. M., 1985 *J. Phys.* **A18**, L809

Barabási, A.-L. and Vicsek, T., 1990 *J. Phys.* **A23**, L729

Barabási, A.-L. and Vicsek, T., 1991 *Phys. Rev.* **A44**

Barabási, A.-L., Szépfalusy, P. and Vicsek, T., 1991a to appear in *Physica A*

Barabási, A.-L., Bourbonnais, R., Jensen, M., Kertész, J., Vicsek, T. and Zhang, Y.-C., 1991 preprint

Barker, P. W. and Ball, R. C., 1990 *Phys. Rev.* **A42**, 6289

Batrouni, C. G. and Hansen, A., 1988 *J. Stat. Phys.* **52**, 747

Ben-Jacob, E. and Garik, P., 1990 *Nature* **343**, 523

Blumenfeld, R. and Ball, R. C., 1991 to appear in *Phys. Rev. A*

Blumenfeld, R. and Aharony, A, 1989 *Phys. Rev. Lett.* **62**, 2977

Bohr, T., Cvitanovic, P. and Jensen, M. H., 1988 *Europhys. Lett.* **6**, 445

Bouchaud, E., Lapasset, G. and Planès, J., 1990 *Europhys. Lett.*, **13**, 73

Bourbonnais, R., Herrmann, H. J. and Vicsek, T., 1991a to appear in *Intl. J. Mod. Phys. C*;

Bourbonnais, R., Kertész, J. and Wolf, D. E., 1991b *J. Physique II.* **1**, 493

Brown, B. F., editor, 1972 *Stress Corrosion in High Strength Steels and in*

Titanium and Aluminium Alloys (Naval Res. Lab., Washington)
Buldyrev, S.V., Havlin, S., Kertész, J., Stanley, H.E. and Vicsek, T., 1991 *Phys. Rev.* **A43**, 7113
Caserta, F., Stanley, H. E., Eldred, W. D., Daccord, G., Hausman, R. E. and Nittmann, J., 1990 *Phys. Rev. Lett.* **64**, 95
Chiarello, R., Panella, V., Krim, J. and Thompson, C., 1991 preprint
Chopard, B., Herrmann, H. J. and Vicsek, T., 1991 to appear in *Nature*
Cieplak, M. and Robbins, M. O., 1988 *Phys. Rev. Lett.* **60**, 2042
Cieplak, M. and Robbins, M. O., 1990 *Phys. Rev.* **B41**, 11508
Coniglio, A. and Zanetti, M., 1990 *Physica* **A163**, 325
Couder, Y., 1988 in *Random Fluctuations and Pattern Growth* edited by Stanley, H. E. and Ostrowsky, N. (Kluwer, Dordrecht) p. 75
Couder, Y., Argoul, F., Arnéodo, A., Maurer, J. and Rabaud, M., 1990 *Phys. Rev.* **A42**, 3499
Dauskardt, R. H., Haubensak, F. and Ritchie, R. O., 1990 *Acta Metall. Mater.* **38**, 143
Eckmann, J.-P., Meakin, P., Procaccia, I. and Zeitak, R., 1989 *Phys. Rev.* **A29**, 3185
Eckmann, J.-P., Meakin, P., Procaccia, I. and Zeitak, R., 1990 *Phys. Rev. Lett* **65**, 52
Family, F. and Vicsek, T., 1985 *J. Phys.* **A18**, L75
Family, F., 1986 *J. Phys.* **A19**, L441
Family, F., Masters, B. R., and Platt, D. E., 1989 *Physica* **D38**, 98
Family, F. and Vicsek, T., 1991 *Dynamics of Fractal Surfaces*, (World Scientific, Singapore)
Feng, S. and Sen, P. N., 1984 *Phys. Rev. Lett.* **52**, 216
Fernandez, J. F. and Albaran, J. M., 1990 *Phys. Rev. Lett.* **64**, 2133
Fernandez, J., Guinea, F. and Louis, E., 1988 *J. Phys.* **A21**, L301
Forrest, B. M. and Tang, L.-H., 1990 *Phys. Rev. Lett.* **64**, 1405.
Frisch, U. and Parisi, G., 1985 in *Turbulence and Predictability in Geophysical Fluid Dynamics and Climate Dynamics* edited by Ghil, M., Benzi, R. and Parisi, G. (North-Holland, Amsterdam)
Fujikawa, H. and Matsushita, M., 1989 *J. Phys. Soc. Jpn.* **58**, 387
Fujikawa, H. and Matsushita, M., 1991 *J. Phys. Soc. Jpn.* **60**, 88
Golubović, L. and Bruinsma, R., 1991 *Phys. Rev. Lett.* **66**, 321

Harris, A. B. and Cohen, M., 1990 *Phys. Rev.* **A41**, 971

Havlin, S., Trus, B. L., Bunde, A. and Roman, H. E., 1989 *Phys. Rev. Lett.* **63**, 1189

Hayakawa, Y. and Matsushita, M., 1989 *Phys. Rev.* **A40**, 2871

Hede, B., Kertész, J. and Vicsek, T., 1991 to appear in *J. Stat. Phys.*

Hentschel, H. G. E. and Family, F., 1991 *Phys. Rev. Lett.* **66**, 1982

Herrmann, H. J., Kertész, J. and de Arcangelis, L., 1989 *Europhys. Lett.* **10**, 147

Herrmann, H. J. and Roux, S., 1990, editors, *Statistical Models for the Fracture in Disordered Media* (North-Holland, Amsterdam)

Herrmann, H. J., 1990 *Correlations and Connectivity: Geometric Aspects of Physics, Chemistry and Biology* edited by Stanley, H. E. and Ostrowsky, N. (Kluwer, Dordrecht)

Hinrichsen, E., Måloy, K. J., Feder, J. and Jøssang, J., 1989 *J. Phys.* **A22**, L271

Horváth, V. K. and Vicsek, T., 1989 *J. Phys.* **A23**, L259

Horváth, V. K. and Herrmann, H. J., 1991 to appear in *Chaos, Solitons and Fractals*

Horváth, V. K., Family, F. and Vicsek, T., 1991a *J. Phys.* **A24**, L25

Horváth, V. K., Family, F. and Vicsek, T., 1991b preprint

Hulin, J. P., Clement, E., Baudet, C., Gouyet, J. F. and Rosso, M., 1988 *Phys. Rev. Lett.* **61**, 333

Huse, D. A., Amar, J. G. and Family, F., 1990 *Phys. Rev.* **A41**, 7075

Hwa, T., Kardar, M. and Paczuski, M., 1991 *Phys. Rev. Lett.* **66**, 441

Jullien, R. and Meakin, P., 1989 *J. Phys.* **A22**, L1115.

Kahng, B., Batrouni, G. G., Redner, S., de Arcangelis, L. and Herrmann, H. J., 1988 *Phys. Rev.* **B37**, 7625

Kardar, M., Parisi, G. and Zhang, Y.-C., 1986 *Phys. Rev. Lett.* **56**, 889

Kardar, M. and Zhang, Y.-C., 1987 *Phys. Rev. Lett.* **58**, 2087

Kardar, M. and Indekeu, J. O., 1990a *Phys. Rev. Lett.* **65**, 662

Kardar, M. and Indekeu, J. O., 1990b *Europhys. Lett.* **12**, 161

Kertész, J. and Vicsek, T., 1986 *J. Phys.* **A19**, L257

Kertész, J. and Wolf, D. E., 1989 *Phys. Rev. Lett.* **62**, 2571

Kertész, J., 1990 in *Statistical Models for the Fracture in Disordered Media*, edited by Herrmann, H. J. and Roux, S., (North-Holland, Amsterdam)

Kessler, D., Koplik, J. and Levine, H., 1988 *Adv. Phys.* **37**, 255

Kim, J. M. and Kosterlitz, J. M., 1989 *Phys. Rev. Lett.* **62**, 2289

Kinzel, W., 1983 in *Percolation Structures and Processes*, edited by Deutscher, G., Zallen, R. and Adler, J. (Adam Hilger, Bristol) p. 425

Klein, W., 1990 *Phys. Rev. Lett.* **65**, 1642

Krim, J., Solina, D. H. and Chiarello, R., 1991 *Phys. Rev. Lett.* **66**, 181

Krug, J. and Spohn, H., 1988 *Phys. Rev.* **A38**, 4271

Krug, J. and Spohn, H., 1990 in *Solids Far from Equilibrium: Growth, Morphology and Defects* edited by Godrèche, C. (Cambridge Univ. Press, Cambridge)

Krug, J., 1991 *J. Physique I.* **1**, 9

Lai, Z.-W. and Das Sarma S., 1991 *Phys. Rev. Lett.* **66**, 2348

Lee, J. and Stanley, H. E., 1988 *Phys. Rev. Lett.* **61**, 2945

Lee, J., Havlin, S., Stanley, H. E. and Kiefer, J.E., 1990 *Phys. Rev.* **A42**, 4832

Lemaire, E. and Van Damme, H., 1989 *C. R. Acad. Sci. Ser. II.* **39**, 859

Li, G., Sander, L. M. and Meakin, P., 1989 *Phys. Rev. Lett.* **63**, 1322

Liu, D. and Plischke, M., 1988 *Phys. Rev.* **B38**, 4781

Louis, E. and Guinea, F., 1987 *Europhys. Lett.* **3**, 871

Mainster, M. A., 1990 *Eye* **4**, 235

Mandelbrot B. B., 1982 *The Fractal Geometry of Nature* (Freeman, San Francisco)

Mandelbrot, B. B., Passoja, D. E. and Paullay, A. J., 1984 *Nature* **308**, 721

Mandelbrot, B. B. and Vicsek, T., 1989 *J. Phys.* **A22**, L377

Mandelbrot, B. B., Evertsz, J. G. and Hayakawa, Y., 1990 *Phys. Rev.* **A42**, 4528

Mandelbrot, B. B. and Evertsz, J. G., 1991 *Nature* **348**, 143

Manna, S. and Vicsek, T., 1991 to appear in *J. Stat. Phys.*

Marconi, U. M.-B. and Zhang, Y.-C., 1990 *J. Stat. Phys.* **61**, 585

Martys, N., Cieplak, M. and Robbins, M. O., 1991 *Phys. Rev. Lett.* **66**, 1058

Masters, B.R., Family, F. and Platt, D. E., 1989 *Invest. Opthalmol. Vis. Sci.* (Suppl) **30**, 391

Matsuyama, T., Sogawa, M. and Nakagawa, Y., 1989 *FEMS Microb. Lett.* **61**, 243.

May, S. E. and Maher. J. V., 1989 *Phys. Rev.* **A40**, 1723

Meakin, P., Ramanlal, P., Sander, L. M. and Ball, R. C., 1986 *Phys. Rev.* **34**, 5091

Meakin, P., 1988 in *Random Fluctuations and Pattern Growth* edited by Stanley, H. E. and Ostrowsky, N. (Kluwer, Dordrecht) p. 174

Meakin, P., Li, G., Sander, L. M., Louis, E. and Guinea, F., 1989 *J. Phys.* **A22**, 1393

Meakin, P. and Jullien, R. 1990, Phys. Rev. A **41**, 983.

Medina, E., Hwa, T. and Kardar, M., 1989 *Phys. Rev.* **A39**, 3053

Nadal, J. P., Derrida, B. and Vannimenius, J., 1984 *Phys. Rev.* **B30**, 376

Nagatani, T., 1988 *Phys. Rev.* **A38**, 2632

Nagatani, T. and Stanley, H. E., 1990 *Phys. Rev.* **A42**, 3512

Nagatani, T., Lee, J. and Stanley, H. E., 1991 *Phys. Rev. Lett.* **66**, 616

Nelkin, M., 1989 *J. Stat. Phys.* **54**, 1

Obert, M., Pfeifer, P. and Sernetz, M., 1990a *J. Bacteriology* **172**, 1180

Obert, M., Sernetz, M., and Neuschulz U., 1990b in *Fractal-90* edited by Mandelbrot, B. B. and Peitgen, H.-O. (North-Holland, Amsterdam)

Ossadnik, P., 1991 to appear in *Physica A*

Pellegrini, Y. P. and Jullien, R., 1990 *Phys. Rev. Lett.* **64**, 1745

Pfeifer, P., Wu, Y. J., Cole, M. W. and Krim, J., 1989 *Phys. Rev. Lett.* **65**, 663

Pfeifer, P., Cole, M. W. and Krim, J., 1990 *Phys. Rev. Lett.* **62**, 1997

Pietronero, L., Erzan, A. and Evertsz, C., 1988a *Phys. Rev. Lett.* **61**, 861

Pietronero, L., Erzan, A. and Evertsz, C., 1988b *Physica* **A151**, 207

Pietronero, L., 1990, editor, *Fractal's Physical Origin and Properties* (Plenum Press, New York)

Plischke, M., Rácz, Z. and Liu, D., 1987 *Phys. Rev.* **B35**, 3485

Procaccia, I. and Zeitak, R., 1988 *Phys. Rev. Lett.* **60**, 2511

Rácz, Z., Siegert, M., Liu, D. and Plischke, M., 1991 *Phys. Rev.* **A43**, 5275

Raz, E., Lipson, S. G. and Polturak, E., 1989 *Phys. Rev.* **A40**, 1088

Roland, C. and Guo, H., 1991 *Phys. Rev. Lett.* **66**, 2104

Rosso, M., Gouyet, J. F. and Sapoval, B., 1986 *Phys. Rev. Lett.* **57**, 3195

Roux, S. and Guyon, E., 1985 *J. Physique Lett.* **46**, L999

Rubio, M. A., Edwards, C. A., Dougherty, A. and Gollub, J. P., 1989 *Phys. Rev. Lett.* **63**, 1685

Sapoval, B., Rosso, M. and Gouyet, J. F., 1985 *J. Physique Lett.* **46**, 149

Das Sarma, S. and Tamborenea, P., 1991 *Phys. Rev. Lett.* **66**, 325

Schwarzer, S., Lee, J., Bunde, A., Havlin, S., Roman, H. E. and Stanley, H. E., 1990 *Phys. Rev. Lett.* **65**, 603

Schwarzer, S., Lee, J., Havlin, S., Stanley, H. E. and Meakin, P., 1991 *Phys. Rev.* **A43**, 1134

Shapiro, J. A., 1987 *J. Bacteriol.* **169**, 142

Sinha, S. K., Sirota, E. B., Garoff, S. and Stanley, H. B., 1988 *Phys. Rev.* **B38**, 2297

Smith, T. G., Marks, W. B., Lange, G. D., Sheriff, W. H. and Neale, E.A., 1989 *J. Neurosci. Methods* **27**, 173

Skjeltorp, A. T., 1988 in *Random Fluctuations and Pattern Growth* edited by Stanley, H. E. and Ostrowsky, N. (Kluwer, Dordrecht) p. 170

Stanley, H. E. and Ostrowsky, N., 1988, editors, *Random Fluctuations and Pattern Growth* (Kluwer, Dordrecht)

Stanley, H. E., 1989 *Physica* **D38**, 330

Stanley, H. E. and Ostrowsky, N., 1990, editors, *Correlations and Connectivity: Geometric Aspects of Physics, Chemistry and Biology* (Kluwer, Dordrecht)

Sun, T., Guo, M. and Grant, M., 1989 *Phys. Rev.* **A40**, 6763

Tang, C., Alexander, A. and Bruinsma, R., 1990 *Phys. Rev. Lett.* **64**, 772

Tang, L.-H., Nattermann, T. and Forrest, B. M., 1991 *Phys. Rev. Lett.* **66**,

Tél, T., Fülöp, Á. and Vicsek, T., 1989 *Physica A*, **159**, 155

Tolman, S. and Meakin, P., 1989 *Phys. Rev.* **A40**, 428

Tolman, S., Meakin, P. and Matsushita, M., 1989 *J. Phys. Soc. Jpn.* **58**, 2721

Van Damme, H., 1989 in *Fractal Approach to Heterogeneous Chemistry* edited by Avnir, D. (John Wiley, New York)

Vicsek, T., Family, F. and Meakin, P., 1990a *Europhys. Lett.* **12**, 217

Vicsek, T., Cserző, M. and Horváth, V. K., 1990b, *Physica* **A167**, 315

Vicsek, T. and Barabási, A.-L., 1991 *J. Phys.* **A24**, Lxx

Vicsek, T. and Wolf, D. E., 1991 to be published

Wolf, D. E. and Villain, J., 1990 *Europhys. Lett.* **13**, 389

Wolf, D. E. and Tang, L.-H., 1990 *Phys. Rev. Lett.* **65**, 159

Yan, H., Kessler, D. and Sander, L. M., 1990 *Phys. Rev. Lett.* **64**, 926

Zhang, Y.-C., 1990a *J. Physique* **51**, 2129

Zhang, Y.-C., 1990b *Physica* **A170**, 1

APPENDICES

Appendix A

Algorithm for growing diffusion-limited aggregates

To generate large DLA clusters one has to use tricks because the original version of the model would require prohibitively large amounts of computer time. In the following, three simple procedures will briefly be described which are helpful in reducing the computational time substantially and allow one to grow – using a mainframe – aggregates consisting of several millions of particles. Of course, a personal computer is much slower and the size of the clusters one can generate in a PC is in the range of a few thousand particles.

i) The first trick is that we release the particles from a circle of radius R_0 which is just a bit larger than the largest distance between the particles already belonging to the aggregate and the origin. This method is justified because the particles released very far from the cluster arrive (for the first time) at different points of a circle of radius larger than R_0 and centred at the origin with the same probability.

ii) Whenever a randomly walking particle leaves the region which is inside the above circle, the distance ΔR of the particle from the circle is determined. Then the next step made by the particle is a jump of length ΔR from its actual position made in a randomly selected direction (the final position is the lattice site closest to the coordinates of the particle after the jump has taken place). Inside the circle of radius R_0 the particle always undergoes a random walk jumping one lattice unit choosing one of the nearest neighbour sites randomly. If the particle making large jumps outside of the circle gets too far (e.g., ten times R_0), it is killed and a new particle is released from the circle R_0.

iii) A further improvement can be achieved using the following algorithm. At the beginning one assigns a quite arbitrary value $l' = l_{max}$ to all lattice sites. (A reasonable choice is in the range of 15–30.) Then, for each site from which the random walker could reach the seed (cluster) by making a jump of length l' lattice units the value assigned to the site is changed to l' if $l' < l_{max}$. In other words, close to the cluster l' is the maximum step length a random walker may take without crossing the arms of the aggregate. Thus the particles jump a distance which is either l' or ΔR (the latter choice is made if $\delta R > l_{max}$). After a new particle has been added to the cluster the l' values in the region where it landed are updated so that the next particle could not make a jump which would result in crossing any parts of the aggregate.

It is straightforward to extend the above method to the off-lattice case by making use of an underlying virtual lattice. In this case the l'-s are assigned to the lattice elements (for the square lattice these are the unit squares bounded by four bonds) and one keeps track of the distances from the cluster and the trajectory of a particle by detecting the lattice elements visited by the off-lattice walk. Below a sample program is given which was written in BASIC for the IBM PC with a colour graphics card. It generates DLA clusters on a square lattice. This short version does not contain trick iii) which is more effective when one grows very large clusters.

```
10 REM DIFFUSION-LIMITED AGGREGATION ON THE SQUARE LATTICE
20 REM SEED PARTICLE AT NO,NO
30 REM NEXT PARTICLE IS RELEASED FROM A CIRCLE AND UNDERGOES
40 REM A RANDOM WALK. IT STICKS TO THE PARTICLE AT THE ORIGIN WHEN
50 REM ARRIVES AT AN ADJACENT SITE. A NEW PARTICLE IS RELEASED ...
60 REM N - LINEAR SIZE OF THE CELL
70 REM N9 - NUMBER OF PARTICLES TO BE DEPOSITED
80 REM A(I,J)=1 IF THERE IS AND A(I,J)=0 IF THERE IS NO PARTICLE
90 REM AT THE SITE I,J
100 OPTION BASE 1
110 SCREEN 2
120 DIM A(90,90),J$(100),B(10,10)
130 CLS
140 N=49
150 C=1
160 N9=50
170 R1=2
180 R2=1
190 W=4
200 NO=(N+1)/2
210 CIRCLE(NO*8,NO*W),3:CIRCLE(NO*8,NO*4),2:CIRCLE(NO*8,NO*4),1
220 R3=R1
230 R4=5*(R3+R2)
240 FOR I1=1 TO N
250 FOR I2=1 TO N
260 A(I1,I2)=0
270 NEXT I2
280 NEXT I1
290 IO=1
300 REM PUT A PARTICLE AT THE ORIGIN
310 A(NO,NO)=1
320 REM             START GROWING
330 RANDOMIZE(.59)
340 FOR I9=2 TO N9
350 F1=6.28319*RND(1)
360 I=INT((R3+R2)*SIN(F1)+NO+.5)
370 J=INT((R3+R2)*COS(F1)+NO+.5)
380 X1=I
390 Y1=J
400 REM JUMP ONTO A CIRCLE OF RADIUS R6 (WHICH IS THE DISTANCE FROM
410 REM THE AGGREGATE) IF R6>2
420 R5=(I-NO)*(I-NO)+(J-NO)*(J-NO)
430 R5=SQR(R5)
440 R6=R5-(R3+R2)
450 IF R6 < 2 THEN GOTO 570
```

```
460 R7=6.28319*RND(1)
470 X1=X1+R6*SIN(R7)
480 Y1=Y1+R6*COS(R7)
490 I=INT(X1+.5)
500 J=INT(Y1+.5)
510 R5=(I-NO)*(I-NO)+(J-NO)*(J-NO)
520 R5=SQR(R5)
530 IF R5 < 20.1 THEN CIRCLE(8*J,4*I),3
540 IF R5 < 20.1 THEN CIRCLE(8*J,4*I),3,0
550 GOTO 760
560 REM                WALK BY STEPS
570 K1=INT(RND(1)*4+1)
580 ON K1 GOTO 590,610,630,650
590 I=I+1
600 GOTO 660
610 I=I-1
620 GOTO 660
630 J=J-1
640 GOTO 660
650 J=J+1
660 REM
670 M1=I-1
680 M2=J-1
690 P1=I+1
700 P2=J+1
710 CIRCLE(8*J,4*I),3: CIRCLE(8*J,4*I),3,0
720 A1=A(M1,J)+A(P1,J)+A(I,M2)+A(I,P2)
730 IF A1 >= .1 THEN GOTO 810
740 X1=I
750 Y1=J
760 R5=(I-NO)*(I-NO)+(J-NO)*(J-NO)
770 R5=SQR(R5)
780 IF R5 > R4 THEN GOTO 350
790 GOTO 400
800 REM FIX NEW PARTICLE
810 A(I,J)=1
820 I0=I0+1
830 IF R5 > R3 THEN R3=R5
840 R4=5*(R3+R2)
850 CIRCLE(J*8,I*W),3
860 NEXT I9
870 END
```

Appendix B

Construction of a simple Hele-Shaw cell

In this Appendix a few details are given for those who are interested in the construction of a Hele-Shaw cell. To build a versatile Hele-Shaw cell is not particularly troublesome but the application of a few tricks makes the experiments easier to carry out. A schematic picture of a possible arrangement is shown in Fig. B.1. The related information is given below.

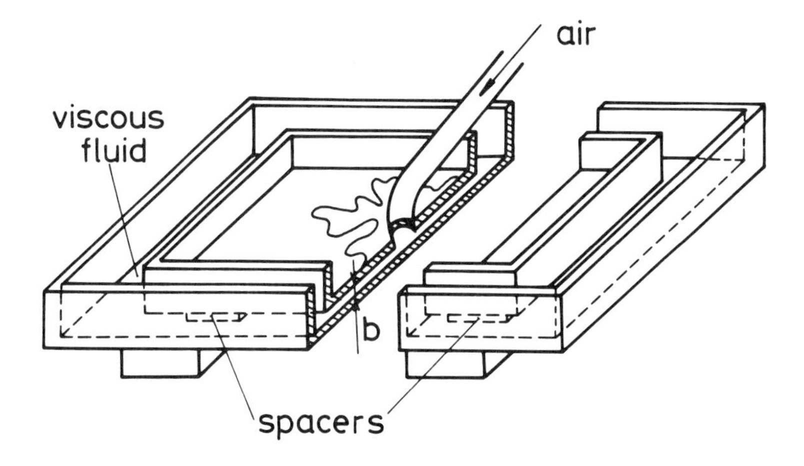

Figure B.1. Schematic picture of a radial Hele-Shaw cell. Its cross-section is also indicated. The distance between the plates is denoted by b.

i) The sizes of the upper and lower plates are 27 × 27 cm and 34 × 34 cm, respectively. They are made of good quality glass of width 5mm.

ii) Air is injected into a viscous fluid through a hole of radius 3 mm drilled at the centre of the upper plate.

iii) In order to prevent the viscous fluid from flowing out, walls are attached to the plates in a manner shown in Fig. B.1.

iv) The distance between the plates is controlled by inserting between them thin metallic strips.

v) At large pressures one needs to clamp the two glass plates together either by using screws or a heavy frame.

The above basic arrangement can be modified in a number of ways and, correspondingly, many types of viscous fingering patterns can be observed. By inserting a third glass plate (in between the two original ones) with a mesh etched on its surface, one is able to study the effects of anisotropy. If small balls are spread randomly on the surface of the third plate the structure of the interface becomes a random fractal, similar to the geometry of diffusion-limited aggregates.

Appendix C

Basic concepts underlying multifractal measures

In Chapter 3 we have discussed a formalism and algorithms which can be used to characterize the properties of fractal measures. Here we describe the *underlying basic concepts* presented (after this book was finished) in a recent paper by Mandelbrot (1988) to be published in *Fluctuations and Pattern Formation* edited by H. E.Stanley and N. Ostrowsky (Kluwer, Dordrecht and Boston).

In example 3.1 a fractal measure is constructed by recursion. This measure fails to have density, and it is not discrete. For example, if one goes from $\epsilon = (1/3)^k$ to $\epsilon = (1/3)^{k+1}$, the sharing of the measure in an original interval of length ϵ among its three parts is usually very uneven. Thus, the values of this measure in the $k \to \infty$ limit can not be described by any distribution function.

To define the appropriate quantity, we fix a given box size ϵ, and for this ϵ plot the corresponding distribution of the measure (number of boxes with the given value of the measure in them) using double logarithmic coordinates. Then a multifractal measure has the property that reducing both logarithmic coordinates by the same factor $\ln \epsilon$, the reduced plots of the distribution *converge to a limit probability distribution* $\rho(\alpha)$. The reduced horizontal logarithmic coordinate is nothing else than α as defined in Section 3.1 while the function $f(\alpha)$ is given by $f(\alpha) = \rho(\alpha) + 1$.

Let us now generalize example 3.1 as done by Mandelbrot 1988, and use this new, multinomial version to demonstrate how one can relate the relevant quantities of the theory in a simple way. Denoting by P_i the weights given to the ith box ($i = 1, 2, ..., b$) at the first step of the recursion we find that

$$\alpha = -\sum_i \varphi_i \log_b P_i \qquad \text{and} \qquad \delta = -\sum_i \varphi_i \log_b \varphi_i,$$

where δ is the box fractal dimension and $\alpha = \log[\mu(\epsilon)]/\log \epsilon$ with

$$\mu(\epsilon) = P_1^{k\varphi_1} ... P_b^{k\varphi_b}.$$

Here φ_i is the relative frequency of the digits i in the b-base development of $x = 0.\eta_1\eta_2...\eta_k$. The set of boxes with the same α is dominated by the term corresponding to the highest dimension. This term maximizes the expression $-\sum_i \varphi_i \log_b \varphi_i$ for given $-\varphi_i \log_b P_i = \alpha$ and $\sum_i \varphi_i = 1$. To calculate φ_i one uses the classical method of Lagrange multipliers in which a multiplier denoted by q ($-\infty < q < \infty$) is introduced. The application of the method gives

$$\varphi_i = \frac{b^{q \log_b P_i}}{\sum_i b^{q \log_b P_i}} = \frac{P_i^q}{\sum_i P_i^q}.$$

In terms of the cumulant generating function $\tau(q) = -\log_b \sum_i P_i^q$ the Lagrange multipliers determine q and $f(\alpha)$ from α by

$$\alpha = -\sum_i \varphi_i \log_b P_i = -\frac{\partial}{\partial q} \log_b \sum_i P_i,$$

$$\max \delta = f(\alpha) = -\frac{\sum_i (q \log_b P_i - \log_b \sum_i P_i^q) P_i^q}{\sum_i P_i^q}.$$

This means that

$$\alpha = \frac{\partial \tau(q)}{\partial q} \qquad \text{and} \qquad f(\alpha) = q\frac{\partial \tau(q)}{\partial q} - \tau = q\alpha - \tau.$$

In the above expressions q, τ and f formally appear as the inverse temperature, the Gibbs free energy, and the entropy, respectively. It is clear that for such *multinomial measures* $\alpha > 0$ and $\delta \geq 0$, therefore, $f(\alpha) \geq 0$.

The latter inequality is, however, not necessarily true for all multifractal distributions. Obviously, if $f(\alpha)$ is viewed as a fractal dimension of the set characterized by the Hölder exponent α, $f(\alpha) < 0$ can not correspond to a fractal dimension in its usual sense. For the $f(\alpha) < 0$ case Mandelbrot (*J. Stat. Phys.* **34** (1984) 895) introduced the term *latent dimension*, expressing that this property of the measure is present, but hidden, and can be observed using high-dimensional cuts from the measure (Mandelbrot 1988).

The above described multinomial measure can easily be genealized to the *random case* (Mandelbrot 1974). Clearly, if the weights P_i are randomly redistributed before each stage of the recursion process, the $f(\alpha)$ of the resulting random multiplicative measure will be the same as that of the deterministic one. As a further step, one can also go from measures embedded into one dimension to multifractals defined on the set of points $\{\vec{x}\}$ in an E-dimensional space, and suppose that $b = B^E$ with positive integers B and E. If $b/B \gg 1$, then the weights in the boxes along a one-dimensional cut from the measure can be regarded as statistically independent. Assuming independence means that the values P_i of the random multiplier M can take on any of the values $m_i' = m_i B^{E-1}$ with the same probability.

Let us now express the amount of measure $\mu(d\vec{x})$ in the B-adic interval of length B^{-k} of the one-dimensional section assuming that this interval is at $x = 0.\eta_1, ...\eta_k$. Since we do not know the exact amount of the measure along the cut (it is normalized to be equal to 1 only in the E-dimensional space), we need a prefactor $\Omega(\eta_1, ...\eta_k)$ which is due to stages of frequency larger than B^k. Thus

$$\mu(d\vec{x}) = \Omega(\eta_1, \dots, \eta_k) M(\eta_1) \dots M(\eta_1, \dots, \eta_k) \dots .$$

Here the multipliers M are identically distributed and independent. The same is valid for the random factors Ω. Introducing $\alpha_L = -(1/k)[\log_B M(\eta_1) + \log_B M(\eta_1, \eta_2) + ...]$ and $\alpha_H = -(1/k) \log_B \Omega$ gives $\alpha = \alpha_L + \alpha_H$. For

$k \to \infty$ α_H becomes negligible, while α_L is the average of k independent random variables. According to the *Cramèr limit theorem*, as $k \to \infty$ the quantity

$$(1/k) \log_b(\text{probability density of } \alpha_L)$$

converges to a limit which is here denoted by $\rho(\alpha)$. However, this $\rho(\alpha)$ is *not universal* in the sense that different multipliers M lead to different $\rho(\alpha)$ distributions. It is easy to show that $f(\alpha) = \rho(\alpha) + \max\ f(\alpha) = \rho(\alpha) +$ dimension of the measure's support.

Finally, a few comments should be made concerning the lognormal character of the distributions considered. If the sum $\log_B M(\eta_1) + \log_B M(\eta_1, \eta_2) + \ldots$ behaved asymptotically as a Gaussian, α would follow a lognormal distribution. In this case $f(\alpha)$ would be a parabola. However, when the distribution is properly normalized, the above sum is not universal and the central limit theorem implies that $\rho(\alpha)$ and $f(\alpha)$ are parabolic *only* close to their maximum. Away from this maximum the behaviour of $\rho(\alpha)$ and $f(\alpha)$ is determined by M and is parabolic only if the multipliers M are lognormal. For more details on the basic concepts of multifractals and the meaning of $f(\alpha) < 0$ see Mandelbrot (1988).

AUTHOR INDEX

SUBJECT INDEX

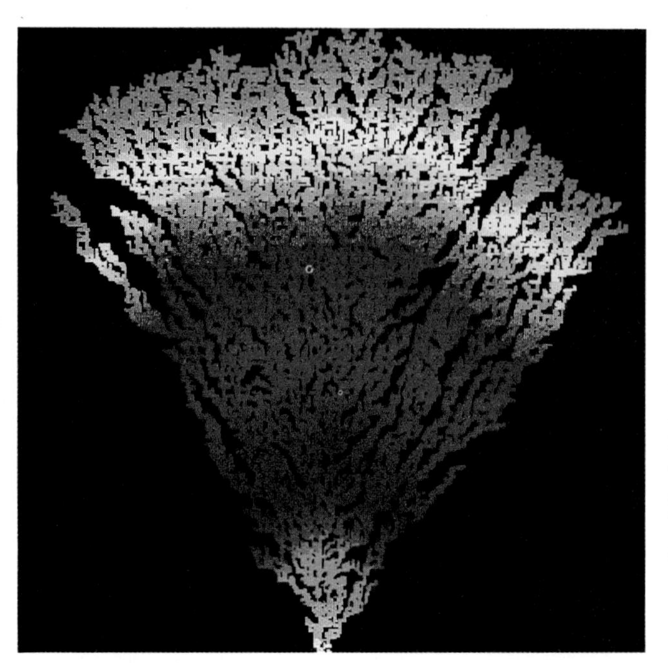

Plate 1

Aggregates obtained in simulations of the two simplest versions of ballistic deposition. i) Irreversible aggregation of vertically falling particles on a single seed (left), the same as i), but along a straight line segment (bottom). The shading and the color coding indicates the elapsed time: the color of a newly added particle depends on the number of already deposited particles. The gradual build up of large scale fluctuations (surface roughening) described by dynamic scaling is demonstrated by the sequence of brighter surfaces shown below *(courtesy of F. Family, Emory University).*

Plate 2
Two related variants of ballistic deposition. i) Example for ballistic deposition on a macroscopic scale (top). This picture shows snow particles sliding down a glass window and forming an aggregate as they stick together on contact *(courtesy of R. Lenormand, Institut Francais du Petrole)*. ii) The bottom picture was generated by a correlated ballistic deposition model in which the particles had a larger probability to be aggregated at positions close to that of the previously added particle *(F. Family and T. Vicsek, Comp. Phys. No. 1 (1990) 44)*. An analogous behaviour is expected to govern the motion of the snow particles which, in the situation shown in the upper picture, are likely to prefer the freshly wetted paths along the window's surface.

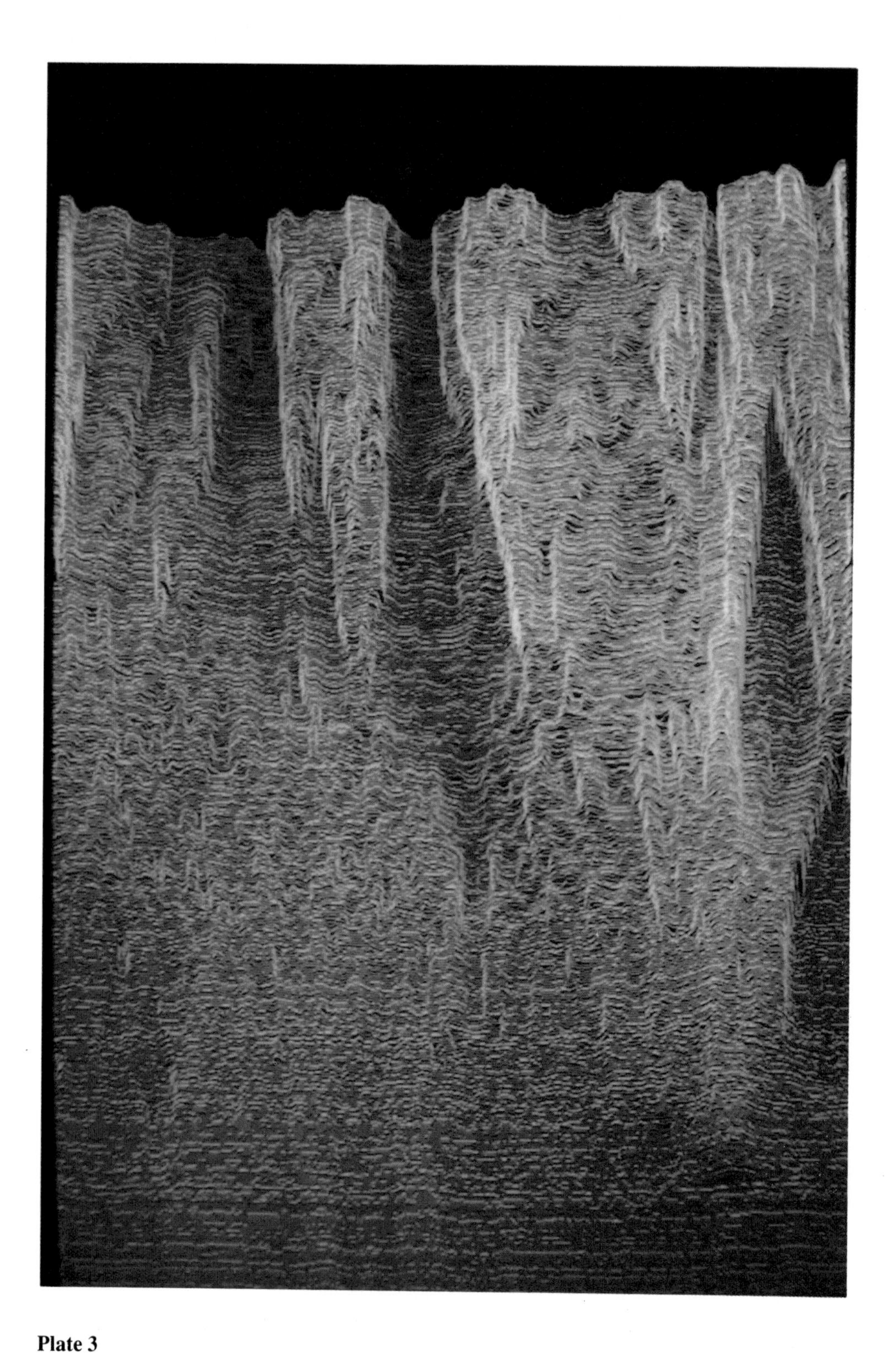

Plate 3

Kinetic roughening in the experiment on the penetration of a wetting fluid into a thin layer of glass beads (Horváth et al 1991a). The flow takes place from bottom to top. The development of the fluid-air interface is demonstrated by plotting the digitized image of the meniscus as it evolves. The three-dimensional appearance of this picture is obtained by making the brightness of the surface at a given point proportional to its distance from the average height.

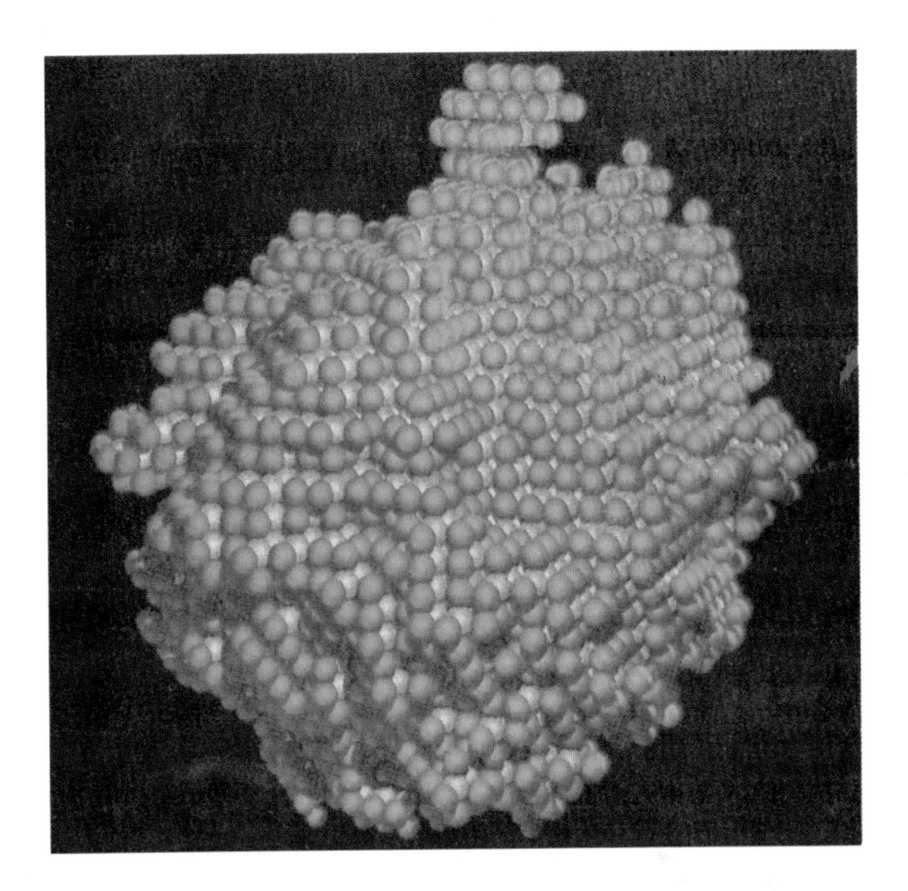

Plate 4
In the growth model introduced by Eden (1961) the aggregates grow on a lattice as particles at randomly selected surface sites are consecutively added to the cluster. The upper picture shows surface (blue) and other (yellow) particles of an Eden cluster generated in three dimensions *(M. T. Batchelor and T. I. Henry, Phys. Lett. 157A (1991) 229)*. Bottom: Two-dimensional Eden cluster-like plants called lichens which are composed of a fungus whose mycelium forms a matrix for algae living in a symbiotic relation with the fungus. Lichens can be found under a wide variety of climates growing on the surface of rocks and trunks.

Plate 5
This rough surface was obtained during the simulation of a physically motivated algorithm designed to model how the self-affine geometry of mountainous regions is created by erosion processes. Blue color indicates the deepest regions on the surface, and as such, these parts of the surface correspond to rivers *(courtesy of H. Takayasu, Kobe University)*. The above picture is a remarkable example for the coexistence of self-affine (the whole surface) and self-similar (rivers) geometries. The system of long valleys and mountain ridges often observed in nature emerges spontaneously in these simulations; a property absent in the various standard methods of generating self-affine functions.

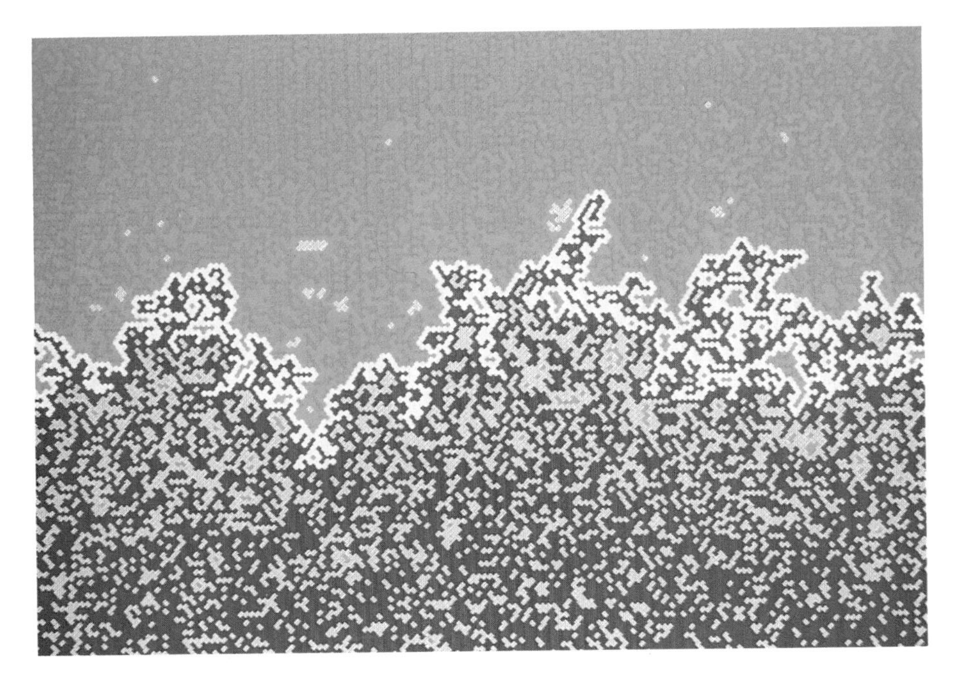

Plate 6a
The yellow dots in this picture *(courtesy of M. Rosso and B. Sapoval, Ecole Polytechnique)* correspond to the so-called diffusion front which is built up by particles diffusing from a line source at the bottom. Some of the particles are disconnected, however, most of them belong to a single large cluster connected to the source line. The border between the unoccupied sites of the lattice (green part) and the region occupied by the above large cluster (blue-pink) is the diffusion front which is a self-similar fractal whose dimension is the same as that of the hull of a percolation cluster.

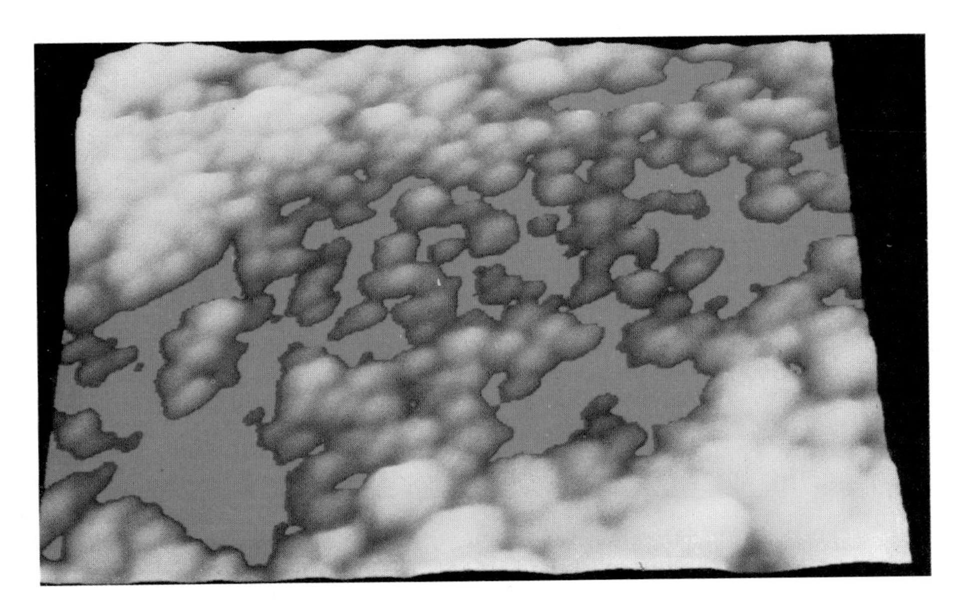

Plate 6b
Image of the surface of a gold electrodeposit obtained using scanning tunneling microscopy *(J. M. Gómez-Rodriguez, R. C. Salvarezza and M. Baró, J. Vac. Sci. Technol. B9 (1991) 495)*. The size of the region shown in this picture is about 1000 nm. The "deep" parts of the image were filled with blue "water" up to a given level to provide data for the fractal analysis of the surface roughness based on the so-called area-perimeter method.

Plate 7
An off-lattice diffusion-limited aggregate *(courtesy of P. Meakin, du Pont)* consisting of 50000 particles. In this model particles released one after another from a distant point are diffusing on the plane and stick rigidly to the growing aggregate when during their random walk arrive at an adjacent site to the cluster. The color coding corresponds to the order in which the particles were added to the cluster.

Plate 8
The distribution of the probability of finding a randomly diffusing particle around a diffusion-limited aggregate (yellow) consisting of 25000 particles *(F. Family and T. Vicsek, Comp. Phys. No. 1 (1990) 44).* This distribution is also analogous to the behaviour of the electrostatic potential around a charged conductor having the shape of the yellow cluster. The system of equipotential lines is shown using a logarithmic scale so that the field drops one order of magnitude between consecutive major layers demonstrating the screening effect of long fjords.

Plate 9

Examples for fractal growth patterns analogous to diffusion-limited aggregates. The upper picture was obtained in a simple experiment in which air (white) was injected through a central whole into a thin layer filled by a viscous liquid (red) and randomly distributed small beads kept between two transparent plates *(F. Family and T. Vicsek, Comp. Phys. No. 1 (1990) 44)*. Bottom: The mineral dendrites shown in this picture *(courtesy of J. M. Garcia-Ruiz, University of Granada)* are chemical deposits (oxides) that formed when at some point in the geological past a thin layer between two limestone plates was penetrated by a supersaturated solution of manganese ions entering through a crack perpendicular to the layer.

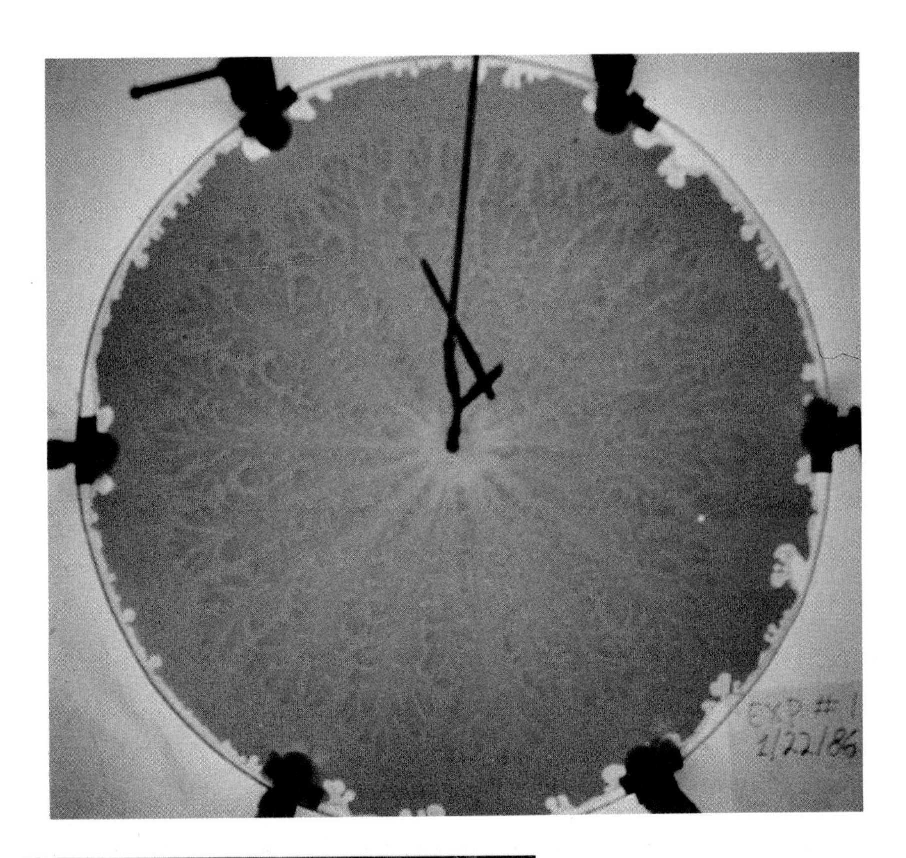

Plate 10
Fractal patterns formed in simple experiments on viscous fingering. In the upper picture the complicated morphology was obtained by injecting the less viscous, miscible green liquid into the layer of more viscous red liquid held between two close circular plates which were kept parallel using spacers and six clamps. The reader can easily repeat the experiment shown in the bottom picture. A thick oil drop (previously made red using food color) is placed on a smooth surface. Next a glass plate is used to flatten the drop and produce a thin layer of viscous fluid. Finally the glass plate is lifted and the air entering from the sides produces intricate figures *(courtesy of E. Ben-Jacob, Tel-Aviv University).*

Plate 11
The presence of anisotropy in experiments on viscous fingering (Plate 10) may result in surprising effects. If a triangular lattice is etched onto the surface of one of the glass plates the growing air bubble advances easier along the grooves and has a hexagonal symmetry (top) closely resembling the geometry of snowflakes *(courtesy of E. Ben-Jacob, Tel Aviv University)*. The strange seven-fold pattern shown in the bottom picture was obtained by injecting air (blue part) into a thin layer of nematic liquid crystal (yellow) which has an orientation dependent shear viscosity.

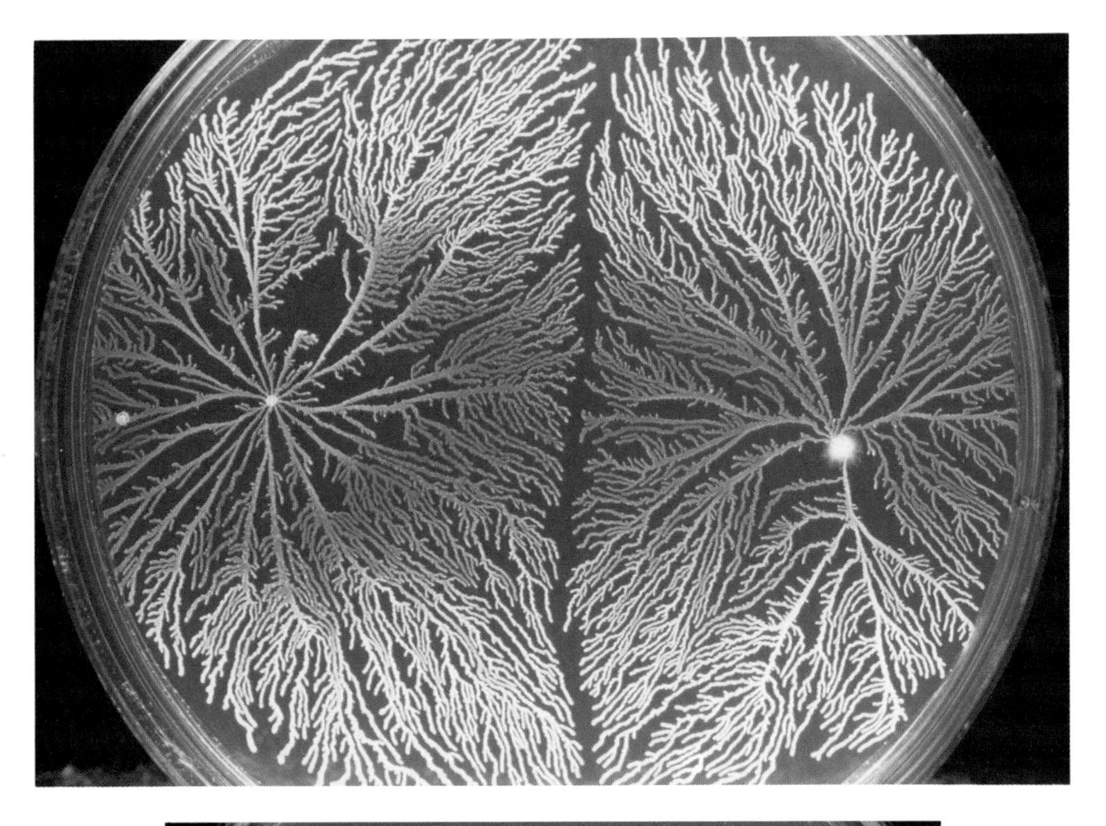

Plate 12
Bacteria colonies growing on the surface of agar plates (layers of gel containing nutrient usually kept in Petri dishes) may develop very complex morphologies depending on the conditions during the experiments. The upper picture was obtained by inoculating the agar's surface at two points with a drop from a culture of *Bacillus Subtilis* (a common kind of bacteria which can be found in our food) and letting the two-dimensional bush-like colonies grow towards each other. If the agar has less nutrient and the humidity in the incubator is kept at a lower level the shape of the bacteria colony (bottom) becomes analogous to a diffusion-limited aggregate *(courtesy of M. Matsushita, Chuo University).*

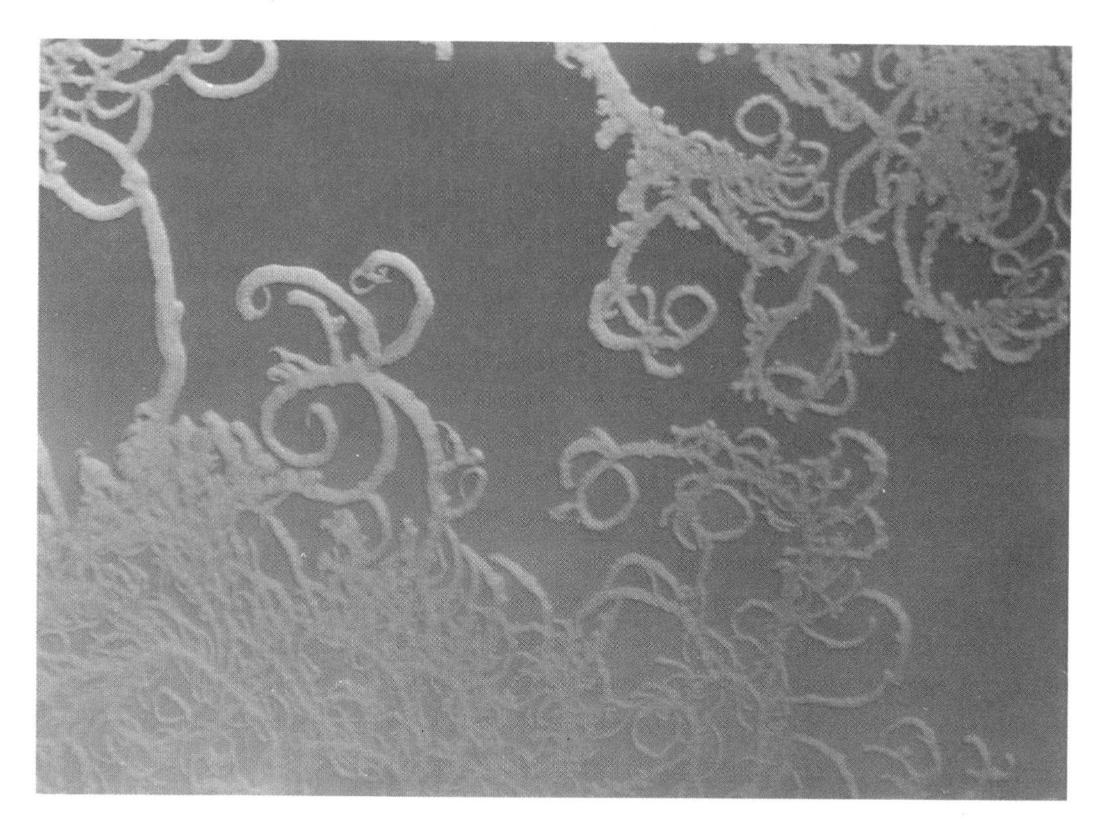

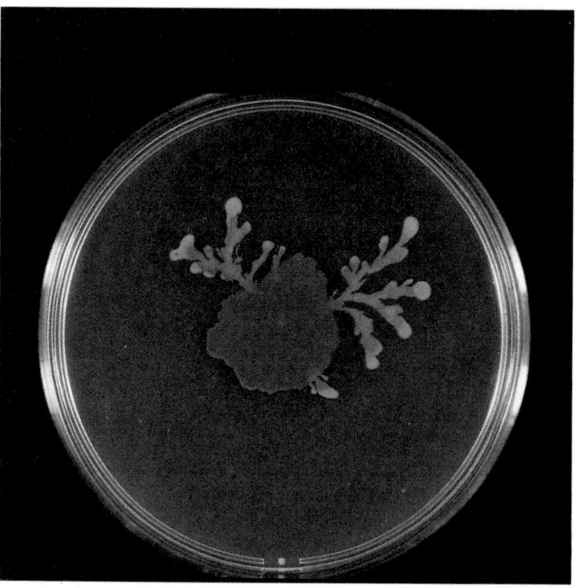

Plate 13
Bacteria colonies (see Plate 12) may exhibit interesting anomalous behaviors. The spiralling growth regime shown in the top was achieved by implementing a curved barrier (not shown) on the surface of the agar plate and in this way forcing the bacteria to locally grow in a symmetry breaking manner. The bacteria apparently adopted the given growth regime and continued growing along spirals of the same direction even far from the barrier *(courtesy of E. Ben-Jacob, Tel-Aviv University)*. The central red part of the bacteria colony shown in the bottom looks like an Eden cluster because the agar was nutrient rich. At some point a new strain appeared as a result of mutation following a different growth habit resulting in the almost white, antlers-like structures *(courtesy of T. Matsuyama, Niigata University)*.

Plate 14
Two plants displaying almost regular fractal structure. The upper picture shows a dried out branch of a bush common in the Middle-East. The bottom structure *(F. Grey and J. K. Kjems, Physica D38 (1989) 154)* looks like a deterministic fractal, but is simply a close up picture of the so-called Minaret cauliflower available at some grocery stores.

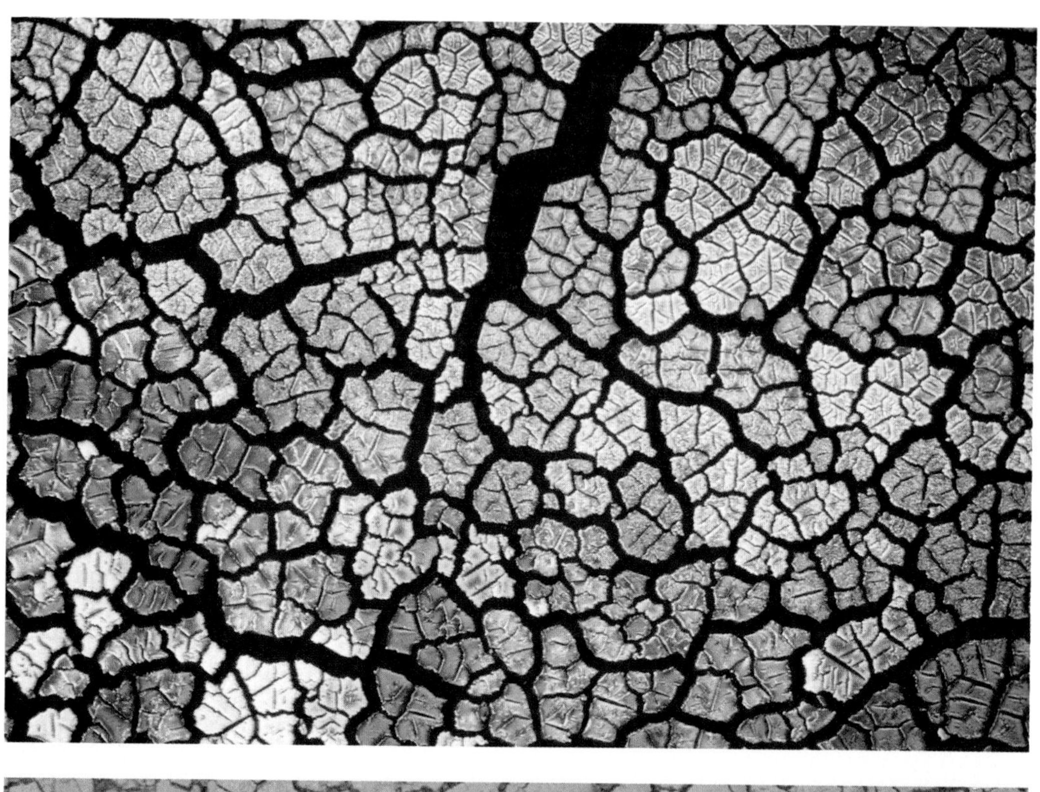

Plate 15
The cracks in the upper picture were formed in a system composed of approximately one million small plastic spheres each of diameter 0.0034 mm. The spheres were originally arranged in a single layer on a wet surface and the cracks were created as a result of shrinking during the process of drying *(Skjeltorp and Meakin, 1988)*. The bottom part of this plate shows a strikingly similar crack network one can observe in a dried out field with clay soil.

Plate 16
This beautiful picture *(courtesy of Y. Furukawa, Hokkaido University)* exhibits a number of features characteristic of most of the virtually infinite variety of snowflakes. It has a well pronounced sixfold symmetry, however, this symmetry is not perfect. The completely filled central part represents a different kind of morphology than the outer dendritic branches. Due to a special effort when taking this picture it is also exceptionally suitable for demonstrating the three-dimensional nature of snowflakes. The snowflakes are crossing different layers of air while falling down in the atmosphere, thus, they undergo randomly changing different physical conditions. Correspondingly, their width and local shape (closely related to the actual growth regime) change in a complicated manner.

Plate 17

Deterministic models of self-similar fractal growth. The sixfold pattern *(courtesy of F. Family, Emory University)* was obtained in an aggregation model simulated on the triangular lattice. In this model the rule determining the addition of new particles to the growing cluster was based on the value of the local temperature field as determined from the solution of the Laplace equation. The fourfold "snowflake" (bottom) can be generated using a simple iterative process based on repeatedly adding the already existing structure to itself at its four corners (the process can be started with a seed configuration of five particles arranged according to a + sign).

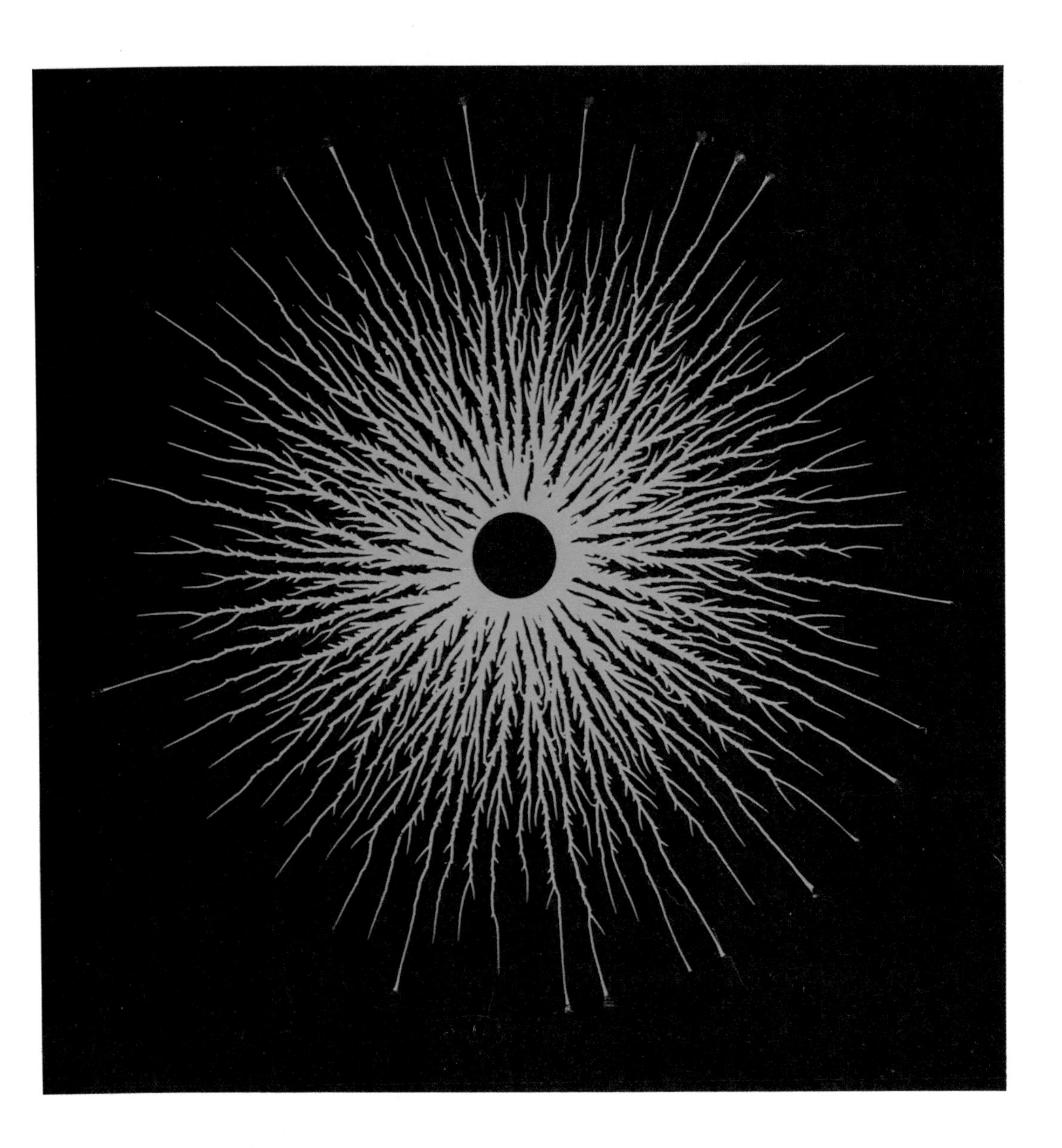

Plate 18
This plate *(courtesy of N. Allen, University of Leeds)* shows a spark or in other words a gaseous discharge pattern (also called dielectric breakdown). To obtain a surface discharge like the one shown here a starter electrode is brought into contact with an insulating (glass) plate surrounded by the other electrodes having a circular shape. Applying a voltage pulse at the starter electrode a propagating discharge pattern can be observed.